CAMBRIDGE TRACTS IN MATHEMATICS

General Editors

H. HALBERSTAM & C. T. C. WALL

---

87 *Exponential diophantine equations*

T. N. SHOREY

Professor of Mathematics
Tata Institute of Fundamental Research, Bombay, India

R. TIJDEMAN

Professor of Number Theory and Analysis
State University at Leiden, The Netherlands

Exponential diophantine equations

The right of the University of Cambridge to print and sell all manner of books was granted by Henry VIII in 1534. The University has printed and published continuously since 1584.

CAMBRIDGE UNIVERSITY PRESS

Cambridge

London New York New Rochelle

Melbourne Sydney

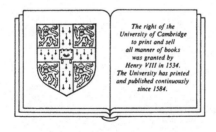

CAMBRIDGE UNIVERSITY PRESS
Cambridge, New York, Melbourne, Madrid, Cape Town, Singapore, São Paulo, Delhi

Cambridge University Press
The Edinburgh Building, Cambridge CB2 8RU, UK

Published in the United States of America by Cambridge University Press, New York

www.cambridge.org
Information on this title: www.cambridge.org/9780521268264

First published 1986
This digitally printed version 2008

A catalogue record for this publication is available from the British Library

Library of Congress Cataloguing in Publication data

Shorey, T. N.
Exponential diophantine equations.
(Cambridge tracts in mathematics; 87)
Bibliography
Includes index.
1. Diophantine analysis. I. Tijdeman, R.
II. Title. III. Series.
QA242.S56 1986 512′.74 86-30143

ISBN 978-0-521-26826-4 hardback
ISBN 978-0-521-09170-1 paperback

To our parents and our wives

Contents

Preface

After the appearance of Baker's fundamental papers 'Linear forms in the logarithms of algebraic numbers' in *Mathematika* in 1966–8, Baker, Coates and others obtained upper bounds for the magnitudes of integer solutions of some polynomial diophantine equations in two unknowns and their p-adic generalisations. The finiteness of the numbers of solutions of these equations had been proved by Thue, Siegel, Mahler and others much earlier. The publication of Baker's papers 'A sharpening of the bounds for linear forms in logarithms' in *Acta Arithmetica* in 1972–5, and van der Poorten's p-adic analogues of it, led to completely new results on exponential diophantine equations such as the work on the Catalan equation by Tijdeman and its p-adic analogue by van der Poorten. Since the numerous publications on exponential diophantine equations are scattered over journals and no thorough introduction is available, we have decided to write a tract on these results.

We were together at the University of Leiden in 1982–3 for one year. A first draft of the manuscript was written during this period. The subsequent work of finalising the manuscript was carried out by correspondence spread over a period of about two years. The stay of one of us (T.N.S.) at the University of Leiden was supported in part by the Netherlands Organisation for the Advancement of Pure Research (Z.W.O.).

We are very grateful to K. Győry for his generous help in preparing the manuscript. Lemma A.16, corollary A.7, theorem 1.4, corollary 1.3, theorem 5.5, theorem 7.2, theorem 7.6 and corollary 7.4 were added or modified on his advice, and he assisted in writing the proofs of these results as well as the changes entailed in other proofs. He read the manuscript with care and brought to our notice several inaccuracies. Without his help, an account of associated literature in the notes of several chapters would have been less complete. In particular the account of decomposable form equations in the notes at the ends of chapters 5 and 7 is due to him.

We are grateful to A. Makowski, A. Pethő, A. Schinzel, C. L. Stewart and B. M. M. de Weger for their remarks on the manuscript. We thank T. Bakker, S. Wassenaar and D. B. Sawant for the excellent work of typing the manuscript and Cambridge University Press for publishing it in Cambridge Tracts in Mathematics.

T.N.S.
R.T.

Introduction

Chapters 1–12 deal with applications of estimates of linear forms in logarithms of algebraic numbers to prove the existence of effectively computable upper bounds for the magnitudes of the solutions of exponential diophantine equations. For the convenience of the readers, we begin with chapters A (on algebraic number theory), B (on linear forms in logarithms) and C (on recurrence sequences) which contain all the required preliminaries.

The simplest exponential diophantine equations are those with fixed bases and variable exponents, the so-called purely exponential equations. Examples of questions leading to such equations are (i) which powers of 2 and powers of 3 differ by 1? and (ii) which powers of 2 are Fibonacci numbers? The equation corresponding to (i) is

$$2^m - 3^n = \pm 1 \quad \text{in non-negative rational integers } m, n \tag{1}$$

and (ii) gives, with $\alpha = \frac{1}{2} + \frac{1}{2}\sqrt{5}$, $\beta = \frac{1}{2} - \frac{1}{2}\sqrt{5}$,

$$\frac{1}{\sqrt{5}}\alpha^m - \frac{1}{\sqrt{5}}\beta^m = 2^n \quad \text{in non-negative rational integers } m, n. \tag{2}$$

Such equations are studied in chapters 1, 3 and 4. Chapter 1 deals with equations $x + y = z$ in algebraic integers x, y, z from a fixed algebraic number field such that the ideal $[xyz]$ is composed of prime ideals from a given finite set. This covers (1) and (2). In chapter 2, a remarkable consequence of Baker (1972) is worked out, namely that, for given non-zero integers A, B, C, D and under suitable conditions, the equation

$$Ax^m + By^m = C \quad \text{in rational integers } m, x, y \quad \text{with } |x| > 1 \tag{3}$$

and, more generally, the equation

$$Ax^m + By^m = Cx^n + Dy^n \quad \text{in rational integers } m, n, x, y \quad \text{with } m > n, \ |x| > 1 \tag{4}$$

implies that m is bounded by an effectively computable number which depends only on A, B, C and D. Note that (3) and (4) are no longer purely exponential equations. Chapter 3 deals with non-degenerate binary recurrence sequences. In chapter 4, recurrence sequences of higher order are investigated. For example, it is proved that elements of a non-degenerate recurrence sequence of order 2 or 3 are distinct after an effectively computable stage. Further it is shown in an effective way that a non-degenerate recurrence sequence of order at most 4 contains only finitely many terms equal to zero.

Chapters 5–8 concern polynomial equations and their p-adic analogues. A polynomial equation is an equation

$$f(x_1,\dots,x_n)=0 \quad \text{in algebraic integers } x_1,\dots,x_n\in K$$

where f is a given polynomial and K is a given algebraic number field. Let $f(X, Y)$ be a binary form (homogeneous polynomial) with rational integer coefficients and with at least three pairwise non-proportional linear factors in its factorisation over the field of complex numbers. For a given non-zero integer k, Thue (1909) proved that the equation

$$f(x, y)=k \quad \text{in rational integers } x, y \tag{5}$$

(now known as Thue's equation) has only finitely many solutions. This is an immediate consequence of Thue's fundamental inequality on the approximations of algebraic numbers by rationals. His argument is non-effective: it fails to provide an explicit bound for the magnitudes of the solutions. Baker (1968*b*), by way of his fundamental researches on linear forms in logarithms, gave an effective version of Thue's theorem (see chapter 5). Consequently it was possible to give effective versions of earlier results on the solutions of superelliptic equations,

$$f(x)=y^m \quad \text{in rational integers } x, y \tag{6}$$

where $m\geqslant 2$ is a given rational integer and f is a given polynomial (see chapter 6). By the natural p-adic analogue of Baker's theory of linear forms in logarithms, it was possible to show in an effective way that equations (5) and (6) have only finitely many solutions in rational numbers with denominators composed of primes from a given finite set (see chapters 7 and 8). By combining results from chapters 2, 5 and 7, we give the necessary conditions under which equations (3) and (4) have only finitely many solutions m, $(n,)x$, y (see chapters 5 and 7).

Equations (3) and (4) are examples of general exponential equations; the equations with a term or factor x^m where $m\geqslant 2$ is a variable rational integer and x is a variable (rational or algebraic) integer. Chapters 9–12 are devoted to such equations. In chapter 9, we again turn to recurrences to show that a non-degenerate binary recurrence sequence has only finitely many perfect powers. Thus there are only finitely many perfect powers in the Fibonacci sequence. A characteristic result of chapter 10 is that if $f(X)$ is a given polynomial with rational integer coefficients and with at least two distinct roots, the equation

$$f(x)=y^m \quad \text{in rational integers} \quad m>1, x, y>1 \tag{7}$$

implies that m is bounded by an effectively computable constant. In combination with the above-mentioned results on equation (6), we obtain general conditions under which (7) has only finitely many solutions. Chapter 11 deals with the Fermat equation

$$x^n+y^n=z^n \quad \text{in rational integers} \quad n>2, \quad x>0, \quad y>0, \quad z>0.$$

Under various conditions, for example when $y-x$ is composed of fixed primes, it is shown that there are only finitely many solutions. Finally, chapter 12 contains a proof that the Catalan equation

$$x^m-y^n=1 \quad \text{in rational integers} \quad m>1, \quad n>1, \quad x>1, \quad y>1$$

implies that m, n, x and y are bounded by an effectively computable constant. There are various results on related equations such as

$$\frac{x^m-1}{x-1}=y^n \quad \text{and} \quad \frac{x^m-1}{x-1}=\frac{y^n-1}{y-1}$$

$$\text{in rational integers} \quad m>2, \quad n>2, \quad x>1, \quad y>1.$$

In all cases we follow the same route as described for equation (7): we apply estimates of linear forms in logarithms to a general exponential equation to show that the exponents are bounded. Then we need to study only finitely many polynomial equations.

The main theme of this tract is the study of exponential diophantine equations. We have therefore dealt only with those polynomial equations which are used in the studies of these exponential equations. Important topics such as the effective results of Baker and Coates on integer points on curves of genus 1, and the effective results of Győry and others on decomposable form equations in several variables, are omitted. For the same reason, Runge's method and the applications of the Thue–Siegel–Roth–Schmidt method to diophantine equations are not included.

Our style is rather leisurely. We have not rushed to give a proof of the most general result in a chapter; we first deal with particular cases so that the reader may take the strain of the proof gradually. However, in certain cases we have proved a generalisation by a method different from the particular. In chapter 5 we follow the original method of Baker, but in chapter 7 we apply results from chapter 1. The proofs in chapter 9 also differ from the proofs of more general results in chapter 10. In most cases, we end with the most far-reaching results available in the literature and, in several instances, we go even further. In order to make the exposition less technical, we have rarely worked out explicit bounds for the magnitudes of solutions of the equations.

Each chapter is in three parts. The first part contains the statements of all the results to be proved in the chapter. The second part, 'Proofs', contains the proofs of these results. The third part, 'Notes', gives an account of the developments related to the results of the chapter. Thus an account of the results of important topics which we could not include in the text is available in the Notes. Lengthy results stated in full in the Notes are often put in italics to indicate where they begin and end.

An important feature of the results in the tract is that upper bounds for (the heights of) the solutions can be effectively computed. We have to add that, in the case of a general exponential equation, the bounds are usually so large that in practice it is not possible to check all the remaining values on a computer to determine all the solutions. In the cases of purely exponential equations and polynomial equations, the situation is different (see Stroeker and Tijdeman, 1982). We speak of 'computable numbers' in place of 'effectively computable numbers', a term which is frequently used in literature. In numerous cases, computable numbers appear which depend not only on integer parameters but also on algebraic numbers, polynomials, number fields, regulators, sets or sequences. However, there are only finitely many algebraic numbers and polynomials of given degree and height. There are only finitely many algebraic number fields of given degree and discriminant. The regulator of an algebraic number field is bounded by its degree and its discriminant. The sets S and $\mathscr{S}$ in the theorems are determined by a finite number of given (algebraic) numbers. The algebraic recurrence sequences are determined by the algebraic recurrence coefficients and the initial values, hence by a finite set of algebraic numbers. Thus, in all cases, it is possible to make the computable numbers depend on a finite set of positive rational integer parameters (degrees, heights, discriminants). By computable number, we always mean a monotonic positive-valued function of these parameters which, moreover, can be calculated by following the proof (compare the remark after corollary A.5). For other conventions, we refer the reader to the list of Notation which follows. Finally we note that a reference with number (1986) which is to appear may be published in 1986, 1987 or later.

Notation

Below we give a list of notation and terms which are often used in the text. References in square brackets on the right-hand side are to the chapter and paragraph where more information can be found.

$A:=B$, $B=:A$	A is defined to be B.
computable constant	A real and positive constant which can be effectively computed by following the given proof.

Sets and sequences

$\mathbb{Z}$	The (rational) integers.
$\mathbb{Z}_+$	The positive elements of $\mathbb{Z}$.
$\mathbb{Q}$	The rational numbers.
$\mathbb{R}$	The real numbers.
$\mathbb{C}$	The complex numbers.
K^*	The non-zero elements of the field K.
algebraic number field	Finite field extension of $\mathbb{Q}$ in $\mathbb{C}$. [A, § 1]
$\mathcal{O}_K$	The algebraic integers of an algebraic number field K.
S	All integers composed of $p_1, \ldots, p_s$, that is, all non-zero rational integers which have no prime divisors different from $p_1, \ldots, p_s$. Here $p_1, \ldots, p_s$ are given prime numbers, not necessarily the first s primes. If $s=0$, then $S=\{-1, 1\}$.

S_+	The positive elements of S.
$\mathcal{S}, \mathcal{S}'$, etc.	An analogue of S in algebraic number fields. [1, Theorems]
$\mathcal{S}$-integers	[7, Notes]
$\mathcal{S}$-units	[1, Theorems]
recurrence sequence	[C, § 1]
Fibonacci sequence	[C, § 2]
Lucas sequence	[C, § 2]
Lehmer sequence	[C, § 2]
fundamental set of units	[A, § 4]
independent set of units	[A, §4]
integral basis	[A, § 3]
for almost all n	For a sequence of positive integers n such that the number of elements of the sequence at most x, divided by x, tends to 1 as $x \to \infty$.

Functions and mappings

(Let K and L be algebraic number fields, a and b in $\mathbb{Z}$ (or in $\mathcal{O}_K$), $n \in \mathbb{Z}$, $x \in \mathbb{R}$, $\alpha \in K$, $\mathfrak{a}$ and $\mathfrak{b}$ ideals in $\mathcal{O}_K$, $\mathfrak{p}$ a prime ideal in $\mathcal{O}_K$)

empty sum	0
empty product	1
0^0	0
$a \mid b$	There is a c in $\mathbb{Z}$ (or in $\mathcal{O}_K$) such that $b = ac$.
$a \nmid b$	It is not true that $a \mid b$.
$a^n \parallel b$	$a^n \mid b$, but $a^{n+1} \nmid b$.
(a, b)	Greatest common divisor of a and b (of the principal ideals generated by a and b). [A, § 2]
$\langle a, b \rangle$	Least common multiple of a and b.
$\phi(n)$	The number of integers a with $(a, n) = 1$ and $1 \leqslant a \leqslant n$.
$\omega(n)$	The number of distinct prime divisors of n.
$P(n)$	The greatest prime factor of n, but $P(0) = P(1) = P(-1) = 1$.

$Q(n)$	The greatest square-free divisor of n, $Q(0)=Q(1)=Q(-1)=1$.
square free	Not divisible by p^2 for any prime p.
m-free	Not divisible by p^m for any prime p.
$\pi(x)$	The number of prime numbers not exceeding x. (By a prime number we shall always mean a positive prime.) We have $$\pi(x) \leqslant \frac{2x}{\log x} \quad \text{for } x \geqslant 2. \qquad \text{(N.1)}$$ (cf. Rosser and Schoenfeld (1962), formula (3.6)).
trivial (function)	(Function) being identically zero.
non-trivial (function)	(Function) not identically zero.
binary form	Homogeneous polynomial in two variables.
height (of polynomial)	Maximal absolute value of the coefficients (of the polynomial).
$\deg(f)$	Degree (of the polynomial f).
non-proportional (functions)	(Two functions) with non-constant ratio.
minimal defining polynomial (of α)	[A, § 1]
degree (of α)	Degree of the minimal defining polynomial of α.
denominator (of α)	Smallest positive integer n such that $n\alpha \in \mathcal{O}_K$.
height (of α)	Height of minimal defining polynomial of α.
$\deg(\alpha)$	Degree of α.
$H(\alpha)$	Height of α.
$\overline{\lvert\alpha\rvert}$	Maximal absolute value of conjugates of α. [A, § 1]
$[L:K]$	The degree of the field extension L/K.
$N_{L/K}(\alpha)$	The norm of α with respect to the field extension L/K. [A, Notes]
$N(\alpha), N_K(\alpha)$	$N_{K/\mathbb{Q}}(\alpha)$, the field norm of α with respect to K. [A, § 1]

d_K	Degree of K.
h_K	Class number of K. [A, § 3]
$\mathscr{D}_K$	Discriminant of K. [A, § 3]
R_K	Regulator of K. [A, § 4]
$\mathscr{a} \mid \mathscr{b}$	There exists an ideal $\mathscr{c}$ in $\mathcal{O}_K$ such that $\mathscr{b} = \mathscr{a}\mathscr{c}$.
$\mathscr{a}^n \parallel \mathscr{b}$	$\mathscr{a}^n \mid \mathscr{b}$, but not $\mathscr{a}^{n+1} \mid \mathscr{b}$.
$(\mathscr{a}, \mathscr{b})$	Greatest common divisor of $\mathscr{a}$ and $\mathscr{b}$. [A, § 2]
$N(\mathscr{a}), N_K(\mathscr{a})$	Norm of $\mathscr{a}$ (with respect to $K/\mathbb{Q}$). [A, § 2]
$[\alpha]$	The principal ideal in K generated by α. (If $\alpha \in \mathcal{O}_K$ then $[\alpha] \subset \mathcal{O}_K$.)
$\mathrm{ord}_{\mathscr{p}}(\alpha)$	[A, § 2]
order (of recurrence sequence)	[C, § 1]
binary recurrence sequence	Recurrence sequence of order 2.
ternary recurrence sequence	Recurrence sequence of order 3.
z-multiplicity (of $\{u_m\}_{m=0}^{\infty}$)	The number of indices m with $u_m = z$.
multiplicity (of $\{u_m\}_{m=0}^{\infty}$)	The supremum of the z-multiplicities taken over all z. [C, § 3]
total multiplicity (of $\{u_m\}_{m=0}^{\infty}$)	The number of pairs (m, n) with $m > n$ and $u_m = u_n$.

Theorems and techniques

unique factorisation theorem for ideals	[A, § 2]
Fermat's theorem for ideal theory	[A, § 2]
Liouville-type argument	[A, § 1]
transferring secondary factors	[1, Proofs]
estimating linear forms in logarithms	[B]
Thue–Siegel–Roth(–Schmidt) method	[Schmidt, 1971*b*]

A. Results from algebraic number theory

---

§**1.** For this chapter we refer to Hecke (1923) and Pollard (1950). Let α be an *algebraic number*. By this we shall always mean that $\alpha \in \mathbb{C}$ and α is algebraic over $\mathbb{Q}$. Then α is a zero of a unique non-zero polynomial of minimal degree (the so-called *minimal defining polynomial*)

$$a_0 x^d + a_1 x^{d-1} + \cdots + a_d$$

where $a_0, a_1, \ldots, a_d \in \mathbb{Z}$ satisfy $a_0 > 0$ and $(a_0, a_1, \ldots, a_d) = 1$. Write

$$H(\alpha) = \max(|a_0|, |a_1|, \ldots, |a_d|)$$

and

$$\deg(\alpha) = d.$$

We call $H(\alpha)$ the *height* of α and $\deg(\alpha)$ the *degree* of α. Notice that $H(\alpha) = H(1/\alpha)$, $\deg(\alpha) = \deg(1/\alpha)$ if $\alpha \neq 0$ and

$$H(\alpha) \leqslant m^d H(m\alpha) \quad (0 < m \in \mathbb{Z}). \tag{A.1}$$

Denote by $\nu = \nu(\alpha)$ the least positive integer such that $\nu\alpha$ is an algebraic integer. The integer ν is called the *denominator* of α. Observe that $a_0\alpha$ is an algebraic integer. Therefore the denominator of α exists and satisfies

$$\nu(\alpha) \leqslant a_0 \leqslant H(\alpha). \tag{A.2}$$

Denote by $\alpha = \alpha_1, \ldots, \alpha_d$ all the conjugates of α. Put

$$\overline{|\alpha|} = \max_{1 \leqslant i \leqslant d} |\alpha_i|.$$

For algebraic numbers α and β, we have

$$\overline{|\alpha + \beta|} \leqslant \overline{|\alpha|} + \overline{|\beta|}, \quad \overline{|\alpha\beta|} \leqslant \overline{|\alpha|}\,\overline{|\beta|}.$$

If $\alpha \neq 0$, observe that

$$|\nu\alpha_1 \cdots \nu\alpha_d| \geq 1. \tag{A.3}$$

Further

$$|\nu\alpha_1 \cdots \nu\alpha_d| \leq \nu^d|\alpha|\,\overline{|\alpha|}^{d-1}. \tag{A.4}$$

Combining (A.4) and (A.3), we have

$$|\alpha| \geq \nu^{-d}\overline{|\alpha|}^{-d+1}. \tag{A.5}$$

This argument for obtaining (A.5) was used by Liouville (1844) to prove his well-known inequality on the approximations of algebraic numbers by rationals. We shall refer to this argument as '*a Liouville-type argument*'. Further we have

Lemma A.1. *Let* α *be an algebraic number. Then*

$$\overline{|\alpha|} \leq \deg(\alpha)H(\alpha). \tag{A.6}$$

Proof. Assume that $|\alpha| > 1$. We have

$$a_0\alpha^d + a_1\alpha^{d-1} + \cdots + a_d = 0.$$

Dividing both sides by α^{d-1}, we have

$$|\alpha| \leq |a_0\alpha| = \left|a_1 + \frac{a_2}{\alpha} + \cdots + \frac{a_d}{\alpha^{d-1}}\right| \leq |a_1| + \cdots + |a_d| \leq dH(\alpha).$$

This inequality is also valid when $|\alpha| \leq 1$. Therefore $|\alpha| \leq dH(\alpha)$. Similarly, we can show that this estimate is valid for all the conjugates of α. □

Corollary A.1. *Let* $\alpha \neq 0$ *be an algebraic number. Then*

$$|\alpha| \geq (\deg(\alpha)H(a))^{-1}. \tag{A.7}$$

Proof. By lemma A.1, we have

$$|\alpha^{-1}| \leq \overline{|\alpha^{-1}|} \leq \deg(\alpha^{-1})H(\alpha^{-1}) = \deg(\alpha)H(\alpha)$$

which implies (A.7). □

On the other hand, we have

Lemma A.2. *Let* $\delta \neq 0$ *be an algebraic integer. Then*

$$H(\delta) \leq (2\overline{|\delta|})^{\deg(\delta)}.$$

Proof. Put $\deg(\delta) = \mu$. Denote by $\delta = \delta_1, \delta_2, \ldots, \delta_\mu$ all the conjugates of δ. Let

$$X^\mu + b_1X^{\mu-1} + \cdots + b_\mu$$

be the minimal polynomial of δ. Then b_j with $1 \leqslant j \leqslant \mu$ is an elementary symmetric function (up to the sign) of $\delta_1, \ldots, \delta_\mu$ of order j. Therefore, since $\overline{|\delta|} \geqslant 1$, we have

$$|b_j| \leqslant \binom{\mu}{j} \overline{|\delta|}^{\,j} \leqslant (2\overline{|\delta|})^\mu \quad (1 \leqslant j \leqslant \mu). \qquad \square$$

Corollary A.2. *Let $v \geqslant 0$ and $d \geqslant 1$ be given. The algebraic integers δ with $\overline{|\delta|} \leqslant v$ and* $\deg(\delta) \leqslant d$ *belong to a computable finite set.*

If β and γ are algebraic numbers of fixed degrees, we shall need bounds for $H(\beta+\gamma)$ and $H(\beta\gamma)$ in terms of $\max(H(\beta), H(\gamma))$.

Lemma A.3. *Let β and γ be algebraic numbers of degrees at most d and heights not exceeding H ($\geqslant 2$). Then*

$$(a)\ \frac{\log H(\beta+\gamma)}{\log H} \leqslant C_1, \quad (b)\ \frac{\log H(\beta\gamma)}{\log H} \leqslant C_2$$

where C_1 and C_2 are computable numbers depending only on d.

Proof. We prove (*a*) first. Note that $\beta+\gamma$ is an algebraic number of degree $\mu \leqslant d^2$. Denote by v_1, v_2 and v_3 the denominators of β, γ and $\beta+\gamma$, respectively. Put

$$\delta = v_3(\beta+\gamma). \tag{A.8}$$

Notice that δ is an algebraic integer of degree μ.

Observe that $v_1 v_2(\beta+\gamma)$ is an algebraic integer. Consequently, by (A.2),

$$v_3 \leqslant v_1 v_2 \leqslant H^2. \tag{A.9}$$

Now it follows from (A.8), (A.1) and (A.9) that

$$H(\beta+\gamma) \leqslant v_3^\mu H(\delta) \leqslant H^{2d^2} H(\delta).$$

Thus it suffices to show that

$$\frac{\log H(\delta)}{\log H} \leqslant C_3 \tag{A.10}$$

for some computable number C_3 depending only on d.

By (A.8), (A.9) and (A.6), we have

$$\|\delta\| \leqslant v_3(\overline{|\beta|} + \overline{|\gamma|}) \leqslant 2dH^3.$$

Then it follows from lemma A.2 that

$$H(\delta) \leqslant (4dH^3)^{\mu} \leqslant (4dH^3)^{d^2}$$

which implies (A.10).

The proof of (*b*) is similar. □

If $\beta \neq 0$ is an algebraic number, then $H(\pm\sqrt{\beta})$ can be bounded similarly.

Lemma A.4. *Let β be an algebraic number such that* $\deg(\beta) \leqslant d$ *and* $H(\beta) \leqslant H$, *where $H \geqslant 2$. Let $\gamma \in \mathbb{C}$ satisfy $\gamma^2 = \beta$. Then*

$$\log H(\gamma) \leqslant C_4 \log H$$

for some computable number C_4 depending only on d.

Proof. Notice that γ is an algebraic number of degree at most $2d$. Denote by ν_4 and ν_5 the denominators of β and γ, respectively. We have

$$(\nu_4\gamma)^2 = \nu_4^2\beta.$$

Therefore $\nu_4\gamma$ is an algebraic integer. Consequently, by (A.2), we have

$$\nu_5 \leqslant \nu_4 \leqslant H. \tag{A.11}$$

Now it follows from (A.1) and (A.11) that

$$H(\gamma) \leqslant \nu_5^{2d} H(\nu_5\gamma) \leqslant H^{2d} H(\nu_5\gamma).$$

Thus it suffices to show that

$$\frac{\log H(\nu_5\gamma)}{\log H} \leqslant C_5 \tag{A.12}$$

for some computable number C_5 depending only on d. By (A.11) and (A.6),

$$\overline{|\nu_5\gamma|} = \nu_5\overline{|\gamma|} \leqslant H\overline{|\beta|}^{1/2} \leqslant (dH^3)^{1/2}.$$

Then it follows from lemma A.2 that

$$H(\nu_5\gamma) \leqslant 2^{2d}(dH^3)^d$$

which implies (A.12). □

If $\beta \neq 0$ is an algebraic integer, then $\overline{|\beta|} \geqslant 1$. Further, an algebraic integer β with $\overline{|\beta|} = 1$ is a root of unity. We strengthen this assertion as follows.

Lemma A.5. *Let β be a non-zero algebraic integer which is not a root of unity. Let* $\deg(\beta) \leqslant d$. *Then*

$$\overline{|\beta|} > 1 + C_6 \tag{A.13}$$

for some computable number $C_6 > 0$ depending only on d.

Proof. Denote by N the number of algebraic integers γ of degree less than or equal to d such that $\overline{|\gamma|} \leqslant 2$. By corollary A.2, we have

$$N < C_7$$

for some computable number C_7 depending only on d. Put

$$C_8 = 2^{1/C_7} - 1.$$

Then

$$(1 + C_8)^N \leqslant 2.$$

Suppose

$$\overline{|\beta|} \leqslant 1 + C_8.$$

Then, for $k = 0, 1, \ldots, N$, we have

$$\overline{|\beta^k|} \leqslant (1 + C_8)^k \leqslant 2.$$

Further,

$$\deg(\beta^k) \leqslant d \quad (0 \leqslant k \leqslant N).$$

Therefore there exist distinct non-negative integers k_1 and k_2 not exceeding N such that $\beta^{k_1} = \beta^{k_2}$. This implies that β is a root of unity, since $\beta \neq 0$. □

Let K be a finite extension of $\mathbb{Q}$ of degree d. From now on we shall assume that K is a subfield of $\mathbb{C}$. All elements of K are algebraic. Such a field K is called an *algebraic number field.* The field K has exactly d distinct embeddings into $\mathbb{C}$. For $\alpha \in K$, we define the *field norm* of α (with respect to K) as

$$N(\alpha) = N_{K/\mathbb{Q}}(\alpha) = \prod_{\sigma} \sigma(\alpha)$$

where the product is taken over all the embeddings of K. The numbers $\sigma(\alpha)$ are called the *field conjugates* of α (with respect to K). If μ denotes the degree of α, then $\mu \mid d$ and the field conjugates of α are the conjugates of α each repeated d/μ times. If $d = \mu$, then $K = \mathbb{Q}(\alpha)$ and α is called a *primitive element* of K over $\mathbb{Q}$. If $\alpha_1, \ldots, \alpha_\mu$ are all the distinct conjugates of α, then

$$N_{\mathbb{Q}(\alpha)/\mathbb{Q}}(\alpha) = \alpha_1 \cdots \alpha_\mu.$$

Further,

$$N_{K/\mathbb{Q}}(\alpha) = (N_{\mathbb{Q}(\alpha)/\mathbb{Q}}(\alpha))^{[K:\mathbb{Q}(\alpha)]}.$$

Observe that

$$|N(\alpha)| \leqslant \overline{|\alpha|}^d$$

and

$$N(\alpha\beta) = N(\alpha)N(\beta).$$

Further we have

$$|N(\alpha)| \geqslant 1$$

whenever $\alpha \neq 0$ is an algebraic integer. Moreover an algebraic integer $\alpha \in K$ satisfies

$$|N(\alpha)| = 1$$

if and only if α is a unit. Recall that α is called a unit if both α and $1/\alpha$ are algebraic integers.

§2. Let K be a finite extension of degree d over $\mathbb{Q}$. Denote by $\mathcal{O}_K$ the ring of all the algebraic integers of K. The letter $\mathfrak{p}$ denotes a prime ideal $\mathfrak{p}$ in $\mathcal{O}_K$. By $\prod_{\mathfrak{p}}$, we shall understand that the product is taken over all the prime ideals $\mathfrak{p}$ in $\mathcal{O}_K$. For $\alpha \in K$, we write $[\alpha]$ for the cyclic module generated by α over $\mathcal{O}_K$. For non-zero $\alpha, \beta \in K$, observe that $[\alpha] = [\beta]$ if and only if α/β is a unit. If $\alpha \in \mathcal{O}_K$, then $[\alpha]$ is called the *principal ideal* generated by α in $\mathcal{O}_K$. If $\mathfrak{a}$ and $[\alpha]$ are ideals in $\mathcal{O}_K$, we write $\mathfrak{a} \mid \alpha$ for $\mathfrak{a} \mid [\alpha]$.

The *unique factorisation theorem for ideals* states that every non-zero ideal $\neq [1]$ in $\mathcal{O}_K$ can be written in one and only one way (except for order) as a product of prime ideals in $\mathcal{O}_K$. Thus every non-zero ideal $\mathfrak{a}$ in $\mathcal{O}_K$ can be written as

$$\mathfrak{a} = \prod_{\mathfrak{p}} \mathfrak{p}^{a_{\mathfrak{p}}} \tag{A.14}$$

where $0 \leqslant a_{\mathfrak{p}} \in \mathbb{Z}$ such that $a_{\mathfrak{p}} = 0$ for all but finitely many $\mathfrak{p}$. Let $\mathfrak{b}$ be a non-zero ideal in $\mathcal{O}_K$. We may write

$$\mathfrak{b} = \prod_{\mathfrak{p}} \mathfrak{p}^{b_{\mathfrak{p}}} \tag{A.15}$$

where $0 \leqslant b_{\mathfrak{p}} \in \mathbb{Z}$ such that $b_{\mathfrak{p}} = 0$ for all but finitely many $\mathfrak{p}$. For all $\mathfrak{p}$, put

$$c_{\mathfrak{p}} = \min(a_{\mathfrak{p}}, b_{\mathfrak{p}}).$$

Define the *greatest common divisor* of $\mathfrak{a}$ and $\mathfrak{b}$ as

$$(\mathfrak{a}, \mathfrak{b}) = \prod_{\mathfrak{p}} \mathfrak{p}^{c_{\mathfrak{p}}}. \tag{A.16}$$

If $(\mathfrak{a}, \mathfrak{b}) = [1]$, then $\mathfrak{a}$ and $\mathfrak{b}$ are called relatively prime. If $\mathfrak{a}_1, \ldots, \mathfrak{a}_l$ are non-zero ideals in $\mathcal{O}_K$, then the greatest common divisor $(\mathfrak{a}_1, \ldots, \mathfrak{a}_l)$ of $\mathfrak{a}_1, \ldots, \mathfrak{a}_l$ can be defined similarly.

Let $\mathfrak{p}$ be a prime ideal in $\mathcal{O}_K$. For a non-zero ideal $\mathfrak{a}$ in $\mathcal{O}_K$, we see from (A.14) that

$$\mathfrak{a} = \mathfrak{p}^{a_\mathfrak{p}} \mathfrak{a}_1$$

where $\mathfrak{a}_1$ is a non-zero ideal in $\mathcal{O}_K$ such that $(\mathfrak{a}_1, \mathfrak{p}) = [1]$. Then we define

$$\mathrm{ord}_\mathfrak{p}(\mathfrak{a}) = a_\mathfrak{p}.$$

Observe that

$$\mathrm{ord}_\mathfrak{p}(\mathfrak{a}\mathfrak{b}) = \mathrm{ord}_\mathfrak{p}(\mathfrak{a}) + \mathrm{ord}_\mathfrak{p}(\mathfrak{b}) \tag{A.17}$$

for all non-zero ideals $\mathfrak{a}, \mathfrak{b}$ in $\mathcal{O}_K$. Let $0 \neq \alpha \in K$. Let v' be a positive rational integer such that $v'\alpha$ is an algebraic integer. Then we define

$$\mathrm{ord}_\mathfrak{p}([\alpha]) = \mathrm{ord}_\mathfrak{p}([v'\alpha]) - \mathrm{ord}_\mathfrak{p}([v']). \tag{A.18}$$

It follows from $v = v(\alpha) \mid v'$ and (A.17) that the right-hand side of (A.18) is equal to

$$\mathrm{ord}_\mathfrak{p}([v\alpha]) - \mathrm{ord}_\mathfrak{p}([v]).$$

Thus $\mathrm{ord}_\mathfrak{p}([\alpha])$ is well defined. We write $\mathrm{ord}_\mathfrak{p}(\alpha)$ for $\mathrm{ord}_\mathfrak{p}([\alpha])$. We have

$$\mathrm{ord}_\mathfrak{p}(\alpha\beta) = \mathrm{ord}_\mathfrak{p}(\alpha) + \mathrm{ord}_\mathfrak{p}(\beta) \tag{A.19}$$

for every non-zero $\alpha, \beta \in K$. To prove this, we write

$$\mathrm{ord}_\mathfrak{p}(\alpha\beta) = \mathrm{ord}_\mathfrak{p}(v(\alpha)v(\beta)\alpha\beta) - \mathrm{ord}_\mathfrak{p}(v(\alpha)v(\beta))$$

and apply (A.17) to obtain (A.19).

Let $0 \neq \alpha \in K$. If α is a unit, then $\mathrm{ord}_\mathfrak{p}(\alpha) = 0$ for every prime ideal $\mathfrak{p}$ in $\mathcal{O}_K$. Conversely, suppose that $\mathrm{ord}_\mathfrak{p}(\alpha) = 0$ for every prime ideal $\mathfrak{p}$ in $\mathcal{O}_K$. Then

$$\mathrm{ord}_\mathfrak{p}(v\alpha) = \mathrm{ord}_\mathfrak{p}(v)$$

for every prime ideal $\mathfrak{p}$ in $\mathcal{O}_K$. Now we apply the unique factorisation theorem for ideals to conclude that $[v\alpha] = [v]$. Consequently α is a unit.

For simplicity, we write $\mathcal{O}$ for $\mathcal{O}_K$. Let $\mathfrak{p}$ be a prime ideal in $\mathcal{O}$. Then there exists a unique positive rational prime p such that $\mathfrak{p} \mid p$. Observe that $\mathbb{Z}/p\mathbb{Z}$ and $\mathcal{O}/\mathfrak{p}$ are fields. Further, the function

$$m + p\mathbb{Z} \to m + \mathfrak{p} \tag{A.20}$$

from $\mathbb{Z}/p\mathbb{Z}$ into $\mathcal{O}/\mathfrak{p}$ is an embedding, since $\mathfrak{p} \cap \mathbb{Z} = p\mathbb{Z}$. Thus $\mathbb{Z}/p\mathbb{Z}$ can be considered as a subfield of $\mathcal{O}/\mathfrak{p}$. Further, $\mathcal{O}/\mathfrak{p}$ is a finite extension of $\mathbb{Z}/p\mathbb{Z}$ of degree

$$f_\mathfrak{p} \leqslant d. \tag{A.21}$$

Thus $\mathcal{O}/\mathfrak{p}$ has $p^{f_{\mathfrak{p}}}$ elements. We define the *norm* of $\mathfrak{p}$ as

$$N(\mathfrak{p}) = N_{K/\mathbb{Q}}(\mathfrak{p}) = p^{f_{\mathfrak{p}}}. \tag{A.22}$$

For every non-zero ideal $\mathfrak{a}$, given by (A.14), in $\mathcal{O}$, we define

$$N(\mathfrak{a}) = N_{K/\mathbb{Q}}(\mathfrak{a}) = \prod_{\mathfrak{p}} (N(\mathfrak{p}))^{a_{\mathfrak{p}}}.$$

Then

$$N(\mathfrak{a}\mathfrak{b}) = N(\mathfrak{a})N(\mathfrak{b}) \tag{A.23}$$

for all non-zero ideals $\mathfrak{a}, \mathfrak{b}$ in $\mathcal{O}$. Further we have

$$N([\alpha]) = |N(\alpha)| \quad (0 \neq \alpha \in \mathcal{O}). \tag{A.24}$$

We have used the same letter N for the norm of an element and the norm of an ideal, but it will be clear from the context what is meant. Since the number of elements in the multiplicative group $(\mathcal{O}/\mathfrak{p})^*$ of non-zero elements of $\mathcal{O}/\mathfrak{p}$ has $N(\mathfrak{p}) - 1$ elements, we have

$$\alpha^{N(\mathfrak{p})-1} \equiv 1 \pmod{\mathfrak{p}}, \quad \text{if } \mathfrak{p} \nmid \alpha. \tag{A.25}$$

This is called *Fermat's theorem for ideal theory.*

We shall apply (A.23), (A.24) and (A.25) to prove the following lemmas.

Lemma A.6. *Let p be a positive rational prime. Then the number of prime ideals in $\mathcal{O}_K$ which divide p does not exceed d.*

Proof. Let $\mathfrak{p}_1, \ldots, \mathfrak{p}_k$ be distinct prime ideals in $\mathcal{O}_K$ dividing p. Then $\mathfrak{p}_1 \cdots \mathfrak{p}_k | p$. Therefore, by (A.23),

$$N(\mathfrak{p}_1 \cdots \mathfrak{p}_k) \,|\, N([p]).$$

By (A.24),

$$N([p]) = N(p) = p^d.$$

By (A.23) and $N(\mathfrak{p}_i) \geqslant p$ for $1 \leqslant i \leqslant k$, we have

$$N(\mathfrak{p}_1 \cdots \mathfrak{p}_k) = N(\mathfrak{p}_1) \cdots N(\mathfrak{p}_k) \geqslant p^k.$$

Therefore $p^k \leqslant p^d$ which implies $k \leqslant d$. □

Lemma A.7. *Let $0 \neq \alpha \in K$ such that $H(\alpha) \leqslant H$, where $H \geqslant 2$. For a prime ideal $\mathfrak{p}$ in $\mathcal{O}_K$, we have*

$$|\mathrm{ord}_{\mathfrak{p}}(\alpha)| \leqslant C_9 \log H \tag{A.26}$$

where C_9 is a computable number depending only on d.

Proof. Let ν be the denominator of α. Recall that

$$\mathrm{ord}_{\wp}(\alpha)=\mathrm{ord}_{\wp}(\nu\alpha)-\mathrm{ord}_{\wp}(\nu).$$

By (A.2) and lemma A.3, it suffices to prove lemma A.7 for $0\neq\alpha\in\mathcal{O}_K$. Then

$$\mathrm{ord}_{\wp}(\alpha)\geqslant 0. \tag{A.27}$$

Further, by (A.24) and (A.6), we have

$$N([\alpha])=|N(\alpha)|\leqslant \overline{|\alpha|}^{d}\leqslant (dH)^{d}. \tag{A.28}$$

On the other hand, it follows from $\wp^{\mathrm{ord}_{\wp}(\alpha)}\,|\,\alpha$, (A.23) and (A.27) that

$$N([\alpha])\geqslant (N(\wp))^{\mathrm{ord}_{\wp}(\alpha)}\geqslant 2^{\mathrm{ord}_{\wp}(\alpha)}. \tag{A.29}$$

Now combine (A.29), (A.28) and (A.27) to obtain (A.26). □

Lemma A.8. *Let p be a positive rational prime and $\wp$ a prime ideal in $\mathcal{O}_K$ dividing p. Let $\alpha\in\mathcal{O}_K$. Suppose that $m\geqslant 2$ is a rational integer satisfying $\alpha^m\neq 1$. Then*

$$\mathrm{ord}_{\wp}(\alpha^m-1)\leqslant C_{10}\log m \tag{A.30}$$

for some computable number C_{10} depending only on d, p and α.

Proof. In view of lemma A.7, we may suppose that α is not a root of unity. Further we may assume $\mathrm{ord}_{\wp}(\alpha^m-1)>0$. Then $\wp\nmid\alpha$. Let s be the least positive integer such that

$$\alpha^s\equiv 1 \pmod{\wp}.$$

Then we see from (A.25), (A.22) and (A.21) that

$$s<N(\wp)\leqslant p^d.$$

Further we observe that

$$\mathrm{ord}_{\wp}(\alpha^m-1)\leqslant \mathrm{ord}_{\wp}((\alpha^s)^m-1).$$

Thus we may suppose that

$$\alpha\equiv 1 \pmod{\wp}. \tag{A.31}$$

We write

$$m=p^{\lambda}m_1,\quad (m_1,p)=1,$$

$$\beta=\alpha^{p^{\lambda}}$$

and

$$\alpha^m-1=\left(\frac{\beta^{m_1}-1}{\beta-1}\right)\left(\frac{\beta-1}{\alpha-1}\right)(\alpha-1). \tag{A.32}$$

We see from (A.31) that $\beta \equiv 1 \pmod{\mathfrak{p}}$. Therefore

$$\frac{\beta^{m_1}-1}{\beta-1} = \beta^{m_1-1} + \cdots + 1 \equiv m_1 \pmod{\mathfrak{p}}.$$

Consequently

$$\operatorname{ord}_{\mathfrak{p}}\left(\frac{\beta^{m_1}-1}{\beta-1}\right) = 0, \tag{A.33}$$

since $(m_1, p) = 1$. Further, (A.31) implies that

$$\operatorname{ord}_{\mathfrak{p}}\left(\frac{\beta-1}{\alpha-1}\right) \leqslant \lambda \operatorname{ord}_{\mathfrak{p}}(p) + \operatorname{ord}_{\mathfrak{p}}(\alpha+1) \leqslant d\lambda + \operatorname{ord}_{\mathfrak{p}}(\alpha+1)$$
$$\leqslant 2d \log m + \operatorname{ord}_{\mathfrak{p}}(\alpha+1). \tag{A.34}$$

Finally, by lemma A.7,

$$\operatorname{ord}_{\mathfrak{p}}(\alpha \pm 1) \leqslant C_{11} \tag{A.35}$$

for some computable number C_{11} depending only on d and α. Now combine (A.32), (A.33), (A.34) and (A.35) to obtain (A.30). □

§3. Let K be a finite extension of degree d over $\mathbb{Q}$. Denote by $\mathcal{O}_K$ the ring of all algebraic integers of K. There exist $w_1, \ldots, w_d \in \mathcal{O}_K$ such that every element of $\mathcal{O}_K$ can be uniquely written as a linear combination of $w_1, \ldots, w_d$ with rational integer coefficients. The set $\{w_1, \ldots, w_d\}$ is called an *integral basis* for K. Put

$$\mathcal{D} = \mathcal{D}_K = \begin{vmatrix} \sigma_1(w_1) & \cdots & \sigma_1(w_d) \\ \text{—} & \cdots & \text{—} \\ \sigma_d(w_1) & \cdots & \sigma_d(w_d) \end{vmatrix}^2$$

where $\sigma_1, \ldots, \sigma_d$ are all the embeddings of K. Observe that $\mathcal{D}$ is independent of the choice of the integral basis, since any two integral bases are connected by a matrix of determinant ± 1. Further notice that $0 \neq \mathcal{D} \in \mathbb{Z}$. We call $\mathcal{D}$ the *discriminant* of K. For $\alpha \in K$, we define the *field discriminant* of α as

$$\mathcal{D}(\alpha) = \mathcal{D}_K(\alpha) = \prod_{1 \leqslant i < j \leqslant d} (\sigma_i(\alpha) - \sigma_j(\alpha))^2.$$

Hence $\mathcal{D}(\alpha) \neq 0$ if and only if α is a primitive element of K. Suppose α is an algebraic integer. Then $\mathcal{D}(\alpha) \in \mathbb{Z}$. Since $\mathcal{D}(\alpha)$ is equal to the determinant $|\sigma_i(\alpha^{j-1})|^2_{\substack{i=1,\ldots,d\\ j=1,\ldots,d}}$, we have $\mathcal{D} | \mathcal{D}(\alpha)$.

Denote by I the set of all non-zero ideals in $\mathcal{O}_K$. We define a relation $\sim$ in I as follows: for non-zero ideals $\mathfrak{a}$, $\mathfrak{b}$ in $\mathcal{O}_K$, $\mathfrak{a} \sim \mathfrak{b}$ if and only if there exist $\alpha, \beta \in \mathcal{O}_K$ such that $\alpha\beta \neq 0$ and $\mathfrak{a}[\alpha] = \mathfrak{b}[\beta]$. It is easy to see that $\sim$ is an equivalence relation. This partitions the set I into equivalence classes. For non-zero ideals $\mathfrak{a}$, $\mathfrak{a}_1$ and $\mathfrak{a}_2$ in $\mathcal{O}_K$, we observe that

$$\mathfrak{a}\mathfrak{a}_1 \sim \mathfrak{a}\mathfrak{a}_2 \Leftrightarrow \mathfrak{a}_1 \sim \mathfrak{a}_2$$

and

$$\mathfrak{a} \sim [1] \Leftrightarrow \mathfrak{a} \text{ is principal.}$$

Further, we have

Theorem A.1 (Minkowski). *In every equivalence class there exists an ideal $\mathfrak{a}$ in $\mathcal{O}_K$ such that*

$$N\mathfrak{a} \leqslant \sqrt{|\mathcal{D}|}.$$

Proof. See Hecke (1923), § 33.

By (A.23) and lemma A.6 there are only finitely many ideals in $\mathcal{O}_K$ of a given norm. Consequently, by theorem A.1, there are only finitely many equivalence classes. The number of equivalence classes is called the *class number h* of K. We have

Lemma A.9. *If $\mathfrak{a}$ is an ideal in $\mathcal{O}_K$ and h is the class number of K, then $\mathfrak{a}^h$ is principal.*

Proof. If $\mathfrak{a} = [0]$, then $\mathfrak{a}^h = [0]$. Therefore we may assume that $\mathfrak{a}$ is non-zero. Choose a set of ideals $\mathfrak{a}_1, \ldots, \mathfrak{a}_h$, one from each equivalence class. Then $\mathfrak{a}\mathfrak{a}_1, \ldots, \mathfrak{a}\mathfrak{a}_h$ fall into distinct equivalence classes. Consequently

$$\mathfrak{a}_1 \cdots \mathfrak{a}_h \sim \mathfrak{a}\mathfrak{a}_1 \cdots \mathfrak{a}\mathfrak{a}_h = \mathfrak{a}^h \mathfrak{a}_1 \cdots \mathfrak{a}_h.$$

This implies that $\mathfrak{a}^h$ is principal. □

Corollary A.3. *If $\mathfrak{a}, \mathfrak{b}$ are non-zero ideals in $\mathcal{O}_K$ and h is the class number of K, then*

$$(\mathfrak{a}^h, \mathfrak{b}^h)$$

is principal.

Proof. Let $\mathfrak{a}$ and $\mathfrak{b}$ be given by (A.14) and (A.15), respectively. Then, by (A.16), we have

$$(\mathfrak{a}^h, \mathfrak{b}^h) = \prod_{\mathfrak{p}} \mathfrak{p}^{hc_{\mathfrak{p}}}$$

and the assertion follows from lemma A.9. □

If $\mathfrak{a}_1, \ldots, \mathfrak{a}_l$ are non-zero ideals in $\mathcal{O}_K$, then it follows similarly that

$$(\mathfrak{a}_1^h, \ldots, \mathfrak{a}_l^h) = [\pi] \tag{A.36}$$

for some non-zero $\pi \in \mathcal{O}_K$. If K is a quadratic field, we have more precise information for the greatest common divisor of two principal ideals in $\mathcal{O}_K$ than corollary A.3 states.

Lemma A.10. *Let r and s be non-zero rational integers. Let α and β be roots of $x^2 - rx - s$. Then the greatest common divisor of the ideals $[\alpha^2]$ and $[\beta^2]$ in the ring of integers of $\mathbb{Q}(\alpha)$ is a principal ideal generated by (r^2, s).*

Proof. Put $l = (r^2, s)$. Then $((r^2+2s)/l, (s/l)^2) = 1$. Further observe that α^2/l and β^2/l satisfy

$$x^2 - \frac{r^2+2s}{l}x + \left(\frac{s}{l}\right)^2.$$

Consequently the ideals $[\alpha^2/l]$ and $[\beta^2/l]$ in the ring of integers of $\mathbb{Q}(\alpha)$ are relatively prime. □

We shall also need the following consequence of theorem A.1 and lemma A.9.

Lemma A.11. *Let $\mathfrak{a}$ be a non-zero ideal in $\mathcal{O}_K$. Then there exists a non-zero ideal $\mathfrak{a}'$ in $\mathcal{O}_K$ such that $N(\mathfrak{a}') \leqslant \sqrt{|\mathscr{D}|}$ and $\mathfrak{a}\mathfrak{a}'$ is principal.*

Proof. By lemma A.9, $\mathfrak{a}^h = \mathfrak{a}\mathfrak{a}^{h-1}$ is principal. By theorem A.1, we can find an ideal $\mathfrak{a}'$ in the equivalence class containing $\mathfrak{a}^{h-1}$ such that $N(\mathfrak{a}') \leqslant \sqrt{|\mathscr{D}|}$. Further observe that $\mathfrak{a}\mathfrak{a}'$ is principal. □

Let $\mathfrak{p}_1, \ldots, \mathfrak{p}_l$ be a finite set of prime ideals in $\mathcal{O}_K$. Put

$$p = \max_{1 \leqslant i \leqslant l} P(N(\mathfrak{p}_i)). \tag{A.37}$$

By lemma A.9, we may write

$$\mathfrak{p}_i^h = [\pi_i] \quad (1 \leqslant i \leqslant l) \tag{A.38}$$

where $\pi_1, \ldots, \pi_l \in \mathcal{O}_K$. Denote by $\mathscr{S}$ the set of all elements α of $\mathcal{O}_K$ such that $[\alpha]$ is exclusively composed of prime ideals $\mathfrak{p}_1, \ldots, \mathfrak{p}_l$. Then we have

Lemma A.12. *Let $\alpha \in \mathscr{S}$. There exist a $\beta \in \mathcal{O}_K$ with $|N(\beta)| \leqslant p^{dhl}$ and a unit $\varepsilon \in \mathcal{O}_K$ such that*

$$\alpha = \varepsilon\beta\pi_1^{a_1} \cdots \pi_l^{a_l}$$

where $a_1, \ldots, a_l$ are non-negative integers.

Proof. Let

$$[\alpha] = \mathfrak{p}_1^{b_1} \cdots \mathfrak{p}_l^{b_l}.$$

For $1 \leqslant i \leqslant l$, write

$$b_i = a_i h + c_i \quad (0 \leqslant c_i < h).$$

Then, by (A.38),

$$[\alpha] = [\pi_1]^{a_1} \cdots [\pi_l]^{a_l} \mathfrak{a} \tag{A.39}$$

where

$$\mathfrak{a} = \mathfrak{p}_1^{c_1} \cdots \mathfrak{p}_l^{c_l}.$$

We see from (A.39) that $\mathfrak{a}$ is a principal ideal. We write

$$\mathfrak{a} = [\beta] \tag{A.40}$$

for some non-zero $\beta \in \mathcal{O}_K$. By (A.24) and (A.23),

$$|N(\beta)| = N(\mathfrak{a}) = (N(\mathfrak{p}_1))^{c_1} \cdots (N(\mathfrak{p}_l))^{c_l}$$

which, together with (A.22) and (A.21), implies that

$$|N(\beta)| \leqslant p^{dhl}.$$

Now the lemma follows from (A.39) and (A.40). □

§ **4.** Let K be a finite extension of degree d over $\mathbb{Q}$. Denote by $\mathcal{O}_K$ the ring of all the algebraic integers of K. Let r_1 and $2r_2$ be the number of conjugate fields of K which are real and non-real, respectively. Further, we shall signify the conjugates of any element α of K by $\alpha^{(1)}, \ldots, \alpha^{(d)}$ with $\alpha^{(1)}, \ldots, \alpha^{(r_1)}$ real and $\alpha^{(r_1+1)}, \ldots, \alpha^{(r_1+r_2)}$ the complex conjugates of $\alpha^{(r_1+r_2+1)}, \ldots, \alpha^{(r_1+2r_2)}$, respectively. Put $r = r_1 + r_2 - 1$. We have

Theorem A.2. (Dirichlet). *There exist units $\eta_1, \ldots, \eta_r \in \mathcal{O}_K$ satisfying*

(*a*) *Every unit $\eta \in \mathcal{O}_K$ can be written as*

$$\eta = \rho \eta_1^{a_1} \cdots \eta_r^{a_r}$$

where $a_1, \ldots, a_r \in \mathbb{Z}$ and $\rho \in \mathcal{O}_K$ is a root of unity.

(*b*) *Let $\rho \in \mathcal{O}_K$ be a root of unity and $b_1, \ldots, b_r \in \mathbb{Z}$. The equation*

$$\eta_1^{b_1} \cdots \eta_r^{b_r} = \rho$$

implies that

$$b_1 = b_2 = \cdots = b_r = 0.$$

Proof. See Hecke (1923, § 34), or Pollard (1950, Ch. XI).

Let $r > 0$. A set of units $\eta_1, \ldots, \eta_r \in \mathcal{O}_K$ satisfying (*b*) of theorem A.2 is called an *independent system of units for K*. If $\eta_1, \ldots, \eta_r$ is an independent system of units for K, then the determinant

$$\det(\log |\eta_j^{(i)}|)_{\substack{i=1,\ldots,r\\ j=1,\ldots,r}} \tag{A.41}$$

is non-zero (cf. Pollard, 1950, p. 137). A set of units $\eta_1, \ldots, \eta_r \in \mathcal{O}_K$ satisfying (*a*) and (*b*) of theorem A.2 is called a *fundamental system of units for K*. Thus every fundamental system of units for K is also an independent system of units for K. Therefore the determinant (A.41) corresponding to a fundamental system of units for K is also non-zero. If $\eta_1, \ldots, \eta_r$ is a fundamental system of units for K, we denote by R the absolute value of the determinant (A.41). Observe that R is independent of the choice of the fundamental system of units for K. We call R the *regulator* of K. Notice that the absolute value of the determinant (A.41) corresponding to an independent system of units for K is at least R.

If $r=0$, then we understand that every independent as well as fundamental system of units for K is the empty set and we put $R=1$. Observe that $r=0$ if and only if either K is $\mathbb{Q}$ or K is an imaginary quadratic field.

For non-zero $\alpha \in \mathcal{O}_K$, put

$$J(\alpha) = \max_{1 \leqslant i \leqslant r} |\log |\alpha^{(i)}||.$$

Hence, for any unit $\eta \in \mathcal{O}_K$,

$$\log \overline{|\eta|} \leqslant dJ(\eta).$$

On applying Minkowski's theorem on successive minima, we have

Lemma A.13. *There exists an independent system $\eta_1, \ldots, \eta_r$ of units for K such that*

$$J(\eta_1) \cdots J(\eta_r) \leqslant R. \tag{A.42}$$

Proof. We may assume $r>0$. Let $\varepsilon_1, \ldots, \varepsilon_r$ be a fundamental system of units for K. For $\boldsymbol{x}=(x_1, \ldots, x_r) \in \mathbb{R}^r$, put

$$l_i(\boldsymbol{x}) = \sum_{j=1}^{r} x_j \log |\varepsilon_j^{(i)}| \quad (1 \leqslant i \leqslant r).$$

Denote by $\mathbb{P}$ the set of all $\boldsymbol{x} \in \mathbb{R}^r$ satisfying

$$|l_i(\boldsymbol{x})| \leqslant 1 \quad (1 \leqslant i \leqslant r).$$

Denote by $\lambda_1, \ldots, \lambda_r$ the successive minima of $\mathbb{P}$. By Minkowski's theorem on successive minima (see Cassels, 1957, p. 154) we have

$$V\lambda_1 \cdots \lambda_r \leqslant 2^r$$

where V is the volume of $\mathbb{P}$. Observe that

$$V = 2^r R^{-1}$$

(see Cassels, 1957, p. 150). Therefore

$$\lambda_1 \cdots \lambda_r \leqslant R. \tag{A.43}$$

For $\boldsymbol{x} \in \mathbb{R}^r$, define

$$F(\boldsymbol{x}) = \max_{1 \leqslant i \leqslant r} |l_i(\boldsymbol{x})|.$$

There exist linearly independent points $\boldsymbol{a}_1, \ldots, \boldsymbol{a}_r \in \mathbb{Z}^r$ such that

$$F(\boldsymbol{a}_i) = \lambda_i \quad (1 \leqslant i \leqslant r) \tag{A.44}$$

(see Cassels, 1959, p. 204). For $1 \leqslant i \leqslant r$, write

$$\boldsymbol{a}_i = (a_{i,1}, \ldots, a_{i,r})$$

and

$$\eta_i = \varepsilon_1^{a_{i,1}} \cdots \varepsilon_r^{a_{i,r}}.$$

Observe that $\eta_1, \ldots, \eta_r$ is an independent system of units for K. Further, notice that

$$J(\eta_i) = F(\boldsymbol{a}_i).$$

Therefore, by (A.44) and (A.43), we have

$$J(\eta_1) \cdots J(\eta_r) = \lambda_1 \cdots \lambda_r \leqslant R.$$ □

Combining (A.42) and (A.13), we obtain

Corollary A.4. *There exists an independent system $\eta_1, \ldots, \eta_r$ of units for K such that*

$$\max_{1 \leqslant i \leqslant r} J(\eta_i) \leqslant C_{12} R \tag{A.45}$$

where C_{12} is a computable number depending only on d.

Corollary A.4 and the inequalities $\log \overline{|\eta_i|} \leqslant dJ(\eta_i)$ with $1 \leqslant i \leqslant r$ imply that $\max_{1 \leqslant i \leqslant r} \overline{|\eta_i|}$ is bounded by a computable number depending only on d and R.

Proof. For $1 \leqslant i \leqslant r$, we have from (A.13) and $N(\eta_i) = \pm 1$ that

$$J(\eta_i) \geqslant C_{13}^{-1}$$

where $C_{13} > 0$ is a computable number depending only on d. Now we apply (A.42) to conclude that

$$J(\eta_i) = J(\eta_1) \cdots J(\eta_r) \left(\prod_{\substack{j=1 \\ j \neq i}}^{r} J(\eta_j) \right)^{-1} \leqslant C_{13}^{r-1} R$$

for $1 \leqslant i \leqslant r$. □

Let $\eta_1, \ldots, \eta_r$ be an independent system of units for K. Let $\alpha, \beta \in \mathcal{O}_K$ such that $\alpha\beta \neq 0$. Then β is called an *associate* of α (with respect to the independent system $\eta_1, \ldots, \eta_r$ of units for K) if

$$\beta = \alpha \eta_1^{a_1} \cdots \eta_r^{a_r}$$

where $a_1, \ldots, a_r \in \mathbb{Z}$. Then we have

Lemma A.14. *Let $\eta_1, \ldots, \eta_r$ be an independent system of units for K. Let $0 \neq \alpha \in \mathcal{O}_K$ with $|N(\alpha)| = m$. Then there exists β such that β is an associate of α and*

$$|\log(m^{-1/d} |\beta^{(j)}|)| \leqslant \sum_{i=1}^{r} \sum_{k=1}^{r} |\log |\eta_i^{(k)}|| \tag{A.46}$$

for $j = 1, \ldots, d$.

Proof. If $r = 0$, then (A.46) holds with $\beta = \alpha$. Thus we may assume $r > 0$. Let Λ be the lattice in $\mathbb{R}^r$ with

$$(\log |\eta_i^{(1)}|, \ldots, \log |\eta_i^{(r)}|) \quad (1 \leqslant i \leqslant r),$$

as basis. The coordinates of every point $\boldsymbol{x} = (x_1, \ldots, x_r) \in \mathbb{R}^r$ can be written as

$$x_j = \sum_{i=1}^{r} u_i \log |\eta_i^{(j)}| \quad (1 \leqslant j \leqslant r),$$

with $u_i \in \mathbb{R}$. There exists $\boldsymbol{\Lambda} = (\Lambda_1, \ldots, \Lambda_r) \in \Lambda$ such that

$$\Lambda_j = \sum_{i=1}^{r} b_i \log |\eta_i^{(j)}|$$

with $b_i \in \mathbb{Z}$ and

$$|x_j + \Lambda_j| \leqslant \sum_{i=1}^{r} |\log |\eta_i^{(j)}|| \quad (1 \leqslant j \leqslant r).$$

Take

$$x_j = \log(m^{-1/d} |\alpha^{(j)}|) \quad (1 \leqslant j \leqslant r),$$

and put

$$\beta = \alpha\eta_1^{b_1}\cdots\eta_r^{b_r}.$$

Then

$$\left|\log(m^{-1/d}|\beta^{(j)}|)\right| \leqslant \sum_{i=1}^{r} \left|\log|\eta_i^{(j)}|\right| \quad (1 \leqslant j \leqslant r). \tag{A.47}$$

Because of complex conjugation, inequalities (A.47) hold for $j=1, \ldots, d$ except, possibly, for $j=r_1+2r_2$ (and $j=r_1+r_2$). Observe that

$$\sum_{j=1}^{d} \log(m^{-1/d}|\beta^{(j)}|)=0.$$

Hence we have

$$\left|\log(m^{-1/d}|\beta^{(j)}|)\right| \leqslant \sum_{i=1}^{r}\sum_{k=1}^{r} \left|\log|\eta_i^{(k)}|\right| \quad (1 \leqslant j \leqslant d). \qquad \square$$

Combining lemma A.14 and corollary A.4 we obtain at once

Lemma A.15. *Let $\eta_1, \ldots, \eta_r$ be an independent system of units for K satisfying* (A.45). *Let $0 \neq \alpha \in \mathcal{O}_K$ with $|N(\alpha)|=m$. Then there exists $\beta \in \mathcal{O}_K$ such that β is an associate of α and*

$$\left|\log(m^{-1/d}|\beta^{(j)}|)\right| \leqslant C_{14}R \quad (1 \leqslant j \leqslant d),$$

where C_{14} is a computable number depending only on d.

We record the following result which is an immediate consequence of lemma A.15.

Corollary A.5. *Let $\eta_1, \ldots, \eta_r$ be an independent system of units for K satisfying* (A.45). *Then every unit $\eta \in \mathcal{O}_K$ can be written as*

$$\eta = \eta'\eta_1^{b_1}\cdots\eta_r^{b_r}$$

where $b_1, \ldots, b_r \in \mathbb{Z}$ and $\overline{|\eta'|}$ is bounded by a computable number depending only on d and R.

Proof. Apply lemma A.15 with $\alpha=\eta$ and $m=1$. $\square$

Remark. By a computable number depending only on d and R, we mean a function of both d and R. It follows from lemma A.15 that, in corollaries A.5 and A.6, this function is monotonic increasing in R.

Combining corollary A.4 and lemma A.15, we have

Corollary A.6. *Let $0 \neq \alpha \in \mathcal{O}_K$ such that $|N(\alpha)| \leqslant M$. There exists a unit $\varepsilon \in \mathcal{O}_K$*

such that $\overline{|\varepsilon\alpha|}$ *is bounded by a computable number depending only on* d, R *and* M.

The remaining results of this chapter will only be applied in chapters 5–8. Theorems A.3 and A.4 enable us to bound the class number and the regulator in terms of the discriminant.

Theorem A.3 (Landau, 1918). *There exists some computable number* C_{15} *depending only on* d *such that*

$$hR < C_{15}|\mathcal{D}|^{1/2}(\log|\mathcal{D}|)^{d-1}. \tag{A.48}$$

Theorem A.4 (Zimmert, 1981). $R > 0.056$.

For certain applications it is important to know that if $K = \mathbb{Q}(\alpha_1, \ldots, \alpha_n)$ then $|\mathcal{D}|$, hence h and R, can be estimated in terms of d and the heights of $\alpha_1, \ldots, \alpha_n$.

Lemma A.16. *There exists an algebraic integer* α *in* K *such that* $K = \mathbb{Q}(\alpha)$ *and* $\overline{|\alpha|} \leqslant |\mathcal{D}|^{1/2}$. *Further, if* $\alpha_1, \ldots, \alpha_n$ *are elements of* K *with heights at most* H *such that* $K = \mathbb{Q}(\alpha_1, \ldots, \alpha_n)$, *then*

$$\max(h, R, |\mathcal{D}|) \leqslant C_{16} \tag{A.49}$$

where C_{16} *is a computable number depending only on* d *and* H.

Proof. As before, we assume that the field conjugates of any $\alpha \in K$ are ordered such that $\alpha^{(1)}, \ldots, \alpha^{(r_1)}$ are real and $\alpha^{(r_1+1)}, \ldots, \alpha^{(r_1+r_2)}$ are the complex conjugates of $\alpha^{(r_1+r_2+1)}, \ldots, \alpha^{(r_1+2r_2)}$, respectively. Let $\{w_1, \ldots, w_d\}$ be an integral basis for K and, for $\boldsymbol{x} = (x_1, \ldots, x_d)$, put

$$\mathcal{L}^{(i)}(\boldsymbol{x}) = w_1^{(i)}x_1 + \cdots + w_d^{(i)}x_d \quad (1 \leqslant i \leqslant d).$$

By virtue of Minkowski's theorem on linear forms (see e.g. Cassels, 1959, p. 73) there exists an $\boldsymbol{x}_0 \in \mathbb{Z}^d$, $\boldsymbol{x}_0 \neq 0$, such that, when $r_1 > 0$,

$$|\mathcal{L}^{(1)}(\boldsymbol{x}_0)| \leqslant |\mathcal{D}|^{1/2}, \quad |\mathcal{L}^{(i)}(\boldsymbol{x}_0)| < 1 \quad (2 \leqslant i \leqslant d)$$

and, when $r_1 = 0$,

$$|\mathcal{L}^{(1)}(\boldsymbol{x}_0) + \mathcal{L}^{(1+r_2)}(\boldsymbol{x}_0)| < \sqrt{2},$$

$$|\mathcal{L}^{(1)}(\boldsymbol{x}_0) - \mathcal{L}^{(1+r_2)}(\boldsymbol{x}_0)| \leqslant \sqrt{(2|\mathcal{D}|)},$$

$$|\mathcal{L}^{(i)}(\boldsymbol{x}_0)| < 1 \quad (i \neq 1, 1+r_2).$$

Put $\alpha = \mathcal{L}^{(1)}(\boldsymbol{x}_0)$. Then α is a non-zero algebraic integer in K with $\overline{|\alpha|} \leqslant$

$|\mathscr{D}|^{1/2}$. Since $\overline{|\alpha|}\geqslant 1$, all the field conjugates of α are distinct. Hence α is a primitive element of K which proves the first assertion.

Put $\alpha_0=1$. For any i we omit α_{i+1} out of the generators if α_{i+1} is an element of the field $K_i=\mathbb{Q}(\alpha_0,\alpha_1,\ldots,\alpha_i)$. We may therefore assume that $n\leqslant d$. We prove by induction on i that there is a primitive element $\theta_i=a_0+a_1\alpha_1+a_2\alpha_2+\cdots+a_i\alpha_i$ of K_i such that $a_j\in\mathbb{Z}$ and $|a_j|\leqslant d^2$ for $j=0,1,\ldots,i$. This assertion is true for $i=0$ by taking $a_0=1$. Suppose that the assertion is true for K_i. If two conjugates of $\theta_i+a\alpha_{i+1}$, $\theta_i^{(\rho)}+a\alpha_{i+1}^{(\rho)}$ and $\theta_i^{(\sigma)}+a\alpha_{i+1}^{(\sigma)}$, say, are equal, then a is uniquely determined. Hence there are at most d^2 non-zero integers a such that two field conjugates of $\theta_i+a\alpha_{i+1}$ with respect to K_{i+1} are equal. Hence we can choose $a_{i+1}\in\mathbb{Z}$ with $|a_{i+1}|\leqslant d^2$ such that all the field conjugates of $\theta_{i+1}=\theta_i+a_{i+1}\alpha_{i+1}$ (with respect to K_{i+1}) are distinct. This implies that θ_{i+1} is a primitive element of K_{i+1} as required. Thus θ_n is a primitive element of K. By lemma A.3 we have $H(\theta_n)\leqslant C_{17}$ where C_{17} as well as C_{18} and C_{19} are computable numbers depending only on d and H. Let t be the denominator of θ_n. Then $t\leqslant C_{17}$. The number $\theta:=t\theta_n$ is an algebraic integer with $K=\mathbb{Q}(\theta)$ such that, by lemmas A.1 and A.3,

$$\overline{|\theta|}\leqslant dH(t\theta_n)\leqslant C_{18}.$$

Hence $0<|\mathscr{D}_K(\theta)|\leqslant C_{19}$. Since $\mathscr{D}$ divides $\mathscr{D}_K(\theta)$, this together with (A.48) implies (A.49). □

The following result is ready-made for our applications.

Corollary A.7. *Let $f\in K[X]$ be a polynomial of degree n. Let H be an upper bound for the heights of the coefficients of f. Let $\beta_1,\ldots,\beta_l$ be zeros of f and put $L=K(\beta_1,\ldots,\beta_l)$. Denote by G the maximum of the heights of $\beta_1,\ldots,\beta_l$ and by d_L, h_L, $\mathscr{D}_L$ and R_L the degree, class number, discriminant and regulator of L, respectively. Then there exists a computable number C_{20} depending only on d, $\mathscr{D}$, n and H such that*

$$\max(d_L,h_L,G,|\mathscr{D}_L|,R_L)\leqslant C_{20}.$$

Proof. By $C_{21},\ldots,C_{27}$ we shall denote computable numbers depending only on d, $\mathscr{D}$, n and H. It is clear that $d_L\leqslant n^n$. Let t be the product of the denominators of the coefficients of f. Let a_0 be the leading coefficient of $tf(X)$. Then we can write

$$a_0^{n-1}tf(X)=(a_0X)^n+a_1(a_0X)^{n-1}+\cdots+a_n$$

where $a_0,a_1,\ldots,a_n$ are algebraic integers in K with $\max_{0\leqslant j\leqslant n}\overline{|a_j|}\leqslant C_{21}$. It follows that

$$\overline{|a_0\beta_i|}^n \leqslant C_{22}\overline{|a_0\beta_i|}^{n-1} \quad (1 \leqslant i \leqslant l).$$

Hence $\overline{|a_0\beta_i|} \leqslant C_{22}$ and, by lemma A.2, $H(a_0\beta_i) \leqslant C_{23}$ for $i = 1, \ldots, l$. Since $H(a_0) = |a_0| \leqslant C_{24}$, we obtain $G \leqslant C_{25}$.

By the first assertion of lemma A.16 there is an algebraic integer α with height at most C_{26} which is a primitive element of K. By applying the second assertion of lemma A.16 to the field $L = \mathbb{Q}(\alpha, \beta_1, \ldots, \beta_l)$ we obtain $\max(h_L, R_L, |\mathscr{D}_L|) \leqslant C_{27}$. □

Notes

In lemma A.5, a remarkably good value of C_6 is due to Dobrowolski (1979). He proved the following. Let $\varepsilon > 0$. There exists a computable number C_{28} depending only on ε such that, if α is an algebraic integer of degree $d > C_{28}$, the inequality

$$\overline{|\alpha|} \leqslant 1 + \frac{2-\varepsilon}{d}\left(\frac{\log\log d}{\log d}\right)^3$$

implies that α is a root of unity (cf. Cantor and Straus (1982), Louboutin (1983)). This improves on earlier results of Schinzel and Zassenhaus (1965), Blanksby (1969), Blanksby and Montgomery (1971) and Stewart (1978). For a non-zero algebraic integer α which is not a root of unity, Schinzel and Zassenhaus (1965) conjectured that

$$\overline{|\alpha|} \geqslant 1 + C_{29}(\deg(\alpha))^{-1}$$

where $C_{29} > 0$ is an absolute constant.

Let K be a finite extension of $\mathbb{Q}$. For $\alpha \in K$ we defined the norm of α with respect to the extension $K/\mathbb{Q}$ in § 1. If L is a finite extension of K and $\alpha \in L$, then the norm $N_{L/K}(\alpha)$ of α with respect to the extension L/K is defined in a similar way (cf. Hecke, 1923, § 38).

Siegel (1969) proved theorem A.3 with

$$C_{15} = 4w2^{-r_1-r_2}\pi^{-r_2}(be(d-1)^{-1})^{d-1}$$

where r_1 and r_2 are as defined at the beginning of § 4, $b = (1 + \frac{1}{2}\log\pi + (r_2/n)\log 2)^{-1}$ and w is the number of roots of unity in K. (Hence w is even and $w = 2$ when $r_1 > 0$.)

For lemmas A.13–A.15 we refer to the original work of Baker (1968*b*), Siegel (1969) and Stark (1973). For further effective aspects of algebraic number theory we refer to the books of Borevich and Shafarevich (1964), Zimmer (1972), Stolarsky (1974) and Narkiewicz (1974).

B. Estimates of linear forms in logarithms

Suppose $\alpha_1, \ldots, \alpha_n$ are non-zero algebraic numbers and let $\log \alpha_1, \ldots, \log \alpha_n$ be any fixed values of the logarithms. If $\log \alpha_1$ and $\log \alpha_2$ are linearly independent over the rationals, then they are linearly independent over the algebraic numbers. This was Hilbert's seventh problem which was solved, independently, by Gelfond (1934) and Schneider (1934). Further, Baker (1966) proved that the linear independence of $\log \alpha_1, \ldots, \log \alpha_n$ over the rationals implies the linear independence of $\log \alpha_1, \ldots, \log \alpha_n$ over the algebraic numbers. Many important generalisations and improvements of this theorem have been obtained. In particular, for rational integers $b_1, \ldots, b_n$, non-trivial lower bounds have been given for the absolute value of the linear form

$$b_1 \log \alpha_1 + \cdots + b_n \log \alpha_n$$

in logarithms. For a survey of the results in this direction, known as the theory of linear forms in logarithms, we refer the reader to a paper of Baker (1977). From this theory, we record the results that we shall use in this tract. We shall refer to these results as *estimating linear forms in logarithms*. We recall our policy that all constants $C_1, C_2, \ldots$ are real and positive.

Let $\alpha_1, \ldots, \alpha_n$ be non-zero algebraic numbers of heights not exceeding $A_1, \ldots, A_n$, respectively. We assume $A_j \geqslant 3$ for $1 \leqslant j \leqslant n$. Put

$$A' = \max_{1 \leqslant j < n} A_j, \quad A = A_n,$$

$$\Omega = \prod_{j=1}^{n} \log A_j, \quad \Omega' = \prod_{j=1}^{n-1} \log A_j,$$

$$K = \mathbb{Q}(\alpha_1, \ldots, \alpha_n), \quad [K : \mathbb{Q}] = d.$$

Then we have

Theorem B.1 (Baker, 1977). *There exist computable absolute constants C_1 and C_2 such that the inequalities*

$$0<|b_1 \log \alpha_1+\cdots+b_n \log \alpha_n|<\exp(-(C_1 nd)^{C_2 n}\Omega \log \Omega' \log B)$$

have no solution in rational integers $b_1, \ldots, b_n$ of absolute values not exceeding B ($\geqslant 2$). It is assumed that the logarithms have their principal values.

We shall use the following formulation of theorem B.1.

Corollary B.1 (Baker, 1977). *There exist computable absolute constants C_3 and C_4 such that the inequalities*

$$0<|\alpha_1^{b_1}\cdots\alpha_n^{b_n}-1|<\exp(-(C_3 nd)^{C_4 n}\Omega \log \Omega' \log B)$$

have no solution in rational integers $b_1, \ldots, b_n$ of absolute values not exceeding B ($\geqslant 2$).

In order to derive corollary B.1 from theorem B.1, we refer the reader to a paper of Shorey *et al.* (1977, p. 66). The transition is trivial if $\alpha_1, \ldots, \alpha_n$ are positive.

Theorem B.2 (Baker, 1973). *There exists a computable number C_5 depending only on n, d and A' such that for any δ with $0<\delta<\frac{1}{2}$, the inequalities*

$$0<|\alpha_1^{b_1}\cdots\alpha_n^{b_n}-1|<(\delta/B')^{C_5 \log A}\,\mathrm{e}^{-\delta n B''}$$

have no solution in rational integers $b_1, \ldots, b_{n-1}$ and b_n ($\neq 0$) satisfying

$$B' \geqslant |b_n|, \quad B'' \geqslant \max_{1\leqslant j\leqslant n} |b_j|.$$

In fact Baker stated theorem B.2 for a linear form $b_1 \log \alpha_1+\cdots+b_n \log \alpha_n$ where all the logarithms have their principal values, but the result for $\alpha_1^{b_1}\cdots\alpha_n^{b_n}-1$ follows as indicated above.

Generalisations in another direction lead to p-adic analogues of such lower bounds. For this theory, which is called the p-adic theory of linear forms in logarithms, the reader is referred to a paper of van der Poorten (1977*a*). We state p-adic analogues of theorems B.1 and B.2 that we use in this tract.

Theorem B.3 (*van der Poorten, 1977a*). *Let $\mathfrak{p}$ be a prime ideal of K lying above a rational prime p. There exist computable absolute constants C_6 and C_7 such* *that*

$$\mathrm{ord}_{\wp}(\alpha_1^{b_1}\cdots\alpha_n^{b_n}-1)\leq (C_6 nd)^{C_7 n}\frac{p^d}{\log p}\,\Omega(\log B)^2$$

for all rational integers $b_1, \ldots, b_n$ with absolute values at most B (≥ 2) such that $\alpha_1^{b_1}\cdots\alpha_n^{b_n}\neq 1$.

Theorem B.4. (van der Poorten, 1977*a*). *Let $\wp$ be a prime ideal of K lying above a rational prime p. There exists a computable number C_8 depending only on n, d and A' such that for any δ with $0<\delta<1$,*

$$\mathrm{ord}_{\wp}(\alpha_1^{b_1}\cdots\alpha_n^{b_n}-1)\leq \max(C_8\log(B'\delta^{-1}p^d)p^d\log A, \delta B/B')$$

for all rational integers $b_1, \ldots, b_{n-1}$ and b_n with $b_n\not\equiv 0 \pmod p$ of absolute values at most B and B', respectively, such that $\alpha_1^{b_1}\cdots\alpha_n^{b_n}\neq 1$.

Putting $b_n=-1$, van der Poorten (1977*a*) derived the following result from theorem B.4.

Corollary B.2. *Let $\wp$ be a prime ideal of K lying above a rational prime p. Suppose that $b_1, \ldots, b_{n-1}$ and $b_n=-1$ are rational integers of absolute values at most B. There exists a computable number C_9 depending only on n, d and A' such that, for every δ with $0<\delta<1$, the inequality*

$$\mathrm{ord}_{\wp}(\alpha_1^{b_1}\cdots\alpha_n^{b_n}-1)>\delta B$$

implies that $\alpha_1^{b_1}\cdots\alpha_n^{b_n}=1$ or

$$B\leq C_9\delta^{-1}p^d\log(\delta^{-1}p^d)\log A.$$

Notes

To find out all the solutions of certain diophantine equations, it is necessary to give explicitly the constants occurring in the theorems of this chapter. The first result in this direction is due to Baker (1968*a*). In fact, Baker proved theorem B.1 and van der Poorten proved theorems B.3, B.4 and corollary B.2 with explicit constants. Other estimates were given by van der Poorten and Loxton (1976), Stewart (1977*b*), Mignotte and Waldschmidt (1978) and Waldschmidt (1980). Loxton (1986) proved a generalisation of theorem B.1 for systems of t linear forms in logarithms in which the factor Ω in the exponent of the upper bound is replaced by $\Omega^{1/t}$. In this direction, the first result is due to Ramachandra (1969). For linear forms with α_is close to 1, see Ramachandra and Shorey (1973), Ramachandra, Shorey and Tijdeman (1975, 1976), Shorey (1974*a*, *b*, 1986*a*) and Waldschmidt (1980). A trivial case of p-adic linear forms in logarithms is already given in lemma A.8.

C. Recurrence sequences

§1. A homogeneous linear recurrence sequence with constant coefficients (*recurrence sequence* for short) is a non-trivial sequence of complex numbers $\{u_m\}_{m=0}^{\infty}$ such that

$$u_{m+k} = v_{k-1}u_{m+k-1} + v_{k-2}u_{m+k-2} + \cdots + v_0 u_m \quad (m=0,1,2,\ldots) \quad \text{(C.1)}$$

for certain complex numbers $v_0, v_1, \ldots, v_{k-1}$ with $v_0 \neq 0$. A recurrence sequence is therefore completely determined by the *initial values* $u_0, \ldots, u_{k-1}$ and the *recurrence coefficients* $v_0, v_1, \ldots, v_{k-1}$. Note that $|u_0| + |u_1| + \cdots + |u_{k-1}| > 0$. A *recurrence of order k* is defined as a sequence of initial values $u_0, u_1, \ldots, u_{k-1}$, not all zero, and a sequence of recurrence coefficients $v_0, v_1, \ldots, v_{k-1}$ with $v_0 \neq 0$. A recurrence generates a recurrence sequence by the recurrence relation (C.1). A recurrence of order 2 is called *binary*; one of order 3 *ternary*.

The *companion polynomial* to a recurrence with coefficients $v_0, v_1, \ldots, v_{k-1}$ is given by

$$G(z) = z^k - v_{k-1}z^{k-1} \cdots - v_0. \quad \text{(C.2)}$$

Let

$$G(z) = \prod_{j=1}^{s} (z - \omega_j)^{\sigma_j}, \quad \text{(C.3)}$$

with distinct numbers $\omega_1, \omega_2, \ldots, \omega_s$, be the factorisation of G. We call $\omega_1, \omega_2, \ldots, \omega_s$ the *roots* of the recurrence. If all roots of G are simple, then we say that the recurrence is *simple*.

A recurrence sequence may satisfy different relations of the form (C.1). Suppose $\{u_m\}_{m=0}^{\infty}$ satisfies two recurrences of order k,

$$u_{m+k} = \sum_{j=0}^{k-1} \lambda_j u_{m+j} = \sum_{j=0}^{k-1} \mu_j u_{m+j} \quad (m=0,1,\ldots).$$

Let $r=\max\{j \mid \lambda_j \neq \mu_j\}$. Then

$$u_{m+r}=\sum_{j=0}^{r-1} -\frac{\lambda_j-\mu_j}{\lambda_r-\mu_r} u_{m+j} \quad (m=0, 1, \ldots),$$

which implies that $\{u_m\}_{m=0}^{\infty}$ satisfies a recurrence of order $r<k$. We conclude that for every recurrence sequence there is a unique recurrence of minimal order. If we speak of the order, the recurrence coefficients, the companion polynomial or the roots of a recurrence sequence or say that a recurrence sequence is simple, this is all meant with respect to this unique recurrence of minimal order.

The following result is fundamental in the theory of recurrence sequences.

Theorem C.1. *(a) Let* $\{u_m\}_{m=0}^{\infty}$ *be a sequence satisfying relation* (C.1) *with* $v_0 \neq 0$. *For* $j=1, 2, \ldots, s$ *let* ω_j *and* σ_j *be determined by* (C.2) *and* (C.3) *where the numbers* $\omega_1, \omega_2, \ldots, \omega_s$ *are distinct. Then there exist uniquely determined polynomials* $f_j \in \mathbb{Q}(u_0, u_1, \ldots, u_{k-1}, v_0, v_1, \ldots, v_{k-1}, \omega_1, \omega_2, \ldots, \omega_s)[z]$ *of degree less than* σ_j $(j=1, 2, \ldots, s)$ *such that*

$$u_m=\sum_{j=1}^{s} f_j(m)\omega_j^m \quad (m=0, 1, \ldots). \tag{C.4}$$

(b) Let $\omega_1, \omega_2, \ldots, \omega_s$ *be distinct complex numbers and* $\sigma_1, \sigma_2, \ldots, \sigma_s$ *positive integers with* $\sum_{j=1}^{s} \sigma_j = k$. *Define* $v_0, v_1, \ldots, v_{k-1}$ *by* (C.3) *and* (C.2). *For* $j=1, 2, \ldots, s$ *let* f_j *be a polynomial of degree less than* σ_j. *Then the sequence* $\{u_m\}_{m=0}^{\infty}$ *defined by* (C.4) *satisfies recurrence relation* (C.1).

Proof. (a) Put

$$u(z)=\sum_{m=0}^{\infty} u_m z^m,$$

$$A(z)=\sum_{i=0}^{k} a_i z^i = 1-\sum_{i=1}^{k} v_{k-i} z^i = \prod_{j=1}^{s} (1-\omega_j z)^{\sigma_j}. \tag{C.5}$$

Then, by (C.1),

$$\begin{aligned} u(z)A(z) &= \sum_{m=0}^{k-1} \sum_{j=0}^{m} a_j u_{m-j} z^m + \sum_{m=k}^{\infty} \sum_{j=0}^{k} a_j u_{m-j} z^m \\ &= \sum_{m=0}^{k-1} \sum_{j=0}^{m} a_j u_{m-j} z^m + \sum_{m=0}^{\infty} \left(u_{m+k} - \sum_{j=1}^{k} v_{k-j} u_{m+k-j} \right) z^{m+k} \\ &= \sum_{m=0}^{k-1} z^m \sum_{j=0}^{m} a_j u_{m-j}. \end{aligned}$$

Put $h_m=\sum_{j=0}^m a_j u_{m-j}$ for $m=0, 1, \ldots, k-1$. By resolution into partial fractions we obtain, from $\sum \sigma_j = k$,

$$u(z)=\frac{\sum_{m=0}^{k-1} h_m z^m}{\prod_{j=1}^{s}(1-\omega_j z)^{\sigma_j}}=\sum_{j=1}^{s}\sum_{i=1}^{\sigma_j}\frac{\beta_{ij}}{(1-\omega_j z)^i} \tag{C.6}$$

for certain numbers $\beta_{ij}\in\mathbb{Q}(u_0, u_1, \ldots, u_{k-1}, v_0, v_1, \ldots, v_{k-1}, \omega_1, \omega_2, \ldots, \omega_s)$ which are uniquely determined. We have $\omega_1\omega_2\cdots\omega_s\neq 0$ in view of $v_0\neq 0$. For $|z|<\min_j|\omega_j|^{-1}$ we have

$$\begin{aligned} u(z) &= \sum_{j=1}^{s}\sum_{i=1}^{\sigma_j}\beta_{ij}\sum_{m=0}^{\infty}\binom{-i}{m}(-\omega_j z)^m \\ &= \sum_{m=0}^{\infty}\left(\sum_{j=1}^{s}\sum_{i=1}^{\sigma_j}\beta_{ij}\binom{m+i-1}{m}\omega_j^m\right)z^m. \end{aligned}$$

Note that the Taylor coefficients are uniquely determined. On comparing coefficients we find (C.4) with f_j defined by

$$f_j(z)=\sum_{i=1}^{\sigma_j}\beta_{ij}\frac{(z+i-1)(z+i-2)\cdots(z+1)}{(i-1)!}\qquad (j=1,2,\ldots,s). \tag{C.7}$$

This proves part (*a*).

(*b*) Define $u(z)$ and $A(z)$ by (C.5) and complex numbers β_{ij} by (C.7). It follows that (C.6) holds for certain numbers $h_0, h_1, \ldots, h_{k-1}$. Hence

$$u(z)A(z)=\sum_{m=0}^{\infty}\sum_{\substack{i=0\\ i\leqslant k}}^{m} a_i u_{m-i} z^m=\sum_{m=0}^{k-1} h_m z^m$$

which implies that the left-hand side is a polynomial of degree less than k. Thus the coefficients of z^{m+k} vanish for $m\geqslant 0$, that is, by (C.5),

$$u_{m+k}-\sum_{i=1}^{k} v_{k-i}u_{m+k-i}=0\quad (m=0, 1, \ldots). \qquad \square$$

It is now easy to characterise all recurrence relations by which a fixed recurrence sequence $\{u_m\}_{m=0}^{\infty}$ can be generated. Let $G(z)=\prod_{j=1}^{s}(z-\omega_j)^{\sigma_j}$ be the companion polynomial to the minimal recurrence relation (C.1) of $\{u_m\}_{m=0}^{\infty}$. Theorem C.1(*a*) yields a representation

$$u_m=\sum_{j=1}^{s} f_j(m)\omega_j^m\quad (m=0, 1, \ldots).$$

Suppose that, moreover,

$$u_{m+l}=\mu_{l-1}u_{m+l-1}+\cdots+\mu_0 u_m\quad (m=0, 1, \ldots).$$

Let

$$g(z)=z^l-\mu_{l-1}z^{l-1}\cdots-\mu_0$$

be the companion polynomial to this recurrence relation. Then theorem C.1(*a*) yields a representation

$$u_m=\sum_{j=1}^{t} f_j^*(m)\omega_j^m \quad (m=0, 1,\ldots),$$

where we define $f_j^*=0$ for those j for which $g(\omega_j)\neq 0$ and $\omega_{s+1},\ldots,\omega_t$ are the roots of g which are not roots of G. We assert that $f_j=f_j^*$ for $1\leqslant j\leqslant s$, $f_j^*=0$ for $s<j\leqslant t$ and that g is divisible by G. To prove this, let

$$g(z)G(z)=\prod_{j=1}^{t}(z-\omega_j)^{\rho_j}=z^{k+l}-\lambda_{k+l-1}z^{k+l-1}\cdots-\lambda_0.$$

By theorem C.1(*b*) the sequence $\{u_m\}_{m=0}^{\infty}$ satisfies the recurrence relation

$$u_{m+k+l}=\sum_{j=0}^{k+l-1}\lambda_j u_{m+j} \quad (m=0, 1,\ldots).$$

By theorem C.1(*a*) applied to this recurrence relation, the polynomials f_j are unique, that is,

$$f_j=f_j^* \quad \text{for } j=1,2,\ldots,s, \quad f_j^*=0 \quad \text{for } j=s+1,\ldots,t.$$

Because of the minimality of the first recurrence and theorem C.1(*b*), the degree of f_j is exactly σ_j-1 for $j=1, 2, \ldots, s$. Hence the zero ω_j of g is of order at least σ_j for $j=1, 2, \ldots, s$. Thus G is a divisor of g.

§2. A recurrence is called *algebraic* (*rational*, *integral*) if all the initial values and recurrence coefficients are algebraic (rational, integral, respectively). If a recurrence is algebraic, then the resulting sequence is algebraic, etc. The converse need not be true. For example, the non-integral recurrence $u_{m+2}=\pi u_{m+1}+(1-\pi)u_m$ $(m=0, 1, \ldots)$ with $u_0=u_1=1$ generates a recurrence sequence of rational integers.

We shall show that, if the elements of a recurrence sequence $\{u_m\}_{m=0}^{\infty}$ belong to a field K, then the recurrence coefficients (of the minimal recurrence relation) of the sequence also belong to K. Let the minimal recurrence relation of $\{u_m\}_{m=0}^{\infty}$ be given by

$$u_{m+k}=v_{k-1}u_{m+k-1}+v_{k-2}u_{m+k-2}+\cdots+v_0u_m \quad (m=0, 1,\ldots). \quad \text{(C.8)}$$

Consider the system of k linear equations in the k variables $x_0, x_1, \ldots, x_{k-1}$,

$$\begin{cases} u_0x_0+u_1x_1+\cdots+u_{k-1}x_{k-1} &= u_k, \\ u_1x_0+u_2x_1+\cdots+u_kx_{k-1} &= u_{k+1}, \\ u_2x_0+u_3x_1+\cdots+u_{k+1}x_{k-1} &= u_{k+2}, \\ \vdots & \\ u_{k-1}x_0+u_kx_1+\cdots+u_{2k-2}x_{k-1} &= u_{2k-1}. \end{cases} \tag{C.9}$$

If the coefficient determinant vanishes, then there exist two distinct solutions $(v_0, v_1, \ldots, v_{k-1})$ and $(\mu_0, \mu_1, \ldots, \mu_{k-1})$. We show by induction on m that

$$u_{m+k}=\mu_{k-1}u_{m+k-1}+\mu_{k-2}u_{m+k-2}+\cdots+\mu_0u_m \quad (m=0, 1, \ldots). \tag{C.10}$$

According to (C.9) this is true for $m<k$. Suppose it has been shown for $m<M$. Then, by (C.8) and the induction hypothesis,

$$\begin{aligned} u_{M+k} &= \sum_{j=0}^{k-1} v_ju_{M+j} = \sum_{j=0}^{k-1} v_j \sum_{i=0}^{k-1} \mu_iu_{(M+j-k)+i} \\ &= \sum_{i=0}^{k-1} \mu_i \sum_{j=0}^{k-1} v_ju_{(M+i-k)+j} = \sum_{i=0}^{k-1} \mu_iu_{M+i}. \end{aligned}$$

This proves (C.10). By an argument given in §1 it follows from (C.8) and (C.10) that the order of $\{u_m\}_{m=0}^{\infty}$ is less than k, a contradiction. Thus the coefficient determinant of (C.9) does not vanish. Then we can apply Cramer's rule to express the solution $v_0, v_1, \ldots, v_{k-1}$ as quotients of sums of products of terms of the recurrence sequence, hence as elements of K.

In particular, a recurrence sequence of algebraic numbers is an algebraic recurrence sequence and a recurrence sequence of rational numbers is a rational recurrence sequence. It follows from a theorem of Fatou that a recurrence sequence of rational integers has a rational integer recurrence. Fatou (1906) proved the following assertion (cf. Pólya and Szegö, 1925, Problem VIII 156).

Let $u(z)$ be a rational function whose Taylor series has rational integer coefficients. Then $u(z)$ can be written in the form $f(z)/g(z)$, where f and g are polynomials with rational integer coefficients and $g(0)=1$.

Let $\{u_m\}_{m=0}^{\infty}$ be a recurrence sequence of rational integers of order k with minimal recurrence relation (C.1). Then, by (C.6) and (C.5) in the notation of

§ 1,

$$u(z)=\sum_{m=0}^{\infty} u_m z^m = H(z)/A(z) = \left(\sum_{m=0}^{k-1} h_m z^m\right)\Big/\left(1-\sum_{i=1}^{k} v_{k-i} z^i\right)$$

is a rational function whose Taylor coefficients are rational integers. We know already that $v_0, v_1, \ldots, v_{k-1}$ hence $h_0, h_1, \ldots, h_{k-1}$ are rational numbers. Let T be the smallest positive integer such that $Th_m \in \mathbb{Z}$ and $Tv_m \in \mathbb{Z}$ for $m=0, 1, \ldots, k-1$. Since A is the reciprocal of the companion polynomial to the minimal recurrence relation, H and A have no common non-constant factor. Hence Fatou's result implies that $TA(0) \mid Th_m$ and $TA(0) \mid Tv_m$ for $m=0, 1, \ldots, k-1$. Since $A(0)=1$, this implies that $v_0, v_1, \ldots, v_{k-1}$ are all rational integers.

Some binary integer sequences are so important that they have special names. The *Fibonacci sequence* is the sequence defined by $u_0=0$, $u_1=1$, $u_{m+2}=u_{m+1}+u_m$ for $m=0, 1, 2, \ldots$. A sequence is called a *Lucas sequence* (*of the first* or *second kind*, respectively) if

$$u_m=\frac{\alpha^m-\beta^m}{\alpha-\beta} \quad \text{for } m=0, 1, 2, \ldots$$

or

$$u_m=\alpha^m+\beta^m \quad \text{for } m=0, 1, 2, \ldots,$$

where $\alpha+\beta$ and $\alpha\beta$ are relatively prime non-zero rational integers and α/β is not a root of unity. Note that the Fibonacci sequence is the Lucas sequence of the first kind with $\alpha, \beta=\frac{1}{2}\pm\frac{1}{2}\sqrt{5}$, and that Lucas sequences are binary rational integer recurrence sequences. Lucas sequences of the first kind satisfy

$$n \mid m \Rightarrow u_n \mid u_m. \tag{C.11}$$

For Lucas sequences of the second kind a similar relation holds. A sequence is called a *Lehmer sequence* (*of the first* or *second kind*, respectively) if

$$u_m=\frac{\alpha^m-\beta^m}{\alpha-\beta} \quad \text{for } m \text{ odd}, \quad u_m=\frac{\alpha^m-\beta^m}{\alpha^2-\beta^2} \quad \text{for } m \text{ even},$$

or

$$u_m=\frac{\alpha^m+\beta^m}{\alpha+\beta} \quad \text{for } m \text{ odd}, \quad u_m=\alpha^m+\beta^m \quad \text{for } m \text{ even},$$

where $(\alpha+\beta)^2$ and $\alpha\beta$ are relatively prime non-zero rational integers and α/β is not a root of unity. Note that Lehmer sequences are Lucas sequences if $\alpha+\beta=\pm 1$. If $\alpha+\beta\neq\pm 1$, Lehmer sequences are rational integer recurrence sequences of order 4 with roots $\pm\alpha$, $\pm\beta$.

§ 3. Several results in this monograph deal with the number of times that a recurrence sequence attains a certain value. The *a-multiplicity* of a sequence $\{u_m\}_{m=0}^{\infty}$ is defined as the number of indices m such that $u_m = a$. The *multiplicity* of a sequence is defined as the supremum of the a-multiplicities taken over all a. The *total multiplicity* of $\{u_m\}_{m=0}^{\infty}$ is defined as the number of pairs (m, n) with $m > n$ such that $u_m = u_n$. The following theorem and its corollary give properties of a recurrence sequence with infinite 0-multiplicity.

Theorem C.2 (Skolem–Mahler–Lech). *If $\{u_m\}_{m=0}^{\infty}$ is a recurrence sequence with infinite 0-multiplicity, then those m for which $u_m = 0$ form a finite union of arithmetic progressions after a certain stage.*

Proof. Lech (1953).

Corollary C.1. *If a recurrence with companion polynomial* (C.3) *generates a sequence with infinite 0-multiplicity, then ω_i/ω_j is a root of unity for some indices i, j with $i \neq j$.*

Proof. In § 1 it was proved that the companion polynomial of a recurrence is divisible by the companion polynomial of the corresponding recurrence of minimal order. Hence we may assume that the recurrence in corollary C.1 is of minimal order. Theorem C.2 implies the existence of positive numbers b and c such that $u_{b+mc} = 0$ for $m = 0, 1, \ldots$. Hence, in terms of the representation (C.4) of u_m,

$$0 = \sum_{j=1}^{s} f_j(b+mc)\omega_j^{b+mc} = \sum_{j=1}^{s} f_j(b+mc)\omega_j^{b}(\omega_j^{c})^m. \tag{C.12}$$

Because of the minimality of the recurrence, the polynomial $f_j(b+xc)\omega_j^b$ in x is non-trivial and of degree $\sigma_j - 1$ for $j = 1, 2, \ldots, s$. Since the generalised power sum at the right-hand side of (C.12) equals 0 for every m, it follows from theorem C.1 that the numbers $\omega_1^c, \omega_2^c, \ldots, \omega_s^c$ are not distinct. Thus $(\omega_i/\omega_j)^c = 1$ for some indices i, j with $i \neq j$. □

A recurrence sequence is called *degenerate* if its companion polynomial has two distinct roots whose ratio is a root of unity and *non-degenerate* otherwise. Every degenerate sequence $\{u_m\}_{m=0}^{\infty}$ can be split into subsequences $\{u_{b+mc}\}_{m=0}^{\infty}$ for $b = 0, 1, \ldots, c-1$, such that each subsequence is either trivial or a non-degenerate recurrence sequence. Here c can be taken as the least common multiple of the orders of those roots of unity which occur as ratio of two distinct roots. It is therefore often sufficient to study the multiplicities of non-degenerate recurrence sequences.

Notes

We do not know a satisfactory introduction to the aspects of recurrence sequences which are relevant for this tract. Some basic concepts and techniques can be found in Lewis (1969). Pólya (1921) wrote an important paper on the prime factors of the numerators and denominators of the terms of a rational recurrence sequence. Loxton and van der Poorten (1977) stated a number of results and conjectures on the rate of growth of the terms of a recurrence sequence and the size of the greatest prime factor of the terms. Stewart (1986) wrote a survey of effective results on the greatest prime factor of terms of recurrence sequences. Results on multiplicities of such sequences can be found in the first part of Tijdeman (1981). See, further, Cerlienco, Mignotte and Piras (1984), LeVeque (1974, §§ B36, B40, B44), Montel (1957) and the notes of chapter 4.

We stress that it is relation (C.4) that makes it possible to apply the Thue–Siegel–Roth–Schmidt method and the theory of linear forms in logarithms to recurrence sequences.

CHAPTER 1

Purely exponential equations

In this chapter, we investigate equations

$$x+y=z$$

in algebraic integers x, y, z from a fixed algebraic number field such that $[xyz]$ is composed of prime ideals from a given finite set.

Let $P \geqslant 3$. Let $p_1, \ldots, p_s$ be given (rational) prime numbers with $s \geqslant 1$ and $0 < p_1 < \cdots < p_s \leqslant P$. Denote by S the set of all rational integers composed of $p_1, \ldots, p_s$. In particular $-1 \in S$, $0 \notin S$, $1 \in S$. Denote by S_+ the set of all positive integers of S and arrange them in the increasing order,

$$n_1 < n_2 < n_3 < \cdots.$$

Then corollary B.1 can be applied to prove

Theorem 1.1 (Tijdeman, 1973). *There exists a computable number C_1 depending only on P such that*

$$n_{i+1} - n_i \geqslant \frac{n_i}{(\log n_i)^{C_1}} \quad \textit{for } n_i \geqslant 3.$$

Theorem 1.1 admits the following consequence which Cassels (1960*b*) derived from a result of Gelfond (1940) on p-adic linear forms in logarithms.

Corollary 1.1. *For a fixed non-zero rational integer k and $x, y \in S_+$, the equation*

$$x - y = k$$

implies that

$$\max(x, y) \leqslant C_2$$

for a certain computable number C_2 depending only on P and k.

An ineffective version of corollary 1.1 is a consequence of a theorem of Thue (1909). See Pólya (1918). For integers $a > 1$, $b > 1$ and $k \neq 0$ it follows from corollary 1.1 that all non-negative integers m and n satisfying

$$a^m - b^n = k$$

are bounded by a computable number depending only on k and $P(ab)$, the greatest prime divisor of ab.

On applying theorem B.3 we obtain the following result.

Theorem 1.2. *Let z be a non-zero rational integer. Suppose $x, y \in S$ with $(x, y, z) = 1$, $|x| < |y|$ and $|y| \geqslant 3$ satisfy*

$$x \equiv y \pmod{z}. \qquad (1)$$

Then there exists a computable absolute constant C_3 such that

$$\log |z| \leqslant (s \log P)^{C_3 s} (\log \log |y|)^2 P(z) \omega(z).$$

Recall that $P(z)$ denotes the greatest prime factor of z and $\omega(z)$ is the number of distinct prime factors of z. It involves no loss of generality to assume $k > 0$ in corollary 1.1. Then corollary 1.1 follows from theorem 1.2 applied to $k \equiv -y (\mathrm{mod}\ x)$. Further, by combining theorems 1.2 and 1.1, we obtain the following generalisation of corollary 1.1.

Corollary 1.2. *Let $x, y \in S$ with $(x, y) = 1$, $|x| < |y|$ and $|y| \geqslant 3$. Then*

$$P(x+y) \geqslant C_4 \left(\frac{\log |y|}{\log \log |y|} \right)^{1/2}$$

where $C_4 > 0$ is a computable number depending only on P. In particular, the equation

$$x + y = z \qquad (2)$$

in x, y, $z \in S$ and $(x, y, z) = 1$, implies that $\max(|x|, |y|, |z|)$ is bounded by a computable number depending only on P.

An ineffective version of the latter assertion follows from a result of Mahler (1933*a*) on the greatest prime factor of a binary form. An effective proof of it is due to Coates (1969, 1970*a*) and Sprindžuk (1969).

Let K be a finite extension of degree d over $\mathbb{Q}$. Denote by $\mathcal{O}_K$ the ring of integers of K. Assume that $\pi_1, \ldots, \pi_s \in \mathcal{O}_K$ are non-zero non-units. Denote by $\mathcal{S}'$ the set of all the products of units of $\mathcal{O}_K$ and of powers of $\pi_1, \ldots, \pi_s$ with non-negative exponents. Then the following analogue of the second part of corollary 1.2 can be derived for algebraic number fields.

Theorem 1.3. *Let $\tau \geqslant 0$. Let $x_1, x_2 \in \mathscr{S}'$ satisfy*

$$\min(\operatorname{ord}_{\mathfrak{p}}(x_1), \operatorname{ord}_{\mathfrak{p}}(x_2)) \leqslant \tau \tag{3}$$

for every prime ideal $\mathfrak{p}$ in $\mathcal{O}_K$. Suppose

$$x_3 = \pm \pi_1^{v_1} \cdots \pi_s^{v_s} \tag{4}$$

where $v_1, \ldots, v_s$ are non-negative integers. If

$$x_1 + x_2 + x_3 = 0, \tag{5}$$

then

$$\max(\overline{|x_1|}, \overline{|x_2|}, \overline{|x_3|}) \leqslant C_5$$

for some computable number C_5 depending only on τ, K and $\mathscr{S}'$.

Next we shall formulate a quantitative result which implies theorem 1.3 and the second part of corollary 1.2 and which will be applied in chapter 7. We need some further notation. Assume that $\mathfrak{p}_1, \ldots, \mathfrak{p}_t$ are distinct prime ideals in $\mathcal{O}_K$. Denote by $\mathscr{S}$ the set of all non-zero elements of $\mathcal{O}_K$ which have no prime ideal divisors different from $\mathfrak{p}_1, \ldots, \mathfrak{p}_t$. Note that in case $t = 0$ the set $\mathscr{S}$ is just the group of units of K. Further, if $\mathfrak{p}_1, \ldots, \mathfrak{p}_t$ are all the prime ideal divisors of $\pi_1, \ldots, \pi_s$, then $\mathscr{S} \supset \mathscr{S}'$. Suppose that the rational primes divisible by $\mathfrak{p}_1, \ldots, \mathfrak{p}_{t-1}$ or $\mathfrak{p}_t$ do not exceed $P (\geqslant 3)$. Let h and R be the class number and regulator of K, respectively. The following result was proved by Győry (1979*a*) even without the factor $(\log \log A)^2$ in the exponent in (7).

Theorem 1.4. *Let α_1, α_2 and α_3 be non-zero elements of $\mathcal{O}_K$ with $\overline{|\alpha_i|} \leqslant A\,(\geqslant 3)$ for $i = 1, 2, 3$. If*

$$\alpha_1 x_1 + \alpha_2 x_2 + \alpha_3 x_3 = 0 \quad \text{for } x_1, x_2, x_3 \in \mathscr{S}, \tag{6}$$

then $x_i = \eta \rho_i$ for some $\eta \in \mathscr{S}$ and $\rho_i \in \mathscr{S}$ $(i = 1, 2, 3)$ such that

$$\max_{i=1,2,3} \overline{|\rho_i|} \leqslant \exp\{(C_6(t+1) \log P)^{C_7(t+1)} P^d \log A (\log \log A)^2\} \tag{7}$$

where C_6 and C_7 are computable numbers such that C_6 depends only on d, h and R, and C_7 only on d.

A non-zero element α of K will be called an $\mathscr{S}$-unit if $\operatorname{ord}_{\mathfrak{p}}(\alpha) = 0$ for all prime ideals $\mathfrak{p}$ apart from $\mathfrak{p}_1, \ldots, \mathfrak{p}_t$. Note that an $\mathscr{S}$-unit is the quotient of two elements from $\mathscr{S}$, and conversely. Denote by $U_{\mathscr{S}}$ the set of $\mathscr{S}$-units of K. It follows from (A.19) that $U_{\mathscr{S}}$ is a multiplicative subgroup of K^* which contains the group of units of K^* as a subgroup.

Let α, β and γ be non-zero elements of $\mathcal{O}_K$. Theorem 1.4 implies that the projective line $\alpha x+\beta y=\gamma z$ has only finitely many points (x, y, z) with $x, y, z \in \mathcal{S}$. An equivalent statement is that the equation

$$\alpha x+\beta y=\gamma \quad \text{in } x, y \in U_{\mathcal{S}} \tag{8}$$

has only finitely many solutions. Equation (8) is called an (inhomogeneous) *$\mathcal{S}$-unit equation* (in two variables). If $t=0$, hence x and y are units of $\mathcal{O}_K$, equation (8) is called a *unit equation.* Ineffective versions of the above-mentioned finiteness assertions concerning (6) and (8) can be deduced from the results of Siegel (1921) (in case $t=0$) and Parry (1950) (in the general case). The first effective variants are due to Baker (1968*b*) in case $t=0$ and Coates (1969, 1970*a*) and Sprindžuk (1969) in the general case.

Let H ($\geqslant 3$) be an upper bound for the heights of α, β and γ and denote by $H(x)$ and $H(y)$ the heights of x and y, respectively. We shall derive the following estimate for the solutions of (8).

Corollary 1.3. *All solutions $x, y \in U_{\mathcal{S}}$ of (8) satisfy*

$$\max\{H(x), H(y)\} \leqslant \exp\{(C_8(t+1)\log P)^{C_9(t+1)} P^d \log H(\log\log H)^2\} \tag{9}$$

where C_8 and C_9 are computable numbers such that C_8 depends only on d, h and R, and C_9 only on d.

This result is due to Kotov and Trelina (1979). The corollary without the factor $(\log\log H)^2$ follows from the independently proved theorem of Győry (1979*a*) mentioned above.

Proofs

Proof of theorem 1.1. Since $n_i \geqslant 3$, we may assume $n_{i+1} \leqslant 2n_i$. Write

$$n_i = p_1^{a_1} \cdots p_s^{a_s}, \quad n_{i+1} = p_1^{b_1} \cdots p_s^{b_s}$$

where a_k and b_k with $1 \leqslant k \leqslant s$ are non-negative integers. Then

$$\frac{n_{i+1}}{n_i} - 1 = p_1^{b_1-a_1} \cdots p_s^{b_s-a_s} - 1.$$

Observe that

$$\max_{1\leqslant k\leqslant s} |b_k - a_k| \leqslant \max_{1\leqslant k\leqslant s} (b_k, a_k) \leqslant 2\log n_{i+1} \leqslant 4\log n_i.$$

Apply corollary B.1 with $n=s, d=1, A_1=A_2=\cdots=A_s=P$ and $B=4\log n_i$

to conclude that

$$\frac{n_{i+1}}{n_i} - 1 \geqslant (\log n_i)^{-c_1}$$

for some computable number c_1 depending only on P. □

Proof of theorem 1.2. We may assume $|z| \geqslant 3$. Write

$$x = \pm p_1^{a_1} \cdots p_s^{a_s}, \quad y = \pm p_1^{b_1} \cdots p_s^{b_s},$$

where a_k and b_k with $1 \leqslant k \leqslant s$ are non-negative integers. For a prime p dividing z, it follows from (1) and $(x, y, z) = 1$ that

$$\operatorname{ord}_p(z) \leqslant \operatorname{ord}_p(x - y) = \operatorname{ord}_p((x/y) - 1). \tag{10}$$

Notice that

$$x/y - 1 = \pm p_1^{a_1 - b_1} \cdots p_s^{a_s - b_s} - 1.$$

Observe that

$$\max_{1 \leqslant k \leqslant s} |a_k - b_k| \leqslant \max_{1 \leqslant k \leqslant s} \max(a_k, b_k) \leqslant 2 \log |y|.$$

We apply theorem B.3 with $n \leqslant s + 1$, $d = 1$, $A_1 = A_2 = \cdots = A_n = P$ and $B = 2 \log |y|$ to obtain

$$\operatorname{ord}_p((x/y) - 1) \leqslant \frac{p}{\log p} (s \log P)^{c_2 s} (\log \log |y|)^2 \tag{11}$$

for some computable absolute constant c_2. Now

$$\log |z| = \sum_{p|z} \operatorname{ord}_p(z) \log p. \tag{12}$$

Hence, by (12), (10), (11) and $\sum_{p|z} p \leqslant P(z)\omega(z)$,

$$\log |z| \leqslant (s \log P)^{c_2 s} (\log \log |y|)^2 P(z) \omega(z).$$ □

Proof of corollary 1.2. We denote by $c_3, \ldots, c_6$ computable numbers depending only on P. We may assume that $|y| \geqslant c_3$, where c_3 is some large constant. Put $z = x + y$. Then corollary 1.1 implies that $|z| \geqslant 2$, hence $P(z) \geqslant 2$. By (N.1) we have $\omega(z) \leqslant \pi(P(z)) \leqslant 2P(z)/\log P(z)$. Further $s \leqslant P$. By theorem 1.2 applied to the congruence $x \equiv -y \pmod z$ we have

$$\log |x + y| \leqslant c_4 (\log \log |y|)^2 (P(z))^2 / \log P(z).$$

On the other hand, we have, by theorem 1.1,

$$\log|x+y| \geqslant \tfrac{1}{2}\log|y|$$

when c_3 is chosen sufficiently large. Thus

$$\frac{\log|y|}{(\log\log|y|)^2} \leqslant 2c_4 \frac{(P(z))^2}{\log P(z)}.$$

We want to get rid of the factor $\log P(z)$. To do this, we apply a standard technique which we shall refer to as *transferring secondary factors to the other side.* A rough estimation gives

$$(\log|y|)^{1/2} \leqslant (P(z))^{c_5}.$$

By taking the logarithm and multiplying on both sides we obtain

$$\frac{1}{2}\frac{\log|y|}{\log\log|y|} \leqslant 2c_4c_5(P(z))^2.$$

Hence

$$P(x+y) \geqslant c_6\left(\frac{\log|y|}{\log\log|y|}\right)^{1/2},$$

which is the first assertion. The second statement is an immediate consequence. □

Proof of theorem 1.4. By $c_7, \ldots, c_{36}$ we shall denote computable positive numbers depending only on d, h and R, and by $e_1, \ldots, e_8$ computable positive numbers depending only on d.

By lemma A.9 we may write $\not{p}_j^h = [\pi_j]$ where $\pi_j \in \mathcal{O}_K$ for $j = 1, \ldots, t$. Further, by corollary A.4 and lemma A.15, we may assume that

$$\overline{|\pi_j|} \leqslant P^{c_7} \quad (j=1,\ldots,t). \tag{13}$$

Hence, by lemma A.2,

$$H(\pi_j) \leqslant P^{c_8} \quad (j=1,\ldots,t). \tag{14}$$

In view of lemmas A.12 and A.15 we may write

$$x_i = \varepsilon_i \gamma_i \pi_1^{u_{i1}} \cdots \pi_t^{u_{it}} \quad (i=1,2,3) \tag{15}$$

where the u_{ij} are non-negative rational integers, the ε_i are units in $\mathcal{O}_K$ and $\gamma_i \in \mathcal{O}_K$ satisfies

$$\overline{|\gamma_i|} \leqslant P^{c_9(t+1)} \quad (i=1,2,3). \tag{16}$$

Put $a_j = \min_i u_{ij}$ and $v_{ij} = u_{ij} - a_j$ for $i = 1, 2, 3$ and $j = 1, \ldots, t$. Put $V =$

$\max_{i,j} v_{ij}$. By permuting $\pi_1, \ldots, \pi_t$ and x_1, x_2, x_3, we may secure $V = v_{11}$. Then $v_{21}v_{31} = 0$. Now by interchanging x_2 and x_3, if necessary, we obtain $v_{31} = 0$. Let r_1 and $2r_2$ be the number of conjugate fields of K which are real and non-real, respectively. We shall signify the conjugates of any element α of K by $\alpha^{(1)}, \ldots, \alpha^{(d)}$ with $\alpha^{(1)}, \ldots, \alpha^{(r_1)}$ real and $\alpha^{(r_1+1)}, \ldots, \alpha^{(r_1+r_2)}$ the complex conjugates of $\alpha^{(r_1+r_2+1)}, \ldots, \alpha^{(r_1+2r_2)}$, respectively. Put $r = r_1 + r_2 - 1$. Let $\eta_1, \ldots, \eta_r$ be an independent system of units for K satisfying (A.45). Then we have, by lemma A.2,

$$\overline{|\eta_l|} \leqslant c_{10} \quad \text{and} \quad H(\eta_l) \leqslant c_{11} \quad (l = 1, \ldots, r). \tag{17}$$

Further, we may write, by corollary A.5,

$$\varepsilon_1/\varepsilon_3 = \varepsilon_1' \eta_1^{w_{11}} \cdots \eta_r^{w_{1r}}, \quad \varepsilon_2/\varepsilon_3 = \varepsilon_2' \eta_1^{w_{21}} \cdots \eta_r^{w_{2r}} \tag{18}$$

with $w_{il} \in \mathbb{Z}$ $(i = 1, 2;\ l = 1, \ldots, r)$ and $\varepsilon_1', \varepsilon_2'$ units in $\mathcal{O}_K$ such that

$$\max(\overline{|\varepsilon_1'|}, \overline{|\varepsilon_2'|}) \leqslant c_{12}. \tag{19}$$

Put $\varepsilon_3' = 1$ and $\gamma_i' = \varepsilon_i' \gamma_i$ for $i = 1, 2, 3$. Then(16) and (19) together with lemma A.2 imply

$$\overline{|\gamma_i'|} \leqslant P^{c_{13}(t+1)} \quad \text{and} \quad H(\gamma_i') \leqslant P^{c_{14}(t+1)} \quad (i = 1, 2, 3). \tag{20}$$

Consequently $x_i = \eta \rho_i$ $(i = 1, 2, 3)$, where $\eta = \varepsilon_3 \pi_1^{a_1} \cdots \pi_t^{a_t}$,

$$\rho_i = \gamma_i' \eta_1^{w_{i1}} \cdots \eta_r^{w_{ir}} \pi_1^{v_{i1}} \cdots \pi_t^{v_{it}} \quad (i = 1, 2, 3) \tag{21}$$

and

$$w_{31} = \cdots = w_{3r} = 0, \tag{22}$$

by definition. It is clear that $\eta \in \mathcal{S}$ and $\rho_1, \rho_2, \rho_3 \in \mathcal{S}$. We shall show that (7) holds which will complete the proof.

By (6) we have

$$\Gamma := -\frac{\alpha_2 \rho_2}{\alpha_3 \rho_3} - 1 = \frac{\alpha_1 \rho_1}{\alpha_3 \rho_3} \neq 0. \tag{23}$$

Hence,

$$\Gamma = -\frac{\gamma_2'}{\gamma_3'} \eta_1^{w_{21}} \cdots \eta_r^{w_{2r}} \pi_1^{v_{21} - v_{31}} \cdots \pi_t^{v_{2t} - v_{3t}} \frac{\alpha_2}{\alpha_3} - 1. \tag{24}$$

We are going to derive an upper bound for $H := \max(V, W)$ where $W = \max_{i,j} |w_{ij}|$. Suppose that

$$H > c_{15}^2 (t+1) \log P \log A \tag{25}$$

for some sufficiently large constant $c_{15} > 1$.

First suppose

$$V \geqslant (c_{15}(t+1) \log P)^{-1} H. \tag{26}$$

We have, by lemmas A.7 and A.2, $\mathrm{ord}_{\not{p}_1}(\alpha_3) \leqslant c_{16} \log A$. Hence, by (23), $v_{11} = V$, $v_{31} = 0$, (26) and (25),

$$\mathrm{ord}_{\not{p}_1}(\Gamma) \geqslant V - c_{16} \log A \geqslant c_{17}(t+1)^{-1}(\log P)^{-1} H \tag{27}$$

where we may take $c_{17} = (2c_{15})^{-1}$ provided that c_{15} is sufficiently large. By lemma A.3 and (20) we have for the heights of $-\gamma_2'/\gamma_3'$ and α_2/α_3,

$$H(-\gamma_2'/\gamma_3') \leqslant P^{c_{18}(t+1)}, \quad H(\alpha_2/\alpha_3) \leqslant A^{e_1}. \tag{28}$$

By applying theorem B.3 to $\mathrm{ord}_{\not{p}_1}(\Gamma)$ and using (24), (28), (17) and (14) we obtain

$$\mathrm{ord}_{\not{p}_1}(\Gamma) \leqslant (c_{19}(t+1) \log P)^{e_2(t+1)} P^d \log A (\log H)^2. \tag{29}$$

We infer from (27) and (29) that

$$\frac{H}{(\log H)^2} \leqslant (c_{20}(t+1) \log P)^{e_3(t+1)} P^d \log A.$$

Hence, by transferring secondary factors to the other side,

$$H \leqslant (c_{21}(t+1) \log P)^{e_4(t+1)} P^d \log A (\log \log A)^2. \tag{30}$$

Now suppose, in place of (26),

$$V < (c_{15}(t+1) \log P)^{-1} H. \tag{31}$$

Notice that $V < H$ and so $H = W$. If $r = 0$, we can take $W = 0$. Thus $r \geqslant 1$ and therefore $d > 1$. In this case, instead of $V = v_{11}$, we may assume that $W = \max_{1 \leqslant l \leqslant r} |w_{1l}|$. This, in view of (22), is possible by permuting ρ_1 and ρ_2. We have, by (21),

$$w_{11} \log |\eta_1^{(k)}| + \cdots + w_{1r} \log |\eta_r^{(k)}| \\ = \log |\rho_1^{(k)}| - \log |\gamma_1'^{(k)}| - \sum_{j=1}^{t} v_{1j} \log |\pi_j^{(k)}| \tag{32}$$

for $k = 1, \ldots, r$. Assume that the right-hand side attains its maximum absolute value when $k = \kappa$ $(1 \leqslant \kappa \leqslant r)$. Consider (32) as a system of r linear equations in r unknowns $w_{11}, \ldots, w_{1r}$. Its determinant E is non-zero. Solving for w_{1l}, it follows from (17) and $|E| \geqslant R$ that

$$W \leqslant c_{22} \left\{ |\log |\rho_1^{(\kappa)}|| + |\log |\gamma_1'^{(\kappa)}|| + \sum_{j=1}^{t} v_{1j} |\log |\pi_j^{(\kappa)}|| \right\}.$$

Hence, by (20), (13) and a Liouville-type argument,

$$|\log |\rho_1^{(\kappa)}|| \geqslant c_{23} W - c_{24}(t+1) \log P - c_{25}(t+1) V \log P.$$

On taking c_{15} large enough we infer from $H=W$, (25) and (31) that

$$|\log|\rho_1^{(\kappa)}||\geqslant c_{26}H.$$

By (21), (20), (13) and (31) we have

$$\begin{aligned}\log|N(\rho_1)|&=\log|N(\gamma_1')|+\sum_{j=1}^{t}v_{1j}\log|N(\pi_j)|\\&\leqslant c_{27}(V+1)(t+1)\log P<2c_{27}H/c_{15}.\end{aligned}$$

By making c_{15} large enough, we secure that $2c_{27}/c_{15}<c_{26}/d$. Hence, if $\log|\rho_1^{(\kappa)}|>0$, we have

$$\sum_{\substack{k=1\\k\neq\kappa}}^{d}\log|\rho_1^{(k)}|<c_{26}\frac{H}{d}-\log|\rho_1^{(\kappa)}|\leqslant-\frac{(d-1)}{d}c_{26}H.$$

Consequently there is a λ with $1\leqslant\lambda\leqslant d$ such that

$$\log|\rho_1^{(\lambda)}|<-c_{26}\frac{H}{d}. \tag{33}$$

Using a Liouville-type argument we deduce

$$\log\left|\frac{\alpha_1^{(\lambda)}}{\alpha_3^{(\lambda)}\rho_3^{(\lambda)}}\right|\leqslant\log A+d\log\left|\overline{\alpha_3\rho_3}\right|. \tag{34}$$

By (21), (20), (22) and (13) we find

$$\log\left|\overline{\rho_3}\right|\leqslant c_{27}(t+1)(V+1)\log P. \tag{35}$$

From (23), (33), (34) and (35) we obtain

$$\log|\Gamma^{(\lambda)}|\leqslant-c_{26}\frac{H}{d}+c_{28}(\log A+(t+1)(V+1)\log P).$$

By taking c_{15} large enough we see from (23), (25) and (31) that

$$0<|\Gamma^{(\lambda)}|<\exp\{-c_{29}H\}. \tag{36}$$

We are going to apply corollary B.1 to $\Gamma^{(\lambda)}$, where we use representation (24) for Γ. On using the estimates (28), (17) and (14) to estimate the heights of the factors, we obtain

$$|\Gamma^{(\lambda)}|\geqslant\exp\{-(c_{30}(t+1)\log P)^{e_5(t+1)}\log A\log H\}. \tag{37}$$

Combining (36) and (37) we find

$$\frac{H}{\log H}<(c_{31}(t+1)\log P)^{e_5(t+1)}\log A.$$

This yields, after transferring the secondary factor to the right,

$$H<(c_{32}(t+1)\log P)^{e_6(t+1)}\log A\log\log A. \tag{38}$$

Collecting (25), (30) and (38) we obtain unconditionally

$$H<(c_{33}(t+1)\log P)^{e_7(t+1)}P^d\log A(\log\log A)^2. \tag{39}$$

By (21), (20), (17) and (13) we have

$$\log\overline{|\rho_i|}\leqslant c_{34}(t+1)(V+1)\log P+c_{35}W\quad(i=1,2,3).$$

By $H=\max(V,W)$ and (39) we may conclude

$$\max_{i=1,2,3}\log\overline{|\rho_i|}\leqslant(c_{36}(t+1)\log P)^{e_8(t+1)}P^d\log A(\log\log A)^2. \qquad\square$$

Proof of corollary 1.3. Let $x, y\in U_{\mathscr{S}}$ be a solution of (8). By c_{37}, c_{38}, c_{39} we shall denote computable positive numbers depending only on d, h and R and by e_9 and e_{10} computable positive numbers depending only on d. By lemma A.9 we have $\mathfrak{p}_j^h=[\pi_j]$ with $\pi_j\in\mathcal{O}_K$ for $j=1,\dots,t$. If $u_1,\dots,u_t$ are sufficiently large rational integers and $\mu=\pi_1^{u_1}\cdots\pi_t^{u_t}$, then

$$\mathrm{ord}_{\mathfrak{p}_j}(\mu x)\geqslant 0,\quad \mathrm{ord}_{\mathfrak{p}_j}(\mu y)\geqslant 0\quad(j=1,\dots,t).$$

By (A.19) we have $\mathrm{ord}_{\mathfrak{p}}(\mu x)=\mathrm{ord}_{\mathfrak{p}}(\mu y)=0$ for all prime ideals $\mathfrak{p}$ different from $\mathfrak{p}_1,\dots,\mathfrak{p}_t$. Hence $\mu x, \mu y, \mu\in\mathscr{S}$. We have, by lemma A.1,

$$\max(\overline{|\alpha|},\overline{|\beta|},\overline{|\gamma|})\leqslant H^{e_9}.$$

Further, (8) implies

$$\alpha(\mu x)+\beta(\mu y)-\gamma\mu=0.$$

By theorem 1.4 there exist $\eta\in\mathscr{S}$ and $\rho_1,\rho_2,\rho_3\in\mathscr{S}$ such that $\mu x=\eta\rho_1$, $\mu y=\eta\rho_2$, $\mu=\eta\rho_3$ and

$$\log\max_{i=1,2,3}\overline{|\rho_i|}\leqslant(c_{37}(t+1)\log P)^{e_{10}(t+1)}P^d\log H(\log\log H)^2.$$

Since $x=\rho_1/\rho_3$ and $y=\rho_2/\rho_3$, we obtain, by lemmas A.3 and A.2,

$$\begin{aligned}\max(H(x),H(y))&\leqslant\exp\{c_{38}\log\max_{i=1,2,3}\overline{|\rho_i|}\}\\&\leqslant\exp\{(c_{39}(t+1)\log P)^{e_{10}(t+1)}P^d\log H(\log\log H)^2\}.\qquad\square\end{aligned}$$

Proof of theorem 1.3. Suppose (5) holds for x_1, x_2, x_3 as specified in theorem 1.3. By $c_{40}, c_{41},\dots,c_{51}$ we shall denote computable positive constants depending only on τ, K and $\mathscr{S}'$. Let $\mathfrak{p}_1, \ldots, \mathfrak{p}_t$ be all the prime ideal divisors of $[\pi_1 \cdots \pi_s]$ in $\mathcal{O}_K$. Denote by $\mathscr{S}$ the set of all elements α of $\mathcal{O}_K$ such that $[\alpha]$ is (exclusively) composed of the prime ideals $\mathfrak{p}_1, \ldots, \mathfrak{p}_t$. Hence $\mathscr{S}' \subset \mathscr{S}$. We have

$$t \leqslant c_{40}, \qquad \max_{j=1,\ldots,t} N(\mathfrak{p}_j) \leqslant c_{41}. \tag{40}$$

Since x_1, x_2, x_3 are in $\mathscr{S}$ and satisfy (5), theorem 1.4 implies the existence of $\eta \in \mathscr{S}$ and $\rho_1, \rho_2, \rho_3 \in \mathcal{O}_K$ such that $x_i = \eta \rho_i$ and

$$\overline{|\rho_i|} \leqslant c_{42} \quad (i = 1, 2, 3). \tag{41}$$

Hence

$$|N(\rho_i)| \leqslant c_{43} \quad (i = 1, 2, 3). \tag{42}$$

By (3) we have $\mathrm{ord}_{\mathfrak{p}_j}(\eta) \leqslant \tau$ for $j = 1, \ldots, t$. Hence $|N(\eta)| \leqslant c_{44}$. This together with (42) gives $|N(x_i)| \leqslant c_{45}$ for $i = 1, 2, 3$. From (4) we obtain $|N(\pi_j)|^{v_j} \leqslant c_{46}$, hence $v_j \leqslant c_{47}$ for $j = 1, \ldots, t$. Thus $\overline{|x_3|} \leqslant c_{48}$. By a Liouville-type argument it follows from (41) that $|\rho_3^{(l)}| \geqslant c_{49}$ for all conjugates $\rho_3^{(l)}$ of ρ_3. Since $\eta = x_3/\rho_3$, we infer $\overline{|\eta|} \leqslant c_{50}$. Thus, by (41),

$$\overline{|x_i|} \leqslant \overline{|\eta|}\,\overline{|\rho_i|} \leqslant c_{51} \quad \text{for } i = 1, 2, 3. \qquad \square$$

Notes

Before it was shown that the method of estimating linear forms in logarithms yields theorem 1.1, some other methods were applied to find lower bounds for $n_{i+1} - n_i$. Størmer (1898) proved that the number of solutions of $n_{i+1} - n_i \leqslant 2$ is finite and that all solutions can effectively be found by solving a finite number of Pell equations (see also Lehmer (1964)). Pólya (1918) noticed that it is a straightforward consequence of a theorem of Thue (1909) on binary forms that $n_{i+1} - n_i \to \infty$. Pólya also remarked that $n_{i+1}/n_i \to 1$ as $i \to \infty$. It follows from the results of Siegel (1921) and Mahler (1933*a*) that for every ε with $0 < \varepsilon < 1$ there is a number N_ε such that $n_{i+1} - n_i > n_i^{1-\varepsilon}$ for $i > N_\varepsilon$. This was observed by Erdös (1965). Tijdeman (1973) proved theorem 1.1 with an explicit value of C_1. This value of C_1 can be improved by using theorem B.1. On the other hand, Tijdeman (1974) showed that there exists a computable number C_{10} depending only on P such that $n_{i+1} - n_i \leqslant n_i/(\log n_i)^{C_{10}}$ for $i \geqslant 3$ and $s > 1$. It is an unsolved problem of Erdös whether $C_{10} = C_{10}(P)$ can be made to increase to infinity when $P \to \infty$.

Størmer's result implies corollary 1.1 for $|k| \leqslant 2$ (see also Skolem (1945*a*)). Skolem (1945*b*) used his method for solving the equation $x - y = k$ in integers $x, y \in S_+$. Particular attention has been given to the special case

$$a^x - b^y = k \quad \text{in positive integers } x, y \tag{43}$$

where a, b, k are fixed integers with $a > 1$, $b > 1$, $k \neq 0$. In this connection there is an interesting conjecture of Skolem (1937, 1938) that if $a^x - b^y \equiv k (\text{mod } m)$ is solvable for every positive integer m then $a^x - b^y = k$ is solvable in integers x, y. Pillai (1931, 1936) showed that (43) has only finitely many solutions and only one solution if k is sufficiently large with respect to a and b. LeVeque (1952) proved that, if $k = 1$, then there is at most one solution, except when $a = 3$, $b = 2$. In the latter case there are exactly two solutions, namely $x = y = 1$ and $x = 2$, $y = 3$. LeVeque indicated how to determine the solution. Cassels (1953) gave a simpler proof of a slightly stronger theorem, dealing with the congruences $a^x \equiv 1 (\text{mod } B)$, $b^y \equiv -1 (\text{mod } A)$ where A, B are the products of the odd divisors of a, b respectively. See also Szymiczek (1965).

Mahler (1933*a*) used his p-adic analogue of the method of Thue–Siegel to prove that $x + y = z$ in integers $x, y, z \in S$ with $(x, y, z) = 1$ has only finitely many solutions. His method is ineffective (see also Schneider (1967)). An effective result for the equation $a^x + b^y = c^z$ was given by Gelfond (1940). Rumsey and Posner (1964) generalised this result to $x + y = c^z$ in $x \in S_+$, $y \in S_+, z \in \mathbb{Z}_+$ where $c > 1$ is some fixed integer. The full effective analogue of Mahler's result was obtained by Coates (1969, 1970*a*) and Sprindžuk (1969). In the latter paper more general equations than (2) were considered. For the greatest square-free divisor of $x + y$, Shorey (1983*c*) proved the following result.

For every $x, y \in S$ with $(x, y) = 1$, $|x| < |y|$ and $\log |y| \geqslant e^e$,

$$\log Q(x + y) \geqslant C_{11} (\log \log |y|)^2 (\log \log \log |y|)^{-1}$$

where $C_{11} > 0$ is a computable number depending only on P.

Theorems B.1 and B.3 can be used for solving equations of types (1) and (2) in practice. Another way of solving such equations is by using congruences and, sometimes, simple algebraic arguments. As early as about 1200 Levi ben Gerson (alias Leo Hebraeus) solved the equations $3^x \pm 1 = 2^y$ in integers x, y and since that time numerous papers with solutions of explicit equations have appeared: see, for example, Pillai (1945), Nagell (1958), LeVeque (1974, § D60), Alex (1976), Brenner and Foster (1982) and Alex and Foster (1983). The classical methods often fail for equations with infinitely many solutions such as $2^x - 2^y = 3^z - 3^w$ in positive integers x, y, z, w. Ellison (1971*a*, 1971*b*) indicated how Baker's method can be used for solving such equations and a detailed account can be found in Stroeker and

Tijdeman (1982). In this paper only estimates in the complex case have been used. An example of the use of p-adic estimates was given by Wagstaff Jr (1979), who solved Nathanson's exponential congruence $5^x \equiv 2 \pmod{3^x}$ in $x \in \mathbb{Z}_+$ (cf. theorem 1.2).

Several equations occurring in the above-mentioned papers are of the form $x_1 + x_2 + \cdots + x_n = 0$ with $x_i \in S$ for $i = 1, \ldots, n$ and $n \geqslant 4$. For such equations congruence methods usually require distinction of many cases. Baker's method is only applicable if n or s is very small. A general result can be obtained by the p-adic analogue of the ineffective Thue–Siegel–Roth–Schmidt method, however. Evertse (1984*b*) improved upon earlier results of Dubois and Rhin (1976) and van der Poorten and Schlickewei (1982) as follows. *Let $c, d \in \mathbb{R}$ with $c > 0$, $0 \leqslant d < 1$. Let $n \in \mathbb{Z}_+$. Then there are only finitely many $(x_1, \ldots, x_n) \in \mathbb{Z}^n$ such that* (i) $x_1 + \cdots + x_n = 0$, (ii) $x_{i_1} + \cdots + x_{i_t} \neq 0$ *for each proper non-empty subset* $\{i_1, \ldots, i_t\}$ *of* $\{1, \ldots, n\}$, (iii) $(x_1, \ldots, x_n) = 1$,

(iv) $$\prod_{k=1}^{n} \left(|x_k| \prod_{p \in S} |x_k|_p \right) \leqslant c \max_{1 \leqslant k \leqslant n} |x_k|^d.$$

Many of the above-mentioned results on rational integers have been generalised to results on algebraic integers from an arbitrary fixed algebraic number field, and some of them even to the elements of an arbitrary finitely generated integral domain over $\mathbb{Z}$. Skolem (1944, 1945*a*) gave an extension of the result of Størmer to equations over algebraic number fields. The general conjecture of Skolem (1937; 1938, p. 56) reads as follows. Let K be an algebraic number field and α_{hij}, β_{hi} non-zero elements of K. If the system of congruences

$$\sum_{h=1}^{g_i} \beta_{hi} \prod_{j=1}^{k} \alpha_{hij}^{x_j} \equiv 0 \pmod{m} \quad (i = 1, 2, \ldots, l)$$

in rational integers $x_1, \ldots, x_k$ is soluble for all moduli m then the corresponding system of equations is soluble in rational integers (cf. Schinzel, 1977). The ineffective analogues of theorem 1.4 and corollary 1.3 are implicitly contained in Siegel (1921) in case $t = 0$ and can be deduced from Parry (1950) in the general case. For an ineffective generalisation see Mahler (1950).

Generalising (8), Lang (1960, 1983) considered the equation

$$\alpha x + \beta y = \gamma \qquad (44)$$

where α, β and γ are fixed elements of an arbitrary field K of characteristic 0, and the unknowns x, y belong to a finitely generated multiplicative subgroup G of K^*. Lang proved that (44) has only finitely many solutions.

In case K is an algebraic number field and G is a group of $\mathscr{S}$-units, Györy (1979*a*), Evertse (1983*b*, Ch. 7; 1984*a*) and Silverman (1983*b*) derived explicit upper bounds for the numbers of solutions of equations of the form (44) which depend only on the degree of K over $\mathbb{Q}$ and the number t of prime ideals generating $\mathscr{S}$. For generalisations to the case considered by Lang, see Evertse and Györy (1985).

Effective results on equations of types (6) and (8) are implicitly contained in Baker (1968*b*) in case $t=0$ and in Coates (1969, 1970*a*) in the general case. See also Sprindžuk (1969). Siegel, Mahler, Parry, Baker, Coates and Sprindžuk were actually interested in the Thue and Thue–Mahler equation, which will be considered in chapters 5 and 7, respectively. Any Thue equation can be reduced to a finite number of appropriate unit equations in two variables. Conversely, any unit equation in two variables can be reduced to a finite number of suitable Thue equations. A similar equivalence holds for the Thue–Mahler equations. We refer to the notes of chapters 5 and 7 for references of papers in which unit equations occur in this context. Explicit bounds for the solutions of equations of types (6) and (8) can be found in Györy (1972, 1973, 1974, 1975, 1976, 1978*a*, 1979*a*, 1980*a*, *b*, *e*), Lang (1978, Ch. VI), Kotov and Trelina (1979) and Sprindžuk (1980; 1982, Ch. VI § 6). Györy (1979*a*) proved theorem 1.4 with (7) replaced by

$$\max_{i=1,2,3} \overline{|\rho_i|} \leqslant \exp\{((C_{12}(t+1))^{C_{13}}\tilde{R}h \log P)^{t+6}P^d \log A\}$$

where $\tilde{R}=\max(R, 1)$ and C_{12}, C_{13} are explicitly given numbers depending only on d.

Let $n \geqslant 2$ be an integer and let $\alpha_1, \dots, \alpha_{n+1}$ be elements of an algebraic number field K such that $\alpha_1 \cdots \alpha_n \neq 0$. As a generalisation of (8), consider the equation

$$\alpha_1 x_1 + \cdots + \alpha_n x_n = \alpha_{n+1} \quad \text{in } x_1, \dots, x_n \in U_{\mathscr{S}}. \tag{45}$$

Equation (45) is called an *$\mathscr{S}$-unit equation* (*in n variables*). Let r_1 and r_2 denote the number of real conjugates and complex conjugate pairs of K, respectively. Under the restriction $r_1 + r_2 + t \leqslant 3$, Vojta (1983) gave effective bounds for the solutions of $\mathscr{S}$-unit equations in three variables. Van der Poorten and Schlickewei (1982) and Evertse (1984*b*) proved some general ineffective finiteness results on $\mathscr{S}$-unit equations in an arbitrary number of variables. As a corollary we have the following result in the style of theorem 1.4.

Apart from multiplication by elements of $\mathscr{S}$, the equation

$$x_1 + \cdots + x_n = 0 \quad \textit{in } x_1, \dots, x_n \in \mathscr{S} \tag{46}$$

has only finitely many solutions such that $x_{i_1} + \cdots + x_{i_k} \neq 0$ *for each proper, non-empty subset* $\{i_1, \ldots, i_k\}$ *of* $\{1, \ldots, n\}$.

Van der Poorten and Schlickewei also obtained a similar finiteness result for the solutions of (46) if $x_1, \ldots, x_n$ are unknowns in some finitely generated multiplicative subgroup of K^* where K is any field of characteristic 0. This result implies Lang's theorem. See also Laurent (1984).

There are several applications of the results given in this chapter. Equations of type (2) occur in the theory of finite groups (see e.g. Brauer (1968) and Alex (1973)). Furthermore, Perelli and Zannier (1982) used a result like theorem 1.2 to give a complete characterisation of all integral-valued arithmetical functions which are periodic modulo p for every large p and take incongruent values modulo p in every period. Equations (5), (6) and (8) play a fundamental role in the theory of diophantine equations. Theorem 1.3 will be applied to prove theorems 9.3′ and 9.5. We shall deal with the consequences of theorem 1.4 to the superelliptic and Thue–Mahler equation in chapters 5, 6 and 7. In the notes of these chapters references can be found to papers dealing with norm form equations, discriminant form equations and index form equations. Equations (5), (6) and (8) also have applications to algebraic number theory. Theorem 1.4 can be used to prove that there are only finitely many algebraic integers (up to translation by rational integers) of given discriminant. This was proved by Birch and Merriman (1972) in an ineffective way and, independently, by Győry (1973) in an effective form. Various extensions of the effective version, for example to algebraic integers with given degree and given relative discriminant over an arbitrary algebraic number field, and applications to algebraic number theory can be found in Győry (1973, 1974, 1976, 1978*a*, *b*, 1980*f*, 1981*c*, 1984*a*) and Trelina (1977*a*). Generalisations to integral elements over an arbitrary finitely generated integral domain over $\mathbb{Z}$ are given in Győry (1982*c*, 1984*a*) and Evertse and Győry (1986*a*). Theorems 1.3 and 1.4 can also be used to obtain irreducibility theorems of Schur-type, see Győry (1972, 1980*b*, 1982*b*). Nagell (1969) calls a unit ε of an algebraic number field exceptional if $1-\varepsilon$ is also a unit. Chowla (1961) and Nagell (1964) proved that every number field has only finitely many exceptional units. This result is an immediate consequence of both theorem 1.3 and theorem 1.4 and corollary 1.3. The best-known bound for exceptional units is due to Győry (1980*a*). Lenstra Jr (1977) used information on exceptional units to find Euclidean number fields of large degree. For further results, references and applications, see Győry (1975, 1980*a*, *b*), Wasén (1977) and Sprindžuk (1982).

Equations (8) and (45) have also been studied and applied in the case of function fields of characteristic 0 in place of number fields. Effective results for the solutions of equation (8) in $\mathscr{S}$-units over function fields were obtained by Schmidt (1978), Mason (1981, 1983, 1984*a*, *b*) and Győry (1983, 1984*a*). Moreover, Mason (1983, 1984*a*, 1986) gave an efficient algorithm for determining all so-called non-trivial solutions of equations (8) and (45) in $\mathscr{S}$-units over function fields. Evertse (1986) derived a good explicit bound for the number of non-trivial solutions of (8) in $\mathscr{S}$-units of a function field.

CHAPTER 2——

Binary recurrence sequences with rational roots

In this chapter we give lower bounds for the absolute value and the greatest prime factor of Ax^m+By^m where A, B, m, x, y are rational integers. As an application we prove, under suitable conditions, that $Ax^m+By^m=Cx^n+Dy^n$ implies that $\max(m,n)$ is bounded by a computable number depending only on A, B, C and D.

Corollary B.1 can be applied to prove:

Theorem 2.1. *For every pair A, B of non-zero rational integers, there exist computable numbers C_1 and C_2 such that*

$$|Ax^m+By^m| \geqslant (\max(|x|,|y|))^{m-C_2\log m} \tag{1}$$

for all rational integers m, x, y with $m \geqslant C_1$ and $|x| \neq |y|$.

An immediate consequence of theorem 2.1 is the following result of Tijdeman (1975).

Corollary 2.1. *If $A\neq 0$, $B\neq 0$, $k\neq 0$, $m\geqslant 0$, $x>1$ and $y\geqslant 0$ are rational integers satisfying*

$$Ax^m+By^m=k, \tag{2}$$

then m is bounded by a computable number depending only on A, B and k.

Van der Poorten (1977*b*) applied theorem 2.1 and theorem B.3 to prove, effectively, that

$$\lim_{m\to\infty} P(Ax^m+By^m)=\infty \tag{3}$$

uniformly in non-zero integers x, y with $|x|\neq|y|$. Further, Stewart (1976, 1982) gave the following quantitative version of this result.

Theorem 2.2. *Suppose that A, B, x and y with $|x| \neq |y|$ are non-zero rational integers. Then*

$$P(Ax^m + By^m) \geqslant C_3(m/\log m)^{1/2} \quad (m \geqslant C_4),$$

where $C_3 > 0$ and C_4 are computable numbers depending only on A and B.

Shorey (1982) applied theorem 2.1 to generalise corollary 2.1 as follows:

Theorem 2.3. *Let $A \neq 0$, $B \neq 0$, C and D be rational integers. Suppose that x, y, m, n with $|x| \neq |y|$ and $0 \leqslant n < m$ are rational integers. There exists a computable number C_5 depending only on A, B, C and D such that the equation*

$$Ax^m + By^m = Cx^n + Dy^n \tag{4}$$

with

$$Ax^m \neq Cx^n \tag{5}$$

implies that

$$m \leqslant C_5.$$

An immediate consequence of theorem 2.3 is the following result.

Corollary 2.2. *Let A and B be non-zero rational integers. Suppose x, y, m, n with $x > y \geqslant 0$, $x > 1$ and $0 \leqslant n < m$ are rational integers. If*

$$Ax^m + By^m = Ax^n + By^n,$$

then m is bounded by a computable number depending only on A and B.

Theorems 2.2 and 2.3 are special cases of the following result.

Theorem 2.4. *Suppose that the assumptions of theorem 2.3 are satisfied. Let λ and μ be non-zero rational integers. There exist computable numbers C_6 and $C_7 > 0$ depending only on A, B, C and D such that, for every $m \geqslant C_6$, the equation*

$$\lambda(Ax^m + By^m) = \mu(Cx^n + Dy^n)$$

with

$$Ax^m Dy^n \neq By^m Cx^n$$

implies that

$$P(\mu) \geqslant C_7\left(\frac{m}{\log m}\right)^{1/2}.$$

Proofs

The constants $c_1, c_2, \ldots$ in the proofs of theorems 2.1 and 2.2 are computable positive numbers depending only on A and B.

Proof of theorem 2.1. Let $A \neq 0$, $B \neq 0$, x and y with $|x| \neq |y|$ be rational integers. We may assume that $xy \neq 0$, otherwise (1) follows immediately. Further, there is no loss of generality in assuming that $x > y > 0$. Then we can find a prime p such that

$$\operatorname{ord}_p(x/y) > 0. \tag{6}$$

If $Ax^m + By^m = 0$, then it follows from (6) that

$$m \leqslant m(\operatorname{ord}_p(x) - \operatorname{ord}_p(y)) = \operatorname{ord}_p(B) - \operatorname{ord}_p(A) =: c_1.$$

Therefore, for $m > c_1$, we have

$$0 \neq |Ax^m + By^m| = |A|x^m \left| -\frac{B}{A}\left(\frac{y}{x}\right)^m - 1 \right|.$$

Apply corollary B.1 with $n = 2$, $d = 1$, $A_1 = \max(|A|, |B|, 3)$, $A_2 = x + 1$ and $B = m + 1$ to conclude that

$$\left| -\frac{B}{A}\left(\frac{y}{x}\right)^m - 1 \right| \geqslant x^{-c_2 \log m}.$$

Hence

$$|Ax^m + By^m| \geqslant x^{m - c_3 \log m} \quad (m > c_1). \qquad \square$$

Proof of corollary 2.1. Let A, B, k, m, x, y be as in corollary 2.1 and suppose that equation (2) is satisfied. There is no loss of generality in assuming that $x > y > 0$. We may suppose that $m \geqslant C_1$. Then, by combining (2) and (1),

$$|k| \geqslant x^{m - C_2 \log m}.$$

Since $x \geqslant 2$, we find that m is bounded by a computable constant depending only on A, B and k. $\square$

Proof of theorem 2.2. There is no loss of generality in assuming that $x > y > 0$ and $(x, y) = 1$. Put

$$W_m = Ax^m + By^m \quad (m = 0, 1, 2, \ldots).$$

We may assume that $m \geqslant c_4$ with c_4 sufficiently large. Then, by (1), $W_m \neq 0$ and

$$\log |W_m| > \frac{m \log x}{2}. \tag{7}$$

Let p be a rational prime dividing W_m. Since $(x, y) = 1$, either $(p, x) = 1$ or

$(p, y)=1$. For simplicity, assume that $(p, x)=1$. Then

$$\operatorname{ord}_p(W_m)\leqslant c_5+\operatorname{ord}_p\left(-\frac{B}{A}\left(\frac{y}{x}\right)^m-1\right).$$

We apply theorem B.3 with $n=2$, $d=1$, $p=p$, $A_1=\max(|A|, |B|, 3)$, $A_2=x+1$ and $B=m$ to obtain

$$\operatorname{ord}_p\left(-\frac{B}{A}\left(\frac{y}{x}\right)^m-1\right)\leqslant c_6(\log m)^2(\log x)\frac{p}{\log p}.$$

Thus we obtain

$$\operatorname{ord}_p(W_m)\leqslant c_7(\log m)^2(\log x)\frac{p}{\log p}$$

if $(p,x)=1$. The above inequality follows similarly when $(p, y)=1$. Consequently it follows from

$$\log|W_m|=\sum_{p\,|\,W_m}\operatorname{ord}_p(W_m)\log p$$

that

$$\log|W_m|\leqslant c_7(\log m)^2(\log x)\sum_{p\,|\,W_m} p.$$

Putting $P=P(W_m)$, we observe

$$\sum_{p\,|\,W_m} p\leqslant P\sum_{p\leqslant P} 1\leqslant\frac{2P^2}{\log P},$$

by formula (N.1). Here notice, by (7), that $P\geqslant 2$. Hence

$$\log|W_m|\leqslant 2c_7(\log m)^2(\log x)\frac{P^2}{\log P}. \tag{8}$$

Now the theorem follows from (7) and (8) by transferring secondary factors. □

Proof of theorem 2.3. Suppose that (4) and (5) are valid. Then $xy\neq 0$. Further, there is no loss of generality in assuming that $x>y>0$. The constants $v_1, v_2, \ldots, v_9$ are computable positive numbers depending only on A, B, C and D. Observe that

$$|Cx^n+Dy^n|\leqslant 2\max(|C|,|D|)x^n. \tag{9}$$

Now it follows from (4), (1) and (9) that

$$m-n\leqslant v_1\log(m+1). \tag{10}$$

In view of (10), we may assume that $n \geqslant v_2$ with v_2 sufficiently large. By (4) and (5),

$$\left(\frac{x}{y}\right)^n = \frac{D - By^{m-n}}{Ax^{m-n} - C} \leqslant v_3$$

which implies that $x/y \leqslant 4/3$, if v_2 is sufficiently large. Consequently

$$r := (x, y) \leqslant x - y \leqslant x/3. \tag{11}$$

From (4), we obtain

$$\left(\frac{x}{r}\right)^n (Ax^{m-n} - C) = \left(\frac{y}{r}\right)^n (D - By^{m-n}).$$

Note that $(x/r, y/r) = 1$ implies that $(x/r)^n$ divides $D - By^{m-n}$. Further, by (4) and (5), $D - By^{m-n} \neq 0$. Consequently, by (10),

$$(x/r)^n \leqslant |D - By^{m-n}| \leqslant x^{v_4 \log m}$$

which implies that

$$x/r \leqslant x^{v_4 (\log m)/n}. \tag{12}$$

Combining (11) and (12),

$$\log x \geqslant (v_4(\log m)/n)^{-1}. \tag{13}$$

By theorem 2.1,

$$|Ax^m + By^m| = r^m \left| A\left(\frac{x}{r}\right)^m + B\left(\frac{y}{r}\right)^m \right| \geqslant r^m \left(\frac{x}{r}\right)^{m - v_5 \log m} = x^m \left(\frac{x}{r}\right)^{-v_5 \log m},$$

which, together with (12), implies that

$$|Ax^m + By^m| \geqslant x^{m - v_6 (\log m)^2/n}. \tag{14}$$

Now it follows from (4), (14) and (9) that

$$x^{m - v_6 (\log m)^2/n} \leqslant v_7 x^n.$$

From $m > n$ and (13), this implies that

$$1 \leqslant m - n \leqslant v_6 \frac{(\log m)^2}{n} + \frac{\log v_7}{\log x} \leqslant v_8 \frac{(\log m)^2}{n}.$$

Consequently

$$n \leqslant v_8 (\log m)^2$$

which, together with (10), gives $m \leqslant v_9$. □

Proof of theorem 2.4. Let A, B, x, y, m, n, λ and μ be as in theorem 2.4 satisfying

$$\lambda(Ax^m+By^m)=\mu(Cx^n+Dy^n) \tag{15}$$

with

$$Ax^mDy^n \neq By^mCx^n. \tag{16}$$

Then $xy \neq 0$. Further, there is no loss of generality in assuming that $x>y>0$. By considering equation (15) with $\lambda(x,y)^{m-n}$ in place of λ, we may assume that $(x,y)=1$. Finally we may assume that $(\lambda,\mu)=1$.

Denote by $v_{10}, v_{11}, \ldots$ computable positive numbers depending only on A, B, C and D. We may suppose that $m \geqslant v_{10}$ with v_{10} sufficiently large. Then $Ax^m+By^m \neq 0$ by theorem 2.1. Let $0<\varepsilon<1$. We suppose that

$$P(\mu) \leqslant \varepsilon\left(\frac{m}{\log m}\right)^{1/2}. \tag{17}$$

We shall arrive at a contradiction for a suitable choice of ε depending only on A, B, C and D.

For a prime p dividing μ, it follows from (15) and $(\lambda,\mu)=1$ that

$$\operatorname{ord}_p(\mu) \leqslant \operatorname{ord}_p(Ax^m+By^m).$$

We apply theorem B.3 and $(x,y)=1$ to conclude that

$$\operatorname{ord}_p(Ax^m+By^m) \leqslant v_{11}(\log m)^2 \log x \frac{p}{\log p}.$$

Hence

$$\log|\mu| \leqslant v_{11}(\log m)^2 \log x \sum_{p|\mu} p.$$

From (N.1), (17) and $\varepsilon<1$,

$$\sum_{p|\mu} p \leqslant P(\mu)\pi(P(\mu)) \leqslant 2\frac{(P(\mu))^2}{\log P(\mu)} \leqslant \frac{6\varepsilon m}{(\log m)^2} \quad (|\mu|>1).$$

Consequently

$$\log|\mu| \leqslant 6\varepsilon v_{11} m \log x \tag{18}$$

which, together with (15), (1) and (9), gives

$$\log|\lambda| \leqslant 6\varepsilon v_{11} m \log x + v_{12} \log m \log x. \tag{19}$$

Observe that

$$|\lambda(Ax^m+By^m)| \geqslant |Ax^m+By^m|. \tag{20}$$

By (15), (20), (1), (18) and (9), we derive

$$m-n \leqslant 6\varepsilon v_{11} m + v_{13} \log m. \tag{21}$$

Rewrite (15) as

$$x^n(\lambda A x^{m-n} - \mu C) = -y^n(\lambda B y^{m-n} - \mu D). \tag{22}$$

In view of (16), equation (15) implies that $\lambda A x^{m-n} - \mu C$ and $\lambda B y^{m-n} - \mu D$ are non-zero. Further, since $(x, y) = 1$, we see from (22) that x^n divides $\lambda B y^{m-n} - \mu D$. Therefore

$$x^n \leqslant |\lambda B y^{m-n} - \mu D|.$$

From (19), (21), (18) and $y < x$, we have

$$\log|\lambda B y^{m-n} - \mu D| \leqslant (18\varepsilon v_{11} m + v_{14} \log m) \log x.$$

Hence

$$n \leqslant 18\varepsilon v_{11} m + v_{14} \log m. \tag{23}$$

This is valid for every ε with $0 < \varepsilon < 1$. Let $\varepsilon = \min((48 v_{11})^{-1}, 2^{-1})$. Then we conclude from (21) and (23) that $m \leqslant 2(v_{13} + v_{14}) \log m$ which implies $m \leqslant v_{15}$. This is not possible if $v_{10} > v_{15}$. □

Notes

Theorems B.1 and B.3 are the best possible with respect to A_n. This feature plays a crucial role in all the results of this chapter, and will appear again in chapters 9–12. It also enabled Stewart (1982) to prove, for non-zero rational integers A, B, x, y with $|x| \neq |y|$,

$$Q(Ax^m + By^m) \geqslant C_8 \frac{m}{(\log m)^2} \quad (m \geqslant C_9),$$

where $C_8 > 0$ and C_9 are computable numbers depending only on A and B. For more results on the square-free divisor and for the results in the direction of theorems 2.1 and 2.2, we refer to chapter 3 and its notes.

Under necessary restrictions, Shorey (1984*a*) showed that equation (4) with $C = A$, $D = B$ implies that m is bounded by a computable number depending only on $P(AB)$. The assertion of theorem 2.3 is also true if $m = n$. Then (4) implies $(A - C)x^m + (B - D)y^m = 0$. Since $A - C \neq 0$ in view of (5), the assertion of theorem 2.3 follows from theorem 2.1.

CHAPTER 3

Binary recurrence sequences

The results of chapter 2 can be put in terms of recurrences. Let m and n be non-negative integers. Consider the binary recurrence

$$u_{m+2}=ru_{m+1}+su_m \quad (m=0,1,2,\ldots)$$

where r and $s\neq 0$ are rational integers satisfying $r^2+4s\neq 0$ and $u_0, u_1\in\mathbb{Z}$ with $|u_0|+|u_1|>0$. Put $T=\max(|u_0|,|u_1|,2)$. Denote by α and β the roots of the companion polynomial z^2-rz-s. Note that α and β are distinct and non-zero. We order α and β such that $|\alpha|\geqslant|\beta|$. We have according to (C.4)

$$u_m=a\alpha^m+b\beta^m \quad (m=0,1,2,\ldots).$$

Here

$$a=\frac{u_0\beta-u_1}{\beta-\alpha} \quad\text{and}\quad b=\frac{u_1-u_0\alpha}{\beta-\alpha}. \tag{1}$$

The results of chapter 2 deal with recurrence sequences for which u_0, u_1, α, β, r, s are rational integers.

We suppose that α/β is not a root of unity and $ab\neq 0$. Hence $\{u_m\}_{m=0}^{\infty}$ is non-degenerate and $|\alpha|>1$. Theorems 3.1, 3.2, 3.3 and 3.6 are formulated in the above notation, but for theorems 3.4 and 3.5 we need some more. Let a_1, a_2, a_3 and a_4 be non-zero algebraic numbers of degrees at most d and heights not exceeding H ($\geqslant 2$). Assume that $A\neq 0$ and B are algebraic numbers of degrees at most d and heights at most H' ($\geqslant 2$). Let λ and μ be non-zero algebraic numbers. For $m=0, 1, 2, \ldots$ put

$$x_m=a_1\lambda^m+a_2\mu^m, \quad y_m=a_3\lambda^m+a_4\mu^m, \tag{2}$$

and set

$$\tau=\max(\overline{|\lambda|}, \overline{|\mu|}).$$

We shall use the above notation and conventions throughout the chapter and without any further reference.

A straightforward application of corollary B.1 yields a good lower bound for $|u_m|$.

Theorem 3.1 (Stewart, 1976, p. 33). *There exist computable numbers C_1 and C_2 depending only on a and b such that*

$$|u_m| \geqslant |\alpha|^{m-C_1 \log m} \quad (m \geqslant C_2). \tag{3}$$

It follows from theorem 3.1 that $u_m = u_n$ implies that $|m-n|$ is small. On combining this with an elementary p-adic argument, we show that the members of $\{u_m\}_{m=0}^{\infty}$ are distinct after a certain stage.

Theorem 3.2 (Parnami and Shorey, 1982). *There exists a computable number C_3 depending only on the sequence $\{u_m\}_{m=0}^{\infty}$ such that*

$$u_m \neq u_n$$

whenever $m \neq n$ and $\max(m, n) \geqslant C_3$.

Another application of corollary B.1 will enable us to derive a lower bound for $|u_m - u_n|$ from theorem 3.2.

Theorem 3.3 (Shorey, 1984*a*). *There exist computable numbers C_4 and C_5 depending only on the sequence $\{u_m\}_{m=0}^{\infty}$ such that*

$$|u_m - u_n| \geqslant |\alpha|^{\max(m,n)}(m+2)^{-C_4 \log(n+2)}$$

whenever $m \neq n$ and $\max(m, n) \geqslant C_5$.

In order to study the behaviour of the greatest prime factor of $u_m/(u_m, u_n)$, we first investigate the difference $Au_m - Bu_n$. Results on this difference are stated in the corollaries of the next two general theorems that include theorems 3.2 and 3.3.

Theorem 3.4 (Shorey, 1984*a*). *Suppose λ/μ is not a root of unity and $\tau > 1$. The equation*

$$x_m = y_n \tag{4}$$

with

$$a_1 \lambda^m \neq a_3 \lambda^n \tag{5}$$

implies that

$$\max(m, n) \leqslant C_6 \log H$$

for some computable number C_6 depending only on d, λ and μ.

The assumption $\tau > 1$ is satisfied if λ and μ are algebraic integers and λ/μ is not a root of unity.

Another application of corollary B.1 will enable us to derive a quantitative version from theorem 3.4.

Theorem 3.5 (Shorey, 1984*a*). *Suppose* $|\lambda|\geqslant|\mu|$, $|\lambda|>1$ *and* λ/μ *is not a root of unity. There exist computable numbers* C_7 *and* C_8 *depending only on* d, λ *and* μ *such that, for all* m,n *with* $m\geqslant n$, $m\geqslant C_7\log(HH')$ *and* $Aa_1\lambda^m\neq Ba_3\lambda^n$, *we have*

$$|Ax_m-By_n|\geqslant|\lambda|^m\,\mathrm{e}^{-C_8 v},$$

where $v=(\log m\log H+\log H')\log(n+2)$.

Putting $a_1=a_3=a$, $a_2=a_4=b$, $\lambda=\alpha$, $\mu=\beta$, $x_m=u_m$ and $y_n=u_n$ in theorem 3.5, we obtain

Corollary 3.1. *There exist computable numbers* C_9 *and* C_{10} *depending only on* d, α *and* β *such that, for all* m,n *with* $m\geqslant n$, $m\geqslant C_9\log(TH')$ *and* $A\alpha^m\neq B\alpha^n$, *we have*

$$|Au_m-Bu_n|\geqslant|\alpha|^m\,\mathrm{e}^{-C_{10}v_1},$$

where $v_1=(\log m\log T+\log H')\log(n+2)$.

We observe that the equations $A\alpha^m=B\alpha^n$ and $A\beta^m=B\beta^n$ with $m\neq n$ cannot hold simultaneously, since α/β is not a root of unity. Thus, if $|\alpha|=|\beta|$, we can interchange α and β, if necessary, to derive the following result from corollary 3.1.

Corollary 3.2. *Suppose* $|\alpha|=|\beta|$. *Then*

$$|Au_m-Bu_n|\geqslant|\alpha|^m\,\mathrm{e}^{-C_{10}v_1}$$

whenever $m>n$ *and* $m\geqslant C_9\log(TH')$.

For given non-zero algebraic numbers A,B and a given sequence $\{u_m\}_{m=0}^{\infty}$ whose companion polynomial has non-real roots, it follows from corollary 3.2 that $|Au_m-Bu_n|\to\infty$, whenever $\max(m,n)$ tends to infinity through non-negative integers m and n with $m\neq n$. This need not be the case with a sequence $\{u_m\}_{m=0}^{\infty}$ whose companion polynomial has real roots. For example, the Fibonacci sequence $\{u_m\}_{m=0}^{\infty}$ satisfies

$$|u_m-\alpha u_{m-1}|=|\beta|^{m-1}=|\alpha|^{-m+1}\quad(m=1,2,\ldots),$$

hence $|u_m-\alpha u_{m-1}|\to 0$ as $m\to\infty$.

By putting $A=B=1$ in corollary 3.2 and recalling that $|\alpha|>1$, we have

Corollary 3.3. *There exist computable numbers* C_{11} *and* C_{12} *depending only on* α *and* β *such that for all pairs* (m,n) *with* $m>n$ *and* $m\geqslant C_{11}\log T$, *we have*

$$|u_m-u_n|\geqslant|\alpha|^m\,\mathrm{e}^{-C_{12}v_2},$$

where

$$\nu_2 = \log m \log T \log(n+2).$$

Corollary 3.3 includes theorem 3.3. Further, corollary 3.3 implies the following refinement of theorem 3.2.

Corollary 3.4. *The equation*

$$u_m = u_n \quad (m \neq n),$$

implies that

$$\max(m, n) \leqslant C_{11} \log T. \tag{6}$$

A simple example shows that (6) is the best possible with respect to T. Let $a^{(n)} = 2^n - 1$, $b^{(n)} = 3^n - 1$ and

$$u_m^{(n)} = a^{(n)} 3^m - b^{(n)} 2^m \quad (m = 0, 1, 2, \ldots).$$

Then $u_n^{(n)} = u_0^{(n)}$ and $0 < \max(|u_0^{(n)}|, |u_1^{(n)}|) \leqslant 2.3^n$ for any n.

Corollary 3.4 states that if $u_m/u_n = 1$ and $m \neq n$ then $\max(m, n)/\log T$ is bounded. By combining theorem 3.4 and corollary B.2, we generalise it as follows. Put

$$\Lambda_{m,n} = \max\left(\frac{\max(m, n)}{\log T}, 2\right), \quad d_1 = [\mathbb{Q}(\alpha) : \mathbb{Q}].$$

Theorem 3.6 (Shorey, 1984*a*). *Let m and n satisfy $m > n \geqslant 0$ and $u_m u_n \neq 0$. There exist computable numbers $C_{13} > 0$ and C_{14} depending only on α and β such that the inequality*

$$P\left(\frac{u_m}{(u_m, u_n)}\right) \leqslant C_{13} \left(\frac{\Lambda_{m,n}}{\log \Lambda_{m,n}}\right)^{1/(d_1+1)}$$

implies that

$$\Lambda_{m,n} \leqslant C_{14}.$$

Since α/β is not a root of unity, the equations $u_m = 0$ and $u_n = 0$ with $m \neq n$ cannot hold simultaneously. Further, by corollary 3.1 with $A = 1$ and $B = 0$, the equation $u_m = 0$ implies that $m \leqslant C_{15} \log T$ for some computable number C_{15} depending only on α and β.

The first part of our next corollary is an immediate consequence of theorem 3.6. For the second part we apply part (i) with the least integer n such that $u_n \neq 0$ (n is either 0 or 1).

Corollary 3.5. *There exist computable positive numbers C_{16}, C_{17} and C_{18}*

depending only on the sequence $\{u_m\}_{m=0}^{\infty}$ *such that*

(i) $$P\left(\frac{u_m}{(u_m, u_n)}\right) \geqslant C_{17}\left(\frac{m}{\log m}\right)^{1/(d_1+1)}$$

whenever $m > n$, $m \geqslant C_{16}$, $u_n \neq 0$;

(ii) $$P(u_m) \geqslant C_{17}\left(\frac{m}{\log m}\right)^{1/(d_1+1)} \qquad (7)$$

whenever $m \geqslant C_{18}$.

Mahler (1934*b*) proved, ineffectively, that $P(u_m) \to \infty$ as $m \to \infty$. Schinzel (1967) gave an effective and quantitative version of Mahler's result. Stewart (1982) proved (7), with constants C_{17} and C_{18} depending only on a and b.

By (C.11) there exist binary integer sequences such that every term divides infinitely many others. The following consequence of corollary 3.5(i) is a result in an opposite direction.

Corollary 3.6. *If* $u_m \mid u_n$ *and* $m > n$, *then m is bounded by a computable number depending only on the sequence* $\{u_m\}_{m=0}^{\infty}$.

Corollary 3.6 includes theorem 3.2.

Finally we give a corresponding result for the greatest square-free factor of a term of a binary sequence.

Theorem 3.7 (Shorey, 1983*c*). *There exist computable positive numbers* C_{19} *and* C_{20} *depending only on the sequence* $\{u_m\}_{m=0}^{\infty}$ *such that*

$$\log Q(u_m) \geqslant C_{19}(\log m)^2(\log\log m)^{-1}$$

whenever $m \geqslant C_{20}$.

Stewart (1983) derived this inequality for the members of Lucas and Lehmer sequences. For other binary sequences his lower bounds are of the order $\log m$ (see Stewart, 1982).

Proofs

Observe that by (1) the heights of a and b do not exceed $C_{21}T^2$ where C_{21} is a computable number depending only on α and β.

Proof of theorem 3.1. Denote by $c_1, c_2, \ldots, c_6$ computable positive numbers depending only on a and b. First we show that

$$u_m \neq 0 \quad (m > c_1). \qquad (8)$$

Suppose that $u_m=0$. Then

$$(\alpha/\beta)^m = -b/a. \tag{9}$$

If α/β is not a unit, there exists a prime ideal $\wp$ in the ring of integers of $\mathbb{Q}(\alpha)$ such that $\operatorname{ord}_\wp(\alpha/\beta)$ is non-zero and hence, by (9),

$$m \leqslant m|\operatorname{ord}_\wp(\alpha/\beta)| \leqslant |\operatorname{ord}_\wp(a)| + |\operatorname{ord}_\wp(b)| \leqslant c_2.$$

Thus we may assume that α/β is a unit. Then, by lemma A.5, we can find a computable absolute constant $c>0$ such that

$$|\alpha/\beta| = \left|\overline{\alpha/\beta}\right| \geqslant 1+c, \tag{10}$$

since α/β is not a root of unity. Combining (10) and (9), we get

$$(1+c)^m \leqslant \frac{|b|}{|a|}$$

which implies that $m \leqslant c_3$. This completes the proof of (8).

For $m>c_1$, we have

$$0 \neq |u_m| = |a\alpha^m + b\beta^m| = |a\alpha^m| \left| -\frac{b}{a}\left(\frac{\beta}{\alpha}\right)^m - 1 \right|.$$

We apply corollary B.1 with $n=3$, $d \leqslant 2$, $A_1=3$, $A_2=c_4$, $A_3=3|\alpha|^2$ and $B=m$ to obtain

$$\left| -\frac{b}{a}\left(\frac{\beta}{\alpha}\right)^m - 1 \right| \geqslant |\alpha|^{-c_5 \log m} \quad (m>c_1).$$

Hence

$$|u_m| \geqslant |\alpha|^{m-c_6 \log m} \quad (m>c_1). \qquad \square$$

Proof of theorem 3.2. Denote by c_7, c_8, c_9, c_{10} computable positive numbers depending only on the sequence $\{u_m\}_{m=0}^{\infty}$. If $|\alpha|>|\beta|$, the assertion follows trivially. Thus we may assume that $|\alpha|=|\beta|$. Then observe that α/β and β/α are conjugate quadratic algebraic numbers in $\mathbb{Q}(\alpha)$ of absolute value 1. Therefore, since α/β is not a root of unity, we see from lemma A.5 that α/β and β/α are not algebraic integers. Thus there exists a prime ideal $\wp$ in the ring of integers of $\mathbb{Q}(\alpha)$ such that $\operatorname{ord}_\wp(\alpha/\beta)>0$.

Let m, n with $m>n$ and $m \geqslant 2$ satisfy

$$u_m = u_n. \tag{11}$$

Re-write (11) as

$$(\alpha/\beta)^n = -\frac{b}{a}\frac{\beta^{m-n}-1}{\alpha^{m-n}-1}.$$

Thus
$$n \leqslant n \operatorname{ord}_{\mathfrak{p}}(\alpha/\beta) \leqslant \operatorname{ord}_{\mathfrak{p}}(b/a) + \operatorname{ord}_{\mathfrak{p}}(\beta^{m-n} - 1).$$

By lemma A.8, we have
$$\operatorname{ord}_{\mathfrak{p}}(\beta^{m-n} - 1) \leqslant c_7 \log m.$$

Consequently
$$n \leqslant c_8 \log m. \tag{12}$$

Now notice that
$$|u_n| \leqslant 2 \max(|a|, |b|)|\alpha|^n. \tag{13}$$

Combining (11), (3) and (13), it follows that
$$m - n \leqslant c_9 \log m. \tag{14}$$

The inequalities (14) and (12) imply that $m \leqslant c_{10}$. If $m < n$, then we can interchange m and n and apply the result proved above. □

Proof of theorem 3.3. Denote by $c_{11}, c_{12}, \ldots, c_{16}$ computable positive numbers depending only on the sequence $\{u_m\}_{m=0}^{\infty}$. For m, n with $m > n$ and $m \geqslant 2$, put
$$\psi = u_m - u_n.$$

If $2|u_n| \leqslant |u_m|$, then
$$|\psi| \geqslant |u_m| - |u_n| \geqslant \frac{|u_m|}{2}$$
and the assertion follows from (3). Thus we may assume that
$$|u_m| < 2|u_n|$$
which, together with (3) and (13), gives
$$m - n \leqslant c_{11} \log m. \tag{15}$$

We may assume that $m \geqslant C_3$ so that, by theorem 3.2, ψ is non-zero. Rewriting ψ, we have
$$0 \neq |\psi| = |a\alpha^n(\alpha^{m-n} - 1) + b\beta^n(\beta^{m-n} - 1)|.$$

Since $a \neq 0$ and α is not a root of unity, we may write
$$0 \neq |\psi| = |a\alpha^n(\alpha^{m-n} - 1)|\Delta$$
where
$$\Delta = \left| -\frac{b}{a}\left(\frac{\beta}{\alpha}\right)^n \frac{\beta^{m-n} - 1}{\alpha^{m-n} - 1} - 1 \right|$$

and Δ is non-zero. We apply corollary B.1 with $n=4$, $d \leqslant 2$, $B=n+2$, $\log A_1 = \log A_2 = \log A_3 = c_{12}$ and, by lemmas A.3 and A.2, $\log A_4 \leqslant c_{13}(m-n)$ which, together with (15), implies that $\log A_4 \leqslant c_{13}c_{11} \log m$. We obtain

$$\Delta \geqslant m^{-c_{14}\log(n+2)}.$$

Further, by $|\alpha|>1$ and (15),

$$|a\alpha^n(\alpha^{m-n}-1)| \geqslant |a|(|\alpha|-1)|\alpha|^n \geqslant |\alpha|^m m^{-c_{15}}.$$

Hence

$$|\psi| \geqslant |\alpha|^m m^{-c_{16}\log(n+2)}.$$

If $m<n$, then we can interchange m and n and apply the result proved above. □

Now we shall prove theorem 3.4. Let λ, μ, a_1, a_2, a_3 and a_4 be non-zero algebraic numbers. Suppose that a_1, a_2, a_3 and a_4 have degrees at most d and heights not exceeding H ($\geqslant 2$). Denote by L the field generated by λ, μ, a_1, a_2, a_3 and a_4 over $\mathbb{Q}$. Let x_m and y_m be given by (2). For $1 \leqslant i \leqslant 4$, we see from (A.6) and (A.7) that

$$\max_\sigma |\sigma(a_i)| \leqslant dH \tag{16}$$

and

$$\min_\sigma |\sigma(a_i)| \geqslant (dH)^{-1} \tag{17}$$

where maximum and minimum are taken over all the embeddings σ of L. Further, for every prime ideal $\wp$ in the ring of integers of L, we observe from lemma A.7 that

$$|\mathrm{ord}_\wp(a_i)| \leqslant k_1 \log H \quad (1 \leqslant i \leqslant 4) \tag{18}$$

for some computable constant k_1 depending only on d. We denote by $k_2, k_3, \ldots$ computable positive constants depending only on d, λ and μ. We apply theorem B.2 to obtain the following estimate for $|x_m|$.

Lemma 3.1. *Suppose λ/μ is not a root of unity. There exist k_2 and k_3 such that, for every δ with $0<\delta<\frac{1}{2}$, we have*

$$|x_m| \geqslant (\max(|\lambda|, |\mu|))^m \exp(-k_2 \log(1/\delta) \log H - 3\delta m) \tag{19}$$

whenever $m \geqslant k_3 \log H$.

Putting $\delta = 1/m$, we obtain

Corollary 3.7. *Suppose λ/μ is not a root of unity. Then*

$$|x_m| \geqslant (\max(|\lambda|, |\mu|))^m \exp(-k_4 \log m \log H) \tag{20}$$

whenever $m \geqslant k_3 \log H$.

Proof of lemma 3.1. We first prove that the equation $x_m = 0$ implies that $m < k_3 \log H$. Suppose $x_m = 0$. Then

$$(\lambda/\mu)^m = -a_2/a_1. \tag{21}$$

If λ/μ is not a unit, there exists a prime ideal $\wp$ in the ring of integers of L such that $\text{ord}_\wp(\lambda/\mu)$ is non-zero. Then, by (21),

$$m \leqslant m|\text{ord}_\wp(\lambda/\mu)| \leqslant |\text{ord}_\wp(a_1)| + |\text{ord}_\wp(a_2)|$$

and the assertion follows from (18). Thus we may assume that λ/μ is a unit. Then, since λ/μ is not a root of unity, we can find an embedding σ of L such that $|\sigma(\lambda/\mu)| > 1$. Further, by taking images under σ on both the sides in (21), we have

$$|\sigma(\lambda/\mu)|^m = |\sigma(a_2)/\sigma(a_1)|$$

and the assertion follows from (16) and (17).

We assume that $m \geqslant k_3 \log H$ so that $x_m \neq 0$. We apply theorem B.2 with $n = 3, d = k_5, \log A' = k_6, \log A = k_7 \log H, B' = 1$ and $B'' = m$ to conclude that

$$\left| -\left(\frac{\mu}{\lambda}\right)^m \frac{a_2}{a_1} - 1 \right| \quad \text{and} \quad \left| -\left(\frac{\lambda}{\mu}\right)^m \frac{a_1}{a_2} - 1 \right|$$

exceed

$$\exp(-k_8 \log(1/\delta) \log H - 3\delta m)$$

for every δ with $0 < \delta < \frac{1}{2}$. Consequently, by (17), we obtain

$$|x_m| \geqslant (\max(|\lambda|, |\mu|))^m \exp(-k_9 \log(1/\delta) \log H - 3\delta m)$$

for every δ with $0 < \delta < \frac{1}{2}$. □

Further we shall prove:

Lemma 3.2. *Suppose λ/μ is not a root of unity. Then* (4) *and* (5) *with $m \geqslant n$ imply that*

$$n \leqslant k_{10}((m-n) + \log H). \tag{22}$$

Proof. Suppose that (4) and (5) with $m \geqslant n$ are valid. Re-write (4) as

$$\lambda^n(a_1\lambda^{m-n} - a_3) = -\mu^n(a_2\mu^{m-n} - a_4). \tag{23}$$

It follows from (5) and (23) that $a_1\lambda^{m-n}-a_3$ and $a_2\mu^{m-n}-a_4$ are non-zero. If λ/μ is not a unit, we can find a prime ideal $\wp$ in the ring of integers of L such that $\mathrm{ord}_\wp(\lambda/\mu)$ is non-zero. Then, by (23),

$$n\leqslant n|\mathrm{ord}_\wp(\lambda/\mu)|\leqslant|\mathrm{ord}_\wp(a_1\lambda^{m-n}-a_3)|+|\mathrm{ord}_\wp(a_2\mu^{m-n}-a_4)|$$

and, by lemmas A.7 and A.3, inequality (22) follows from (18). Thus we may suppose that λ/μ is a unit. Then, since λ/μ is not a root of unity, we can find an embedding σ of L such that $|\sigma(\lambda/\mu)|>1$. Further, by taking images under σ on both sides of (23), we have

$$|\sigma(\lambda/\mu)|^n=\left|\frac{\sigma(a_2\mu^{m-n}-a_4)}{\sigma(a_1\lambda^{m-n}-a_3)}\right|.$$

Now inequality (22) follows from (16) and a Liouville-type argument. □

Corollary 3.8. *Put* $k_{11}=2(k_{10}+1)$. *Suppose* λ/μ *is not a root of unity and* $m\geqslant n$. *Assume that* (4) *and* (5) *are satisfied. Then*

$$m-n\leqslant k_{11}^{-1}m \tag{24}$$

implies that

$$m\leqslant 2k_{10}\log H.$$

Proof. By (22) and (24),

$$n\leqslant k_{11}^{-1}k_{10}m+k_{10}\log H$$

which, together with (24), implies that

$$m\leqslant k_{11}^{-1}(k_{10}+1)m+k_{10}\log H=2^{-1}m+k_{10}\log H.$$

Hence $m\leqslant 2k_{10}\log H$. □

Proof of theorem 3.4. Assume that (4) and (5) are satisfied. There is no loss of generality in assuming that $m\geqslant n$. Further there exists an embedding σ of L such that

$$\tau=\max(|\sigma(\lambda)|,|\sigma(\mu)|).$$

By considering the equation $\sigma(x_m)=\sigma(y_n)$ in place of (4), there is no loss of generality in assuming that $\max(|\lambda|,|\mu|)>1$. Write

$$\log\max(|\lambda|,|\mu|)=k_{12}. \tag{25}$$

We may assume that $m\geqslant k_{13}\log H$ with k_{13} sufficiently large. Let $k_{13}>\max(k_3,2k_{10})$ so that the assertion of lemma 3.1 is valid and, by corollary 3.8,

$$m-n>k_{11}^{-1}m. \tag{26}$$

Further, by (16) and (25),

$$|y_n| \leqslant 2dH\, e^{k_{12}n}. \tag{27}$$

Now it follows from (4), (19) with $\delta = \min(k_{12}/6k_{11}, \frac{1}{4})$, (25) and (27) that

$$m - n \leqslant (2k_{11})^{-1}m + k_{14}\log H. \tag{28}$$

Combining (26) and (28), we obtain $m < 2k_{11}k_{14}\log H$. □

Proof of theorem 3.5. Suppose that $Aa_1\lambda^m \neq Ba_3\lambda^n$ with $m \geqslant n$. Put

$$\psi_1 = Ax_m - By_n.$$

We assume that $m \geqslant k_{15}\log(HH')$ with k_{15} sufficiently large. Let $k_{15} > k_3$ so that the assertion of corollary 3.7 is valid. We see from (A.6) and (A.7) that, for $B \neq 0$,

$$\max(|A|, |B|) \leqslant dH', \quad \min(|A|, |B|) \geqslant (dH')^{-1}. \tag{29}$$

If $|Ax_m| \geqslant 2|By_n|$, then

$$|\psi_1| \geqslant |Ax_m| - |By_n| \geqslant \frac{|Ax_m|}{2}$$

and the theorem follows from (29) and (20). Thus we may assume that

$$|Ax_m| < 2|By_n|. \tag{30}$$

Further, by lemma A.1,

$$|y_n| \leqslant 2dH|\lambda|^n. \tag{31}$$

Now it follows from (30), (20), (31), (29) and $\max(|\lambda|, |\mu|) \geqslant |\lambda| > 1$ that

$$m - n \leqslant k_{16}(\log m \log H + \log H'). \tag{32}$$

If k_{15} is sufficiently large, it follows from theorem 3.4 that ψ_1 is non-zero. Further, re-writing ψ_1, we obtain

$$0 \neq |\psi_1| = |\lambda^n(Aa_1\lambda^{m-n} - Ba_3) + \mu^n(Aa_2\mu^{m-n} - Ba_4)|.$$

Since $Aa_1\lambda^{m-n} - Ba_3 \neq 0$, we may write

$$0 \neq |\psi_1| = |\lambda^n(Aa_1\lambda^{m-n} - Ba_3)|\Delta_1$$

where

$$\Delta_1 = \left| -\left(\frac{\mu}{\lambda}\right)^n \frac{Aa_2\mu^{m-n} - Ba_4}{Aa_1\lambda^{m-n} - Ba_3} - 1 \right|$$

and Δ_1 is non-zero. We apply corollary B.1 with $n = 3$, $d = k_{17}$, $B = n + 2$, $\log A_1 = \log A_2 = k_{18}$ and $\log A_3 \leqslant k_{19}((m-n) + \log(HH'))$ which, together with (32), implies that $\log A_3 \leqslant k_{20}(\log m \log H + \log H')$. We obtain

$$\Delta_1 \geqslant e^{-k_{21}\nu}.$$

Further, by (32), (16), (29) and a Liouville-type argument, we obtain

$$|\lambda^n(Aa_1\lambda^{m-n} - Ba_3)| \geqslant |\lambda|^m \exp(-k_{22}(\log m \log H + \log H')).$$

Hence

$$|\psi_1| \geqslant |\lambda|^m e^{-k_{23}\nu}. \qquad \square$$

Proof of theorem 3.6. For an integer x in $\mathbb{Q}(\alpha)$, denote by $[x]$ the ideal generated by x in the ring of integers of $\mathbb{Q}(\alpha)$. By lemma A.10, we have

$$([\alpha^2], [\beta^2]) = [l]$$

where l is a positive rational integer. In fact, $l = (r^2, s)$. Put

$$\alpha_1 = \alpha^2/l, \quad \beta_1 = \beta^2/l.$$

Then α_1 and β_1 are non-zero algebraic integers such that the ideals $[\alpha_1]$ and $[\beta_1]$ are relatively prime. Further, observe that $|\alpha_1| \geqslant |\beta_1|$, α_1/β_1 is not a root of unity and α_1, β_1 are roots of a quadratic monic polynomial with rational integer coefficients. Consequently, we find that $|\alpha_1| > 1$. For $m' = 0, 1, 2, \ldots$ and $\delta' = 0, 1$, write

$$u_{2m'+\delta'} = l^{m'} v_{2m'+\delta'} \tag{33}$$

where

$$v_{2m'+\delta'} = a\alpha^{\delta'}\alpha_1^{m'} + b\beta^{\delta'}\beta_1^{m'}.$$

Let m and n be non-negative integers such that $m > n$ and $u_m u_n \neq 0$. Write

$$u_m/u_n = B_1/A_1 \tag{34}$$

where $A_1 > 0$ and B_1 are relatively prime non-zero integers. Then

$$B_1 = \pm \frac{u_m}{(u_m, u_n)}.$$

Further, write $m = 2m_1 + \delta_1$, $n = 2n_1 + \delta_2$ where $\delta_1, \delta_2 \in \{0, 1\}$. Observe that $m_1 \geqslant n_1$, since $m > n$. By (34) and (33),

$$A_1 l^{m_1 - n_1} v_m = B_1 v_n.$$

Cancelling the common factors of $A_1 l^{m_1 - n_1}$ and B_1, we can find non-zero rational integers A_2, B_2 with $(A_2, B_2) = 1$ and

$$P(B_2) \leqslant P(B_1) \tag{35}$$

such that

$$A_2 v_m = B_2 v_n. \tag{36}$$

We apply theorem 3.4 with $a_1 = A_2 a\alpha^{\delta_1}$, $a_2 = A_2 b\beta^{\delta_1}$, $a_3 = B_2 a\alpha^{\delta_2}$, $a_4 = B_2 b\beta^{\delta_2}$, $\lambda = \alpha_1$, $\mu = \beta_1$, $x_{m_1} = A_2 v_m$, $y_{n_1} = B_2 v_n$ and

$$\log H \leqslant c_{17}(\log |A_2 B_2| + \log T)$$

where c_{17} and the subsequent symbols $c_{18}, c_{19}, \ldots, c_{25}$ are computable positive constants depending only on α and β. We see that $\tau > 1$, since $|\alpha_1| > 1$. If $a_1 \lambda^{m_1} = a_3 \lambda^{n_1}$, then, by (36), $a_2 \mu^{m_1} = a_4 \mu^{n_1}$ and, consequently, we find that $(\alpha/\beta)^m = (\alpha/\beta)^n$, which is not possible since α/β is not a root of unity and $m \neq n$. Further, λ/μ is not a root of unity. Thus all the assumptions of theorem 3.4 are satisfied. Hence, by theorem 3.4, we conclude that

$$m \leqslant 2m_1 + 1 \leqslant c_{18}(\log |A_2 B_2| + \log T) \tag{37}$$

with $c_{18} > 1$.

We assume that $m > c_{19} \log T$ with c_{19} sufficiently large. Let $c_{19} > 2c_{18}$. Then, by (37),

$$m < 2c_{18} \log |A_2 B_2|. \tag{38}$$

By (36) and lemma 3.1 with $\delta = (12c_{18})^{-1}$ and $m > n$,

$$\log |A_2| \leqslant \log |B_2| + c_{20} \log T + (4c_{18})^{-1} m$$

which, together with (38), implies that

$$m < c_{21} \log |B_2| \quad (c_{21} > 1), \tag{39}$$

if c_{19} is sufficiently large. Write $P = P(B_2)$. By (39), we find $P \geqslant 2$. For a prime p dividing B_2, it follows from (36) that

$$\operatorname{ord}_p(B_2) \leqslant \operatorname{ord}_p(v_m),$$

since A_2 and B_2 are relatively prime. Further,

$$\log |B_2| = \sum_{p|B_2} \operatorname{ord}_p(B_2) \log p \leqslant \log P \sum_{p \leqslant P} \operatorname{ord}_p(v_m). \tag{40}$$

Thus, by (39) and (40), we can find a prime $p_0 \leqslant P$ such that

$$\operatorname{ord}_{p_0}(v_m) > (c_{21} \pi(P) \log P)^{-1} m$$

which, by (N.1), gives

$$\operatorname{ord}_{p_0}(v_m) > (2c_{21} P)^{-1} m.$$

Let $\wp_0$ be a prime ideal in the ring of integers of $\mathbb{Q}(\alpha)$ dividing p_0. Then

$$\operatorname{ord}_{\wp_0}(v_m) > (2c_{21} P)^{-1} m. \tag{41}$$

We see from $([\alpha_1], [\beta_1]) = [1]$ that $\wp_0$ is prime to at least one of the ideals

$[\alpha_1]$ and $[\beta_1]$. For simplicity, we assume that $\wp_0$ and $[\alpha_1]$ are relatively prime. Put

$$\Delta_2 = \text{ord}_{\wp_0}\left(-\left(\frac{\beta}{\alpha}\right)^m\left(\frac{a}{b}\right)^{-1}-1\right).$$

Then, by lemma A.7,

$$\text{ord}_{\wp_0}(v_m) \leqslant \Delta_2 + c_{22}\log T. \tag{42}$$

By (42) and (41), we find that

$$\Delta_2 > (2c_{21}P)^{-1}m - c_{22}\log T.$$

We may assume that $m > 4c_{21}c_{22}P\log T$; otherwise the theorem follows from (35). Then

$$\Delta_2 > (4c_{21}P)^{-1}m.$$

Now we apply corollary B.2 with $p = p_0 \leqslant P$, $n = 3$, $d = d_1$, $\log A' = c_{23}$, $\log A = c_{24}\log T$, $B = m$ and $\delta = (4c_{21}P)^{-1}$ to Δ_2. We obtain

$$m \leqslant c_{25}P^{d_1+1}\log P\log T$$

which, together with (35), completes the proof of theorem 3.6. □

Proof of theorem 3.7. By lemma A.10, we have

$$([\alpha^2], [\beta^2]) = [l]$$

where l is a positive rational integer. By considering integral binary recurrences $\{l^{-m}u_{2m}\}_{m=0}^{\infty}$ and $\{l^{-m}u_{2m+1}\}_{m=0}^{\infty}$ separately, we may assume that $([\alpha], [\beta]) = [1]$. Denote by $c_{26}, c_{27}, \ldots$ computable positive numbers depending only on the sequence $\{u_m\}_{m=0}^{\infty}$. We may assume that $m \geqslant c_{26}$ with c_{26} sufficiently large. Then $|u_m| > 1$. We may suppose that

$$\log P(u_m) \leqslant (\log m)^2;$$

otherwise the assertion follows immediately. Let $p_1, \ldots, p_s$ be all the prime factors of u_m such that $p_i > m^{1/4}$ for $1 \leqslant i \leqslant s$. It suffices to show that

$$s > c_{27}(\log m)(\log\log m)^{-1}.$$

Let $0 < \varepsilon < 1$. We assume that

$$s \leqslant \varepsilon(\log m)(\log\log m)^{-1}$$

and we shall arrive at a contradiction for a suitable value of ε depending only on the sequence $\{u_m\}_{m=0}^{\infty}$.

For a prime p, let $\mathfrak{p}$ be a prime ideal in the ring of integers of $\mathbb{Q}(\alpha)$ dividing p. We apply theorem B.3 and $([\alpha],[\beta])=[1]$ to conclude that

$$\operatorname{ord}_p(u_m)\leqslant\operatorname{ord}_{\mathfrak{p}}(u_m)\leqslant c_{28}\frac{p^{d_1}}{\log p}(\log m)^2$$

where $d_1=[\mathbb{Q}(\alpha):\mathbb{Q}]$. Consequently

$$\sum_{p\leqslant m^{1/4}}\operatorname{ord}_p(u_m)\log p\leqslant c_{28}(\log m)^2\sum_{p\leqslant m^{1/4}}p^{d_1}.$$

Further, since $d_1=1$ or 2,

$$\sum_{p\leqslant m^{1/4}}p^{d_1}\leqslant m^{3/4}.$$

Hence

$$\sum_{p\leqslant m^{1/4}}\operatorname{ord}_p(u_m)\log p\leqslant c_{28}m^{3/4}(\log m)^2.$$

Thus we may write

$$u_m=a\alpha^m+b\beta^m=U_m p_1^{a_1}\cdots p_s^{a_s}\tag{43}$$

where $a_1,\ldots,a_s$ are positive integers and $0\neq U_m\in\mathbb{Z}$ with

$$\log|U_m|\leqslant c_{28}m^{3/4}(\log m)^2.$$

Further notice that, for some $c_{29}>1$,

$$a_i\leqslant c_{29}m\quad(1\leqslant i\leqslant s).$$

We put

$$\Lambda=a^{-1}\alpha^{-m}p_1^{a_1}\cdots p_s^{a_s}U_m-1.\tag{44}$$

Observe that $\Lambda\neq 0$.

Suppose that $|\alpha|>|\beta|$. Then it follows from (44) and (43) that

$$0<|\Lambda|<c_{30}^{-m}\quad\text{with } c_{30}>1.\tag{45}$$

We apply corollary B.1 with $n=s+3\leqslant\varepsilon(\log m)(\log\log m)^{-1}+3$, $d\leqslant 2$, $\log A_1=\log A_2=c_{31}$, $\log A_3=\cdots=\log A_{n-1}=(\log m)^2$, $\log A_n=c_{28}m^{3/4}(\log m)^2$ and $B=c_{29}m$ to conclude that

$$|\Lambda|\geqslant\exp(-m^{3/4+c_{32}\varepsilon}(\log m)^{c_{33}}).\tag{46}$$

Let $\varepsilon=\min(1/8c_{32},\frac{1}{2})$. Then it follows from (46) and (45) that $m\leqslant c_{34}$ which is not possible if $c_{26}>c_{34}$.

Thus we may assume that $|\alpha|=|\beta|$. Then we employ an elementary p-adic argument from the proof of theorem 3.2 to find out a prime ideal $\mathfrak{p}$ in the ring of integers of $\mathbb{Q}(\alpha)$ such that $\operatorname{ord}_{\mathfrak{p}}(\beta/\alpha)>0$. Since $([\alpha],[\beta])=[1]$, we see that $\mathfrak{p}\mid\beta$ and $\mathfrak{p}\nmid\alpha$. Now it follows from (43) and (44) that

$$m\leqslant c_{35}+\operatorname{ord}_{\mathfrak{p}}(\Lambda). \tag{47}$$

We apply theorem B.3, with the same parameters as used for corollary B.1 to derive (46), to conclude that

$$\operatorname{ord}_{\mathfrak{p}}(\Lambda)\leqslant m^{3/4+c_{36}\varepsilon}(\log m)^{c_{37}}. \tag{48}$$

Put $\varepsilon=\min(1/8c_{36},\frac{1}{2})$. Then, by combining (47) and (48), we see that $m\leqslant c_{38}$ which is not possible if $c_{26}>c_{38}$. □

Notes

In these notes $\{u_m\}_{m=0}^{\infty}$ will denote a non-degenerate binary recurrence sequence of rational integers with distinct roots α,β unless it is explicitly stated that we deal with a Lehmer sequence.

Mahler (1934*b*) showed, by a p-adic generalisation of the Thue–Siegel theorem, that $|u_m|\to\infty$ as $m\to\infty$. Lower bounds for $|u_m|$ in the special cases of Lucas or Lehmer sequences were given by Schinzel (1962*a*), Townes (1962) and P. Chowla, S. Chowla, Dunton and Lewis (1963). The first general effective lower bound was given by Schinzel (1967), who employed his version of a p-adic theorem of Gelfond (see also Mahler (1966) and P. Chowla (1969)). Stewart's proof of theorem 3.1 can be found in Stewart (1976, p. 33) and in Shorey and Stewart (1983). Kiss (1979) has given completely explicit estimates for $|u_m|$.

It is a trivial consequence of the fact $|u_m|\to\infty$ as $m\to\infty$ that $\{u_m\}_{m=0}^{\infty}$ attains every value only finitely many times. Ward conjectured that the multiplicity of $\{u_m\}_{m=0}^{\infty}$ is at most 5. Partial results in this direction were obtained by P. Chowla, S. Chowla, Dunton and Lewis (1959), S. Chowla, Dunton and Lewis (1961), Laxton (1967) and Alter and Kubota (1973). By using Skolem's method, Kubota (1977*a*) established Ward's conjecture and proved that the multiplicity of $\{u_m\}_{m=0}^{\infty}$ is in fact at most 4. Improved results and simplified proofs were given by Beukers (1980). His results imply that the multiplicity is at most 3 with essentially only one exception which has multiplicity 4, namely the sequence defined by $u_0=1$, $u_1=-1$, $u_{m+2}=-u_{m+1}-2u_m$ for $m\geqslant 2$. Multiplicities of Lucas sequences were studied by Kubota (1977*a*) and Beukers (1980).

The first lower bounds for $P(u_m)$ were obtained as by-products of results on primitive divisors of Lucas and Lehmer sequences. Let A and B be non-zero integers of an algebraic number field K. A prime ideal $\mathfrak{p}$ of K is called a *primitive divisor* of A^m-B^m if $\mathfrak{p}\mid[A^m-B^m]$ but $\mathfrak{p}\nmid[A^n-B^n]$ for $0<n<m$. Zsigmondy (1892) and Birkhoff and Vandiver (1904) showed that if v and w

are coprime non-zero rational integers with $v \neq \pm w$, then $v^m - w^m$ has a primitive divisor for $m > 6$. This result was improved by Schinzel (1962*b*). Schinzel (1974) improved upon earlier work of Postnikova and Schinzel (1968) by showing that if $([A, B]) = [1]$ and A/B is not a root of unity, then $A^m - B^m$ has a primitive divisor for all $m > C_{22}$, where C_{22} is a computable number depending only on the degree d of A/B. The proof depends on theorem B.1. Further, Stewart (1976, Ch. V; 1977*b*) showed that C_{22} can be taken to be equal to $\max(2(2^d - 1), e^{452} d^{67})$. He used a refinement of theorem B.1. By taking $A = \alpha$, $B = \beta$ in the definitions of Lucas and Lehmer sequences as given in chapter C it follows that every Lucas number u_m with $m > e^{452} 2^{67}$ and every Lehmer number u_m with $m > e^{452} 4^{67}$ has a prime factor which does not divide u_n for $0 < n < m$. Stewart proved the stronger result that there are only finitely many Lucas and Lehmer sequences whose mth term, $m > 6, m \neq 8$, 10 or 12, does not possess a primitive divisor and that these sequences may be explicitly determined. This improves upon results of Carmichael (1913), Ward (1955*a*), Durst (1959) and Schinzel (1968, 1974). Since primitive factors of Lucas and Lehmer numbers have the property that they are $\pm 1 \pmod m$, we see that $P(u_m) \geqslant m - 1$ for $m > e^{452} 4^{67}$ for all Lucas and Lehmer sequences $\{u_m\}_{m=0}^{\infty}$. Stewart (1976, p. 57) conjectured that, for any Lucas or Lehmer sequence with α, β real, $P(u_m) > C_{23}(\phi(m))^2$ for all m, where C_{23} is a computable positive number. However, the much weaker assertion $P(u_m)/m \to \infty$ is still open. Stewart (1975) proved that $P(v^m \pm w^m)/m$ tends to infinity as $m \to \infty$ and $\omega(m) \leqslant \kappa \log\log m$ and $0 < \kappa < 1/\log 2$. Stewart (1977*a*) and Shorey and Stewart (1981) generalised this result to all Lucas and Lehmer sequences. In fact it follows from the results of Stewart (1975, 1977*a*), Erdös and Shorey (1976) and Shorey and Stewart (1981) that $P(v^m \pm w^m) > C_{24} m \log m$ for $m \geqslant 3$ where $C_{24} > 0$ is a computable number depending only on $P(vw)$ and $\omega(m)$, and that, in the case of a Lucas or Lehmer sequence, $P(u_m) > C_{25} m \log m$ for $m \geqslant 3$ where $C_{25} > 0$ is a computable number depending only on α, β and $\omega(m)$, and, moreover, $P(u_m) > m(\log m)^2/(\log\log m)^2$ for almost all m. Erdös and Shorey (1976) applied estimates of linear forms in logarithms and Brun's sieve to show that $P(2^p - 1) \geqslant p(\log p)^2/(\log\log p)^3$ for almost all primes p. Györy, Kiss and Schinzel (1981) proved the following result. If u_m is the mth term of a Lucas sequence ($m > 4$) or Lehmer sequence ($m > 6$), then $|\alpha|$, $|\beta|$ and $|u_m|$ are bounded by computable numbers depending only on $P(u_m)$ and $\omega(u_m)$. Györy (1982*a*) computed explicit bounds. As a consequence he showed that there exists a computable absolute constant C_{26} such that every Lucas and Lehmer number u_m with $m > 6$ and $|u_m| > C_{26}$ satisfies $P(u_m) > \frac{1}{2}(\log\log |u_m|)^{1/3}$. For more information on primitive divisors, we refer to Schinzel (1974) and Stewart (1976, 1977*a*, *b*, 1982).

We consider again sequences $\{u_m\}_{m=0}^{\infty}$ as described at the beginning of these notes. Pólya (1921) proved that there exist infinitely many primes which divide some term u_m. Mahler (1934*b*) proved that $P(u_m) \to \infty$ as $m \to \infty$. The first general lower bound for $P(u_m)$ was given by Schinzel (1967), namely

$$P(u_m) > C_{27} m^{C_{28}}$$

where $C_{27} > 0$ is a computable constant depending only on the sequence, and C_{28} is an absolute positive constant. Stewart (1976, Ch. 3) gave the first proof of corollary 3.5(ii). As noted in the text, a proof of corollary 3.5(ii) where C_{17} depends on a and b only is given in Stewart (1982). In both publications he further proved that for almost all integers m

$$P(u_m) > \varepsilon(m) m \log m,$$

where $\varepsilon(m)$ is any real-valued function such that $\varepsilon(m) \to 0$ as $m \to \infty$. See also Pethö (1985) and papers of Pethö and de Weger to appear in *Math. Comp.*

Shorey (1983*a*) applied theorem 3.2 to obtain estimates for the greatest prime factor and the number of distinct prime factors of the product of blocks of consecutive terms in binary recurrence sequences.

Stewart (1982) also showed that

$$Q(u_m) > C_{29} \left(\frac{m}{(\log m)^2} \right)^{1/d},$$

where $C_{29} > 0$ is a computable constant depending only on a and b. In Stewart (1983) he proved that for Lucas and Lehmer numbers

$$Q(u_m) > m^{C_{30}(d(m) \log m)/(q(m) \log \log m)},$$

where $d(m)$ denotes the divisor function, $q(m) = 2^{\omega(m)}$ and C_{30} is a positive computable number depending only on the sequence. This result implies the assertion of theorem 3.7 for Lucas and Lehmer numbers. Further, Stewart (1983) proved that, for Lucas and Lehmer numbers, the inequality

$$Q(u_m) > m^{(\log m)^{1+\log 2-\varepsilon}} \quad (\varepsilon > 0),$$

is valid for almost all m. Shorey (1983*c*) proved the inequality of theorem 3.7 with $Q(u_m)$ replaced by

$$Q\left(\frac{[u_m, u_n]}{(u_m, u_n)} \right)$$

for all $m > n$ with $u_m u_n \neq 0$.

Lewis and Turk (1985) showed that the restriction $\tau > 1$ in theorem 3.4 can be removed if a_1, a_2 are fixed and $a_3 = a' a_1, a_4 = a' a_2$ where a' is a fixed algebraic number. Thus they proved: if $a' \neq 0$, then $x_m = a'x_n$ with $m \neq n$ implies that $\max(m, n) \leqslant C_{31}$ where C_{31} is a computable number depending only on $\{x_m\}_{m=0}^{\infty}$ and a'.

Ramanujan's function $\tau(n)$ satisfies a binary recurrence

$$\tau(p^{m+1}) = \tau(p)\tau(p^m) - p^{11}\tau(p^{m-1}) \quad (m = 1, 2, \ldots).$$

Ram Murty, Kumar Murty and Shorey (1986) applied the theory of linear forms in logarithms, via this relation, to show that, for an odd integer a, the equation $\tau(n) = a$ implies that $\log n \leqslant (2|a|)^{C_{32}}$ where C_{32} is a computable absolute constant.

CHAPTER 4

Recurrence sequences of order 2, 3 and 4

Let $\{u_m\}_{m=0}^{\infty}$ be a recurrence sequence of algebraic numbers. Let the minimal recurrence relation of $\{u_m\}_{m=0}^{\infty}$ be given by

$$u_{m+k} = v_{k-1}u_{m+k-1} + v_{k-2}u_{m+k-2} + \cdots + v_0 u_m \quad (m=0, 1, \ldots). \tag{1}$$

It follows from chapter C that $v_0, v_1, \ldots, v_{k-1}$ are algebraic numbers with $v_0 \neq 0$. Put $K = \mathbb{Q}(u_0, u_1, \ldots, u_{k-1}, v_0, v_1, \ldots, v_{k-1})$. Then $u_m \in K$ for all m. Write the companion polynomial of (1) as

$$G(z) = z^k - v_{k-1}z^{k-1} \cdots - v_0 = \prod_{j=1}^{s} (z - \omega_j)^{\sigma_j}, \tag{2}$$

where $\omega_1, \omega_2, \ldots, \omega_s$ are distinct complex numbers and $\sigma_1, \sigma_2, \ldots, \sigma_s$ positive integers. Without loss of generality we may assume

$$|\omega_1| \geqslant |\omega_2| \geqslant \cdots \geqslant |\omega_s| > \omega_{s+1} := 0. \tag{3}$$

Define r by

$$|\omega_1| = |\omega_2| = \cdots = |\omega_r| > |\omega_{r+1}|. \tag{4}$$

By theorem C.1 and the minimality of the recurrence there exist non-trivial polynomials f_j with coefficients in $K(\omega_1, \omega_2, \ldots, \omega_s)$ and of degree $\sigma_j - 1$ for $j = 1, 2, \ldots, s$ such that

$$u_m = \sum_{j=1}^{s} f_j(m)\omega_j^m \quad (m=0, 1, 2, \ldots). \tag{5}$$

We shall use the above notation throughout the chapter without further reference. Furthermore, m and n will denote non-negative integers.

We consider the equation

$$u_m = u_n \tag{6}$$

in integers m, n with $m>n$. If $r=1$, a direct application of theorem B.1 yields the following result of Mignotte (1979).

Theorem 4.1. *Assume* $r=1, |\omega_1|>1$. *Equation* (6) *with* $m>n$ *implies that* m *is bounded by a computable number depending only on the sequence* $\{u_m\}_{m=0}^{\infty}$.

In case $r=2$ a direct application of theorem 3.5 gives an analogous result:

Theorem 4.2. *Assume* $r=2, |\omega_1|>1, \omega_1/\omega_2$ *is not a root of unity. Equation* (6) *with* $m>n$ *implies that* m *is bounded by a computable number depending only on the sequence* $\{u_m\}_{m=0}^{\infty}$.

If $s=2$, we can use theorem 3.5 to replace $|\omega_1|>1$ by $|\overline{\omega_1}|>1$ in theorem 4.2. The analogous result for $r=3$ requires a more complicated argument involving corollary B.2 and an argument due to Beukers (see Beukers and Tijdeman, 1984). Theorems 4.3–4.8 are contained in Mignotte, Shorey and Tijdeman (1984).

Theorem 4.3. *Assume* $r=3, |\omega_1|>1$ *and at least one of the numbers* ω_1/ω_2, ω_2/ω_3 *and* ω_3/ω_1 *is not a root of unity. Equation* (6) *with* $m>n$ *implies that* m *is bounded by a computable number depending only on the sequence* $\{u_m\}_{m=0}^{\infty}$.

Theorem 4.3 is contained in the following result which is an extension of theorem 3.3.

Theorem 4.4. *Assume* $r=3, |\omega_1|>1$ *and at least one of the numbers* ω_1/ω_2, $\omega_2/\omega_3, \omega_3/\omega_1$ *is not a root of unity. There exist computable numbers* C_1 *and* C_2 *depending only on the sequence* $\{u_m\}_{m=0}^{\infty}$ *such that*

$$|u_m-u_n|\geqslant|\omega_1|^m\exp(-C_1(\log m)^2\log(n+2)) \tag{7}$$

whenever $m\geqslant C_2$ *and* $m>n$.

Theorem 4.4 will be deduced from the following result corresponding to the case $n=0$.

Theorem 4.5. *Assume* $r=3$ *and at least one of the numbers* ω_1/ω_2, ω_2/ω_3, ω_3/ω_1 *is not a root of unity. Then there exist computable numbers* C_3 *and* C_4 *depending only on the sequence* $\{u_m\}_{m=0}^{\infty}$ *such that*

$$|u_m|\geqslant|\omega_1|^m\exp(-C_3(\log m)^2) \tag{8}$$

whenever $m\geqslant C_4$.

Theorem 4.5 and corollary 3.7 imply the following result on 0-multiplicity.

Corollary 4.1. *Assume $r \leqslant 3$ and ω_2/ω_1 is not a root of unity. The equation $u_m = 0$ implies that m is bounded by a computable number depending only on the sequence $\{u_m\}_{m=0}^{\infty}$.*

We cannot prove an analogous result for $r = 4$, but we have

Theorem 4.6. *Assume $s \leqslant 4$ and $\{u_m\}_{m=0}^{\infty}$ is a non-degenerate recurrence sequence of real algebraic numbers. The equation $u_m = 0$ implies that m is bounded by a computable number depending only on the sequence $\{u_m\}_{m=0}^{\infty}$.*

The proofs of theorems 4.3, 4.4, 4.5 and 4.6 depend on the following two results.

Theorem 4.7. *Let A_1, A_2, A_3 be non-zero algebraic numbers of degrees at most d and of heights at most H ($\geqslant 2$). Let $\gamma_1, \gamma_2, \gamma_3$ be non-zero algebraic numbers such that at least one of the numbers $\gamma_1/\gamma_2, \gamma_2/\gamma_3, \gamma_3/\gamma_1$ is not a root of unity. Then the equation*

$$A_1\gamma_1^m + A_2\gamma_2^m + A_3\gamma_3^m = 0 \tag{9}$$

implies that $m \leqslant C_5 \log H$ for some computable number C_5 depending only on $\gamma_1, \gamma_2, \gamma_3$ and d.

Theorem 4.8. *Let A_1, A_2, A_3 be as in theorem 4.7. Let $\gamma_1, \gamma_2, \gamma_3$ be algebraic numbers with $|\gamma_1| = |\gamma_2| = |\gamma_3|$. Let $m \geqslant 2$. Then either $A_1\gamma_1^m + A_2\gamma_2^m + A_3\gamma_3^m = 0$ or*

$$|A_1\gamma_1^m + A_2\gamma_2^m + A_3\gamma_3^m| \geqslant |\gamma_1|^m m^{-C_6 \log H}$$

for some computable number C_6 depending only on $\gamma_1, \gamma_2, \gamma_3$ and d.

Proofs

The proofs of the theorems depend on the following result.

Lemma 4.1. *Let $f(z)$ be a non-constant polynomial with algebraic coefficients and ω a non-zero algebraic number. The equation*

$$f(m)\omega^m = f(n)\omega^n \quad (m > n) \tag{10}$$

implies that m is bounded by a computable number depending only on f and ω.

Proof. Suppose that m and n with $m > n$ satisfy (10). Denote by $b_1, b_2, \ldots$ computable positive numbers depending only on f and ω. We may assume that $m \geqslant b_1$ with b_1 sufficiently large. Denote by L the field generated over $\mathbb{Q}$ by ω and the coefficients of f. Write $v = \deg f \geqslant 1$.

Suppose ω is a root of unity. Then $\omega^\mu = 1$ for some positive integer μ. Consequently, by (10),

$$g(m) = g(n) \tag{11}$$

where

$$g(x) = (f(x))^\mu.$$

Observe that $g(x)$ is a polynomial of degree $\rho = \mu\nu \geqslant 1$. By (11), we see that $\rho \geqslant 2$ and

$$m^{\rho-1} \leqslant \frac{m^\rho - n^\rho}{m-n} \leqslant b_2 m^{\rho-2}$$

which implies that $m \leqslant b_2$.

Thus we may assume that ω is not a root of unity. We first prove that

$$m - n \leqslant b_3 \log m. \tag{12}$$

If ω is not a unit, there exists a prime ideal $\mathfrak{p}$ in the ring of integers of L such that $\operatorname{ord}_{\mathfrak{p}}(\omega)$ is non-zero. Counting the power of the prime ideal $\mathfrak{p}$ on both sides in (10), we obtain (12) from lemma A.7. Suppose ω is a unit. Then, since ω is not a root of unity, there exists an embedding σ of L such that $|\sigma(\omega)| > 1$. Further, by taking images under σ on both the sides in (10), we have

$$|\sigma(\omega)|^{m-n} = \left|\frac{\sigma(f(n))}{\sigma(f(m))}\right|$$

and inequality (12) follows from a Liouville-type argument.

Re-writing (10), we have

$$\omega^{m-n} - 1 = \frac{f(n) - f(m)}{f(m)}. \tag{13}$$

Observe that

$$|f(n) - f(m)| \leqslant b_4 (m-n) m^{\nu-1}$$

and, by taking b_1 large enough,

$$|f(m)| \geqslant b_5 m^\nu.$$

Thus we obtain from (13) and (12), since ω is not a root of unity,

$$0 < |\omega^{m-n} - 1| \leqslant b_6 m^{-1} \log m.$$

Now apply corollary B.1 with $n = 1$, $d = b_7$, $\log A_1 = b_8$ and, by (12), $B = m - n + 1 \leqslant 2b_3 \log m$ to conclude that

$$|\omega^{m-n} - 1| > (\log m)^{-b_9}.$$

Consequently $m \leqslant b_6 (\log m)^{b_9+1}$ which implies that $m \leqslant b_{10}$. □

Corollary 4.2. *Let $f(z)$ be a non-trivial polynomial with algebraic coefficients and ω a non-zero algebraic number. Suppose that ω is not a root of unity. Then (10) implies that m is bounded by a computable number depending only on f and ω.*

Corollary 4.2 with '$|\omega|>1$' in place of 'ω not a root of unity' is sufficient for our purpose.

Proof. In view of lemma 4.1, we may assume that deg $f=0$. Then equation (10) implies that ω is a root of unity. □

The constants $c_1, c_2, \ldots$ in the proofs of theorems 4.1–4.6 and corollary 4.1 are computable positive numbers depending only on the sequence $\{u_m\}_{m=0}^{\infty}$.

Proof of theorem 4.1. Suppose that equation (6) with $m>n$ is valid. We may assume that $m \geqslant c_1$ with c_1 sufficiently large. Then, by corollary 4.2 with $f=f_1$ and $\omega=\omega_1$, we see that $s \geqslant 2$ and $f_1(m)\omega_1^m \neq f_1(n)\omega_1^n$. Further

$$0<|f_1(m)\omega_1^m - f_1(n)\omega_1^n| \leqslant m^{c_2} \max(1, |\omega_2|^m)$$

which, by taking c_1 large enough, implies that

$$0<\left|\omega_1^{n-m}\frac{f_1(n)}{f_1(m)}-1\right| \leqslant m^{c_3}\max(|\omega_1|^{-m}, |\omega_1/\omega_2|^{-m}). \tag{14}$$

We apply corollary B.1 with $n=2, d=c_4, \log A_1=c_5, \log A_2=c_6 \log m$ and $B=m$ to conclude that

$$\left|\omega_1^{n-m}\frac{f_1(n)}{f_1(m)}-1\right| > \exp(-c_7(\log m)^2). \tag{15}$$

It follows from (15), (14), $|\omega_1|>1$ and $|\omega_1/\omega_2|>1$ that $m \leqslant c_8(\log m)^2$ which implies that $m \leqslant c_9$. □

Proof of theorem 4.2. Suppose that equation (6) with $m>n$ is valid. We may assume that $m \geqslant c_{10}$ with c_{10} sufficiently large. Then, by corollary 4.2 with $f=f_1$ and $\omega=\omega_1$, we obtain

$$f_1(m)\omega_1^m \neq f_1(n)\omega_1^n. \tag{16}$$

In the notation of corollary 3.7, put $a_1=f_1(m)$, $a_2=f_2(m)$, $\log H = c_{22}\log m, \lambda=\omega_1, \mu=\omega_2$ and $x_m=f_1(m)\omega_1^m+f_2(m)\omega_2^m$. Observe that λ/μ is not a root of unity. By taking c_{10} large enough, we see that $a_1a_2 \neq 0$. Hence, by corollary 3.7,

$$|x_m| \geqslant |\omega_1|^m \exp(-c_{11}(\log m)^2)$$

which, together with $r=2$, implies that

$$|u_m| \geqslant |\omega_1|^m \exp(-c_{12}(\log m)^2). \tag{17}$$

Further observe that

$$|u_n| \leqslant m^{c_{13}}|\omega_1|^n. \tag{18}$$

It follows from (6), (17), (18) and $|\omega_1|>1$ that

$$m-n \leqslant c_{14}(\log m)^2.$$

Consequently, by taking c_{10} large enough, we obtain

$$n \geqslant 2^{-1}m \geqslant 2^{-1}c_{10}.$$

In the notation of theorem 3.5, put $a_1=f_1(m)$, $a_2=f_2(m)$, $a_3=f_1(n)$, $a_4=f_2(n)$, $\log H=c_{27}\log m$, $A=B=1$, $H'=2$, $\lambda=\omega_1$, $\mu=\omega_2$ and

$$x_m=f_1(m)\omega_1^m+f_2(m)\omega_2^m, \quad y_n=f_1(n)\omega_1^n+f_2(n)\omega_2^n.$$

Observe, by taking c_{10} sufficiently large, that $a_1a_2a_3a_4\neq 0$. Further notice that $|\lambda|=|\mu|>1$, λ/μ is not a root of unity and, by (16), $Aa_1\lambda^m\neq Ba_3\lambda^n$. Hence, by theorem 3.5, we conclude that

$$|x_m-y_n| \geqslant |\omega_1|^m \exp(-c_{15}(\log m)^3). \tag{19}$$

In particular $x_m\neq y_n$ and consequently, by (6), we see that $s>2$. Further, it follows from (6) that

$$|x_m-y_n| \leqslant m^{c_{16}}\max(1,|\omega_3|^m). \tag{20}$$

Now it follows from (19), (20), $|\omega_1|>|\omega_3|$ and $|\omega_1|>1$ that $m\leqslant c_{17}(\log m)^3$. Hence $m\leqslant c_{18}$. □

The constants e_1, e_2, ... in the proofs of theorems 4.7 and 4.8 are computable positive numbers depending only on $\gamma_1,\gamma_2,\gamma_3$ and d.

Proof of theorem 4.7. Suppose that (9) is valid. By interchanging the indices of γ_1,γ_2 and γ_3, there is no loss of generality in assuming that γ_1/γ_2 is not a root of unity. Further, we may assume that $m\geqslant e_1\log H$ with e_1 sufficiently large. Denote by L the field generated over $\mathbb{Q}$ by $A_1, A_2, A_3, \gamma_1, \gamma_2$ and γ_3. Observe that $[L:\mathbb{Q}]\leqslant e_2$.

Assume that γ_1/γ_2 is not a unit. Then there exists a prime ideal $\mathfrak{p}$ in the ring of integers of L such that $\mathrm{ord}_{\mathfrak{p}}(\gamma_1/\gamma_2)$ is non-zero. By permuting the indices of γ_1 and γ_2, we may assume that $\mathrm{ord}_{\mathfrak{p}}(\gamma_1/\gamma_2)>0$. Then, by (9), we see that $A_2\gamma_2^m+A_3\gamma_3^m\neq 0$ and

$$m\leqslant \mathrm{ord}_{\mathfrak{p}}((\gamma_1/\gamma_2)^m)=\mathrm{ord}_{\mathfrak{p}}(A_2A_1^{-1})+\mathrm{ord}_{\mathfrak{p}}\left(-\left(\frac{\gamma_3}{\gamma_2}\right)^m\frac{A_3}{A_2}-1\right). \tag{21}$$

We apply corollary B.2 with $n=3, d=e_3, \delta=\frac{1}{2}, A'=e_4, A=H^{e_5}$ and $B=m$ to conclude that

$$\operatorname{ord}_{\wp}\left(-\left(\frac{\gamma_3}{\gamma_2}\right)^m \frac{A_3}{A_2}-1\right) \leqslant m/2, \tag{22}$$

if e_1 is sufficiently large. Consequently we see from (21), (22), lemma A.3 and lemma A.7 that $m \leqslant e_6 \log H$ which is not possible if $e_1 > e_6$.

Thus we may suppose that γ_1/γ_2 is a unit. Then, since γ_1/γ_2 is not a root of unity, there exists an embedding σ of L such that $|\sigma(\gamma_1/\gamma_2)|>1$. Therefore $|\sigma(\gamma_1)|>|\sigma(\gamma_2)|$. By taking images under σ on both sides of (9), we may assume that $|\gamma_1|>|\gamma_2|$. By (9), $A_1\gamma_1^m + A_3\gamma_3^m \neq 0$. Further, we may write

$$0 \neq |A_1\gamma_1^m + A_3\gamma_3^m| = |A_1\gamma_1^m| \left|-\left(\frac{\gamma_3}{\gamma_1}\right)^m \frac{A_3}{A_1} - 1\right|.$$

We apply theorem B.2 with $n=3, \delta=\min(\frac{1}{6}\log|\gamma_1/\gamma_2|, \frac{1}{4}), d=e_7, \log A'=e_8$, $\log A = e_9 \log H$, $B'=1$ and $B''=m$ to conclude that

$$\left|-\left(\frac{\gamma_3}{\gamma_1}\right)^m \frac{A_3}{A_1} - 1\right| \geqslant \left|\frac{\gamma_1}{\gamma_2}\right|^{-m/2} H^{-e_{10}}.$$

Thus, by (A.7),

$$|A_1\gamma_1^m + A_3\gamma_3^m| \geqslant |\gamma_1|^m \left|\frac{\gamma_1}{\gamma_2}\right|^{-m/2} H^{-e_{11}}. \tag{23}$$

Further, by lemma A.1,

$$|A_2\gamma_2^m| \leqslant dH|\gamma_2|^m. \tag{24}$$

By (9), (23) and (24),

$$\left|\frac{\gamma_1}{\gamma_2}\right|^{m/2} \leqslant H^{e_{12}}$$

which, together with $|\gamma_1/\gamma_2|>1$, implies that $m \leqslant e_{13} \log H$. □

Proof of theorem 4.8. Without loss of generality we may assume $|A_2| \leqslant |A_1| \leqslant |A_3|$ and $A_1\gamma_1^m + A_2\gamma_2^m + A_3\gamma_3^m \neq 0$. We may also assume that $\gamma_1, \gamma_2, \gamma_3$ are distinct; otherwise the assertion follows from corollary B.1. Put

$$v_m = -\frac{A_1}{A_3}\left(\frac{\gamma_1}{\gamma_3}\right)^m - \frac{A_2}{A_3}\left(\frac{\gamma_2}{\gamma_3}\right)^m - 1 \quad (m=1,2,\ldots). \tag{25}$$

Then v_m is of the form

$$v_m = a_1\alpha_1^m - a_2\alpha_2^m - 1, \tag{26}$$

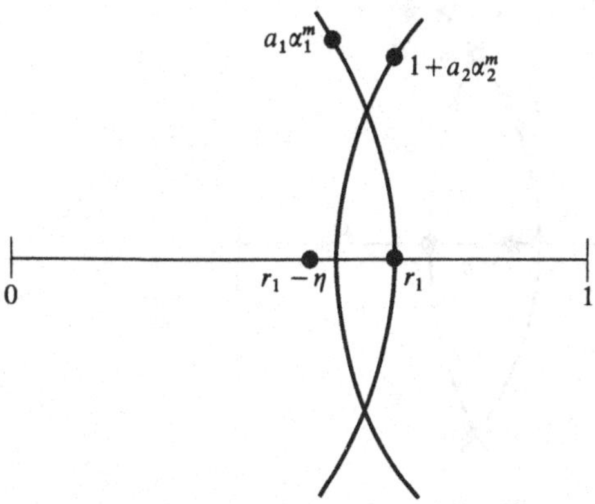

Fig. 4.1

with $a_1, a_2, \alpha_1, \alpha_2$ algebraic, $0<|a_2|\leqslant|a_1|\leqslant 1$, whereas α_1 and α_2 are distinct and $|\alpha_1|=|\alpha_2|=1$. Note that $v_m\neq 0$ and, by lemma A.3, the heights of a_1 and a_2 do not exceed $H^{e_{14}}$. Let η be any positive number with $\eta<\frac{1}{2}$. We distinguish two cases:

Case (*i*). $|a_1|+|a_2|\leqslant 1+\eta$ (see fig. 4.1). Put $|a_1\alpha_1^m|=|a_1|=r_1$. Let $|v_m|<\frac{1}{2}$. We have

$$\mathrm{Re}(1+a_2\alpha_2^m)\geqslant r_1-\eta, \quad r_1-\mathrm{Re}(a_1\alpha_1^m)\leqslant|v_m|+\eta.$$

Hence

$$(\mathrm{Im}(a_1\alpha_1^m))^2=r_1^2-(\mathrm{Re}(a_1\alpha_1^m))^2\leqslant 2r_1(|v_m|+\eta).$$

It follows that

$$|r_1-a_1\alpha_1^m|\leqslant|r_1-\mathrm{Re}(a_1\alpha_1^m)|+|\mathrm{Im}(a_1\alpha_1^m)|\leqslant(|v_m|+\eta)+2\sqrt{(|v_m|+\eta)}\leqslant 3\sqrt{(|v_m|+\eta)}.$$

Thus

$$|v_m|\geqslant\min(\tfrac{1}{2}, \tfrac{1}{9}|r_1-a_1\alpha_1^m|^2-\eta). \tag{27}$$

Case (*ii*). $|a_1|+|a_2|>1+\eta$ (see fig. 4.2). Put $r_1=|a_1|, r_2=|a_2|$. Note that $z_1:=a_1\alpha_1^m$ is on the circle $|z|=r_1$ and that $z_2:=a_2\alpha_2^m+1$ is on the circle $|z-1|=r_2$. These circles intersect in two points, z_0 and $\overline{z_0}$. We may assume that $(\mathrm{Im}\, z_0)(\mathrm{Im}\, z_1)\geqslant 0$. Because of symmetry it is no restriction to assume that $\mathrm{Im}\, z_0>0$. Put $z_0=x+iy$. By $x^2+y^2=r_1^2$, $(1-x)^2+y^2=r_2^2$ and $y\leqslant r_2\leqslant r_1$, we have

$$1=x+(1-x)=\sqrt{(r_1^2-y^2)}+\sqrt{(r_2^2-y^2)}$$
$$=r_1\sqrt{\left(1-\frac{y^2}{r_1^2}\right)}+r_2\sqrt{\left(1-\frac{y^2}{r_2^2}\right)}\geqslant r_1\left(1-\frac{y^2}{r_1^2}\right)+r_2\left(1-\frac{y^2}{r_2^2}\right).$$

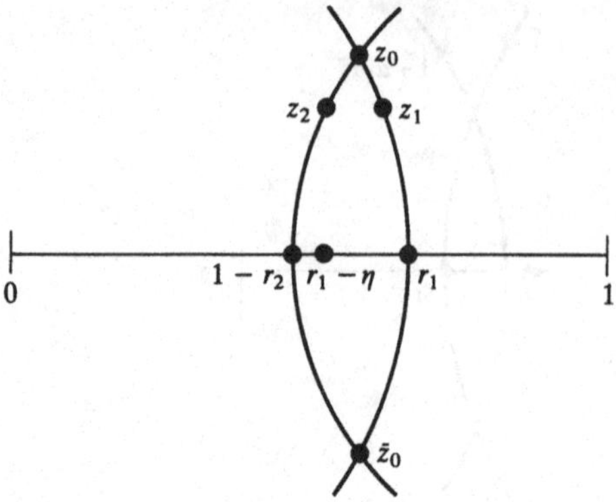

Fig. 4.2

We infer, by $r_2 \geqslant 1+\eta-r_1 \geqslant \eta$,

$$\eta < r_1+r_2-1 \leqslant \frac{y^2}{r_1}+\frac{y^2}{r_2} \leqslant \frac{2y^2}{r_2} \leqslant \frac{2y^2}{\eta}. \tag{28}$$

We shall derive a lower bound for $|v_m|=|z_1-z_2|$. Observe that

$$|z_1-1|^2=r_1^2+1-2\,\mathrm{Re}\,z_1,$$

$$|z_2-1|^2=|z_0-1|^2=r_1^2+1-2\,\mathrm{Re}\,z_0.$$

Hence

$$2|\mathrm{Re}\,z_1-\mathrm{Re}\,z_0|=\big||z_1-1|^2-|z_2-1|^2\big| \leqslant |z_1-z_2|(|z_1|+|z_2|+2)$$

which implies

$$|\mathrm{Re}\,z_1-\mathrm{Re}\,z_0| \leqslant 3|z_1-z_2|. \tag{29}$$

From $|z_0|=|z_1|$ and (29), we obtain

$$|(\mathrm{Im}\,z_1)^2-(\mathrm{Im}\,z_0)^2|=|(\mathrm{Re}\,z_1)^2-(\mathrm{Re}\,z_0)^2| \leqslant 6|z_1-z_2|. \tag{30}$$

On the other hand, by $\mathrm{Im}\,z_1 \geqslant 0$ and $\mathrm{Im}\,z_0=y>0$,

$$|(\mathrm{Im}\,z_1)^2-(\mathrm{Im}\,z_0)^2| \geqslant |\mathrm{Im}\,z_1-\mathrm{Im}\,z_0|\,y. \tag{31}$$

On combining (31), (30) and (28), we see that

$$|\mathrm{Im}\,z_1-\mathrm{Im}\,z_0| \leqslant \frac{6\sqrt{2}}{\eta}|z_1-z_2|. \tag{32}$$

By (29) and (32),

$$|z_1-z_0| \leqslant \left(3+\frac{9}{\eta}\right)|z_1-z_2|,$$

or, equivalently,

$$|v_m| \geq \frac{\eta}{3\eta+9}|z_0 - a_1\alpha_1^m|. \tag{33}$$

Subsequently we show that z_0 is an algebraic number of degree at most $e_{15}H^{e_{16}}$. By $z_0\overline{z_0}=a_1\overline{a_1}$ and $(1-z_0)(1-\overline{z_0})=(1-a_2)(1-\overline{a_2})$ we have $z_0+\overline{z_0}=a_2+\overline{a_2}-a_2\overline{a_2}+a_1\overline{a_1}$. Thus

$$(z-z_0)(z-\overline{z_0})=z^2-(a_2+\overline{a_2}-a_2\overline{a_2}+a_1\overline{a_1})z+a_1\overline{a_1}.$$

This proves that z_0 is algebraic indeed and that it can be expressed in terms of a_1 and a_2 by using only sums, products and square roots.

On using lemmas A.3 and A.4 for the heights of sums, products and square roots of algebraic numbers it becomes clear that we may assume without loss of generality that the heights of the numbers $A_3, a_1, a_2, r_1 = (a_1\overline{a_1})^{1/2}, r_1-1, r_1^2/9, a_1/r_1, a_2/(r_1-1), z_0, z_0-1, a_1/z_0$ and $a_2/(z_0-1)$ are at most $e_{17}H^{e_{18}}$. By a Liouville-type argument these numbers are in absolute values at least $e_{19}H^{-e_{18}}$. We use these estimates without reference.

Case (i). If $a_1\alpha_1^m = r_1 = 1$, then $v_m = -a_2\alpha_2^m$ and hence

$$|v_m| = |a_2||\alpha_2|^m = |a_2| \geq e_{19}H^{-e_{18}} \geq \exp(-e_{20}\log m \log H). \tag{34}$$

If $a_1\alpha_1^m = r_1 < 1$, then $v_m = r_1 - 1 - a_2\alpha_2^m$. On applying corollary B.1 we obtain

$$|v_m| = |r_1 - 1|\left|\alpha_2^m\frac{a_2}{r_1-1} - 1\right| \geq e_{19}H^{-e_{18}}\exp(-e_{21}\log m \log H)$$

$$\geq \exp(-e_{22}\log m \log H). \tag{35}$$

If $a_1\alpha_1^m \neq r_1$, then we find by applying corollary B.1,

$$\frac{1}{9}|r_1 - a_1\alpha_1^m|^2 = \frac{r_1^2}{9}\left|\alpha_1^m\frac{a_1}{r_1} - 1\right|^2 \geq 2\exp(-e_{23}\log m \log H)$$

where e_{23} is so large that the right-hand side is less than 1. Put $\eta = \exp(-e_{23}\log m \log H)$. Then $\eta < \frac{1}{2}$ and, by (27),

$$|v_m| \geq \eta \geq \exp(-e_{24}\log m \log H). \tag{36}$$

Case (ii). If $a_1\alpha_1^m = z_0$, then $v_m = z_0 - 1 - a_2\alpha_2^m$, hence, by corollary B.1,

$$|v_m| = |z_0 - 1|\left|\alpha_2^m\frac{a_2}{z_0-1} - 1\right| \geq \exp(-e_{25}\log m \log H). \tag{37}$$

If $a_1\alpha_1^m \neq z_0$, then, by (33), $\eta<1$ and corollary B.1,

$$|v_m| \geqslant \frac{\eta|z_0|}{12}\left|\alpha_1^m \frac{a_1}{z_0}-1\right| \geqslant \exp(-e_{26}\log m \log H). \tag{38}$$

By (34)–(38) and $|\gamma_1|=|\gamma_3|$ we deduce

$$|A_1\gamma_1^m+A_2\gamma_2^m+A_3\gamma_3^m|=|A_3||v_m||\gamma_1|^m \geqslant \exp(-e_{27}\log m \log H)|\gamma_1|^m. \quad \square$$

Proof of theorem 4.5. We may assume that $m \geqslant c_{19}$ with c_{19} sufficiently large. Put

$$\Lambda=f_1(m)\omega_1^m+f_2(m)\omega_2^m+f_3(m)\omega_3^m.$$

We apply theorem 4.7 with $A_i=f_i(m)$, $\gamma_i=\omega_i$ for $1\leqslant i\leqslant 3$ and $\log H=c_{20}\log m$ to conclude that

$$\Lambda \neq 0$$

if c_{19} is sufficiently large. Then, by theorem 4.8 with the same choice of the parameters, we obtain

$$|\Lambda| \geqslant |\omega_1|^m \exp(-c_{21}(\log m)^2). \tag{39}$$

Further observe that

$$|u_m| \geqslant |\Lambda|-\delta \tag{40}$$

where $\delta=m^{c_{22}}|\omega_4|^m$. Now estimate (8) follows from (40), (39) and $|\omega_1|>|\omega_4|$. $\square$

Proof of theorem 4.4. Let $m>n$. We may assume that $m>c_{23}$ with c_{23} sufficiently large. Let $c_{23}>C_4$ so that estimate (8) is valid. We may assume that

$$|u_m|<2|u_n|; \tag{41}$$

otherwise

$$|u_m-u_n| \geqslant |u_m|-|u_n| \geqslant \frac{|u_m|}{2}$$

and (7) follows from (8). Further, observe that inequality (18) is valid. Now it follows from (41), (8), (18) and $|\omega_1|>1$ that

$$m-n \leqslant c_{24}(\log m)^2. \tag{42}$$

Then $n \geqslant 2$. For $1\leqslant i\leqslant 3$, put

$$B_i=f_i(m)\omega_i^{m-n}-f_i(n)$$

and

$$\Lambda_1 = B_1\omega_1^n + B_2\omega_2^n + B_3\omega_3^n.$$

By taking c_{23} sufficiently large, it follows from corollary 4.2 and $\min(|\omega_1|, |\omega_2|, |\omega_3|) > 1$ that $B_1B_2B_3 \neq 0$. We apply theorem 4.7 with $A_i = B_i, \gamma_i = \omega_i$ for $1 \leqslant i \leqslant 3$ and, by (42), $\log H \leqslant c_{25}(\log m)^2$ to conclude that

$$\Lambda_1 \neq 0$$

if c_{23} is large enough. Then, by theorem 4.8 with $m = n$, $A_i = B_i, \gamma_i = \omega_i$ for $i = 1, 2, 3$ and $\log H \leqslant c_{25}(\log m)^2$, we obtain

$$|\Lambda_1| \geqslant |\omega_1|^n \exp(-c_{26}(\log m)^2 \log n)$$

which, together with (42), implies that

$$|\Lambda_1| \geqslant |\omega_1|^m \exp(-c_{27}(\log m)^2 \log n). \tag{43}$$

Observe that

$$|u_m - u_n| \geqslant |\Lambda_1| - \delta_1 \tag{44}$$

where $\delta_1 \leqslant m^{c_{28}} \max(1, |\omega_4|^m)$. Now inequality (7) follows from (44), (43) and $|\omega_1| > \max(1, |\omega_4|)$. □

Proof of corollary 4.1. If $r = 3$, then it follows from theorem 4.5 that $m < C_4$. Suppose $r \leqslant 2$. If $s = 1$, then $u_m = f_1(m)\omega_1^m$ with f_1 non-trivial, $\omega_1 \neq 0$, and the assertion is obvious. Since we may replace u_m by $v^m u_m$ for any integer m, and hence ω_i by $v\omega_i$ for $i = 1, 2, \ldots, s$, it involves no loss of generality to assume that $\omega_1, \ldots, \omega_s$ are algebraic integers. Then, since ω_2/ω_1 is not a root of unity, we see from lemma A.5 that $\max(\overline{|\omega_1|}, \overline{|\omega_2|}) > 1$. Therefore we may assume $|\omega_1| > 1$. Apply corollary 3.7 with $\lambda = \omega_1$, $\mu = \omega_2$, $a_1 = f_1(m)$, $a_2 = f_2(m)$, $H = c_{29} \log m$. Hence

$$|f_1(m)\omega_1^m + f_2(m)\omega_2^m| \geqslant |\omega_1|^m \exp(-c_{30}(\log m)^2)$$

for $m \geqslant c_{31}$. If $s > 2$, then the contribution of the other terms is of smaller order. Hence,

$$|u_m| \geqslant |\omega_1|^m \exp(-c_{32}(\log m)^2)$$

for $m \geqslant c_{33}$. Since $u_m = 0$, we have $m < c_{33}$. □

Proof of theorem 4.6. By corollary 4.1 we may assume $r = 4$. By § 2 of chapter C the sequence $\{u_m\}_{m=0}^{\infty}$ is induced by a recurrence of real algebraic numbers. By considering the sequence $\{v^m u_m\}_{m=0}^{\infty}$ where v is a positive integer such that $v\omega_1, \ldots, v\omega_4$ are algebraic integers, we see that there is no loss of generality in assuming that the coefficients of the companion polynomial to $\{u_m\}_{m=0}^{\infty}$ are real algebraic integers. Consequently $\omega_1,\ldots,\omega_4$ are algebraic integers. Put $L=\mathbb{Q}(\omega_1,\ldots,\omega_4)$ and denote by h the class number of L. For an integer $m\geqslant 0$ let

$$u_m=0. \tag{45}$$

Since the companion polynomial to $\{u_m\}_{m=0}^{\infty}$ has real coefficients and none of the quotients ω_i/ω_j $(1\leqslant i<j\leqslant 4)$ is ± 1, it follows that $\omega_1,\omega_2,\omega_3,\omega_4$ are non-real. Further, by permuting the indices of $\omega_1,\omega_2,\omega_3,\omega_4$, there is no loss of generality in assuming that

$$\omega_1=\overline{\omega_3},\quad \omega_2=\overline{\omega_4}. \tag{46}$$

Suppose ω_1/ω_3 is a unit. Since ω_1/ω_3 is not a root of unity, there exists an embedding σ of L such that $|\sigma(\omega_1)|>|\sigma(\omega_3)|$. Further, (45) implies that $\sigma(u_m)=0$. Now we apply corollary 4.1 to $\{\sigma(u_m)\}_{m=0}^{\infty}$ to conclude that m is bounded by a computable number depending only on the sequence $\{u_m\}_{m=0}^{\infty}$.

Thus we may assume that ω_1/ω_3 is not a unit. Further, we see from (A.36) that

$$([\omega_1^h],[\omega_2^h],[\omega_3^h],[\omega_4^h])=[\pi]$$

where π is an algebraic integer in L. For $1\leqslant i\leqslant 4$, put

$$W_i=\omega_i^h\pi^{-1}. \tag{47}$$

Notice that W_1, W_2, W_3, W_4 are algebraic integers in L satisfying

$$([W_1],[W_2],[W_3],[W_4])=[1]. \tag{48}$$

It follows from (47), (46) and $|\omega_3|=|\omega_4|$ that

$$W_1W_3=W_2W_4. \tag{49}$$

Since ω_1/ω_3 is not a unit, it follows from (47) that W_1/W_3 is not a unit. We know that W_1 and W_3 are algebraic integers. Hence W_1W_3 is not a unit. Thus there exists a prime ideal $\mathfrak{p}$ in the ring of integers of L such that $\mathfrak{p}\mid W_1W_3$. Consequently, by (49), $\mathfrak{p}\mid W_2W_4$. By permuting the indices of W_1, W_3 and W_2, W_4, there is no loss of generality in assuming that

$$\mathfrak{p}\mid W_3,\quad \mathfrak{p}\mid W_4. \tag{50}$$

Write $m=nh+q$ with $0\leqslant q<h$ and $p_i(z)=f_i(z)\omega_i^q$ for $1\leqslant i\leqslant 4$. We may assume that $n\geqslant c_{34}$ with c_{34} sufficiently large; otherwise the assertion of the theorem follows. Dividing both sides of (45) by π^n, we obtain

$$p_1(m)W_1^n+p_2(m)W_2^n=-p_3(m)W_3^n-p_4(m)W_4^n.$$

By counting the power of prime ideal $\mathfrak{p}$ on both sides, it follows from (50) and lemma A.7 that

$$n \leqslant c_{35} \log m + \operatorname{ord}_{\mathfrak{p}}(\Delta) \tag{51}$$

where

$$\Delta = p_1(m) W_1^n + p_2(m) W_2^n.$$

By (47) and ω_1/ω_2 not a root of unity, we see that W_1/W_2 is not a root of unity. Further, by taking c_{34} large enough, it follows from corollary 4.1 that $\Delta \neq 0$.

In view of (48) and (50), we find that either $\mathfrak{p} \nmid W_1$ or $\mathfrak{p} \nmid W_2$. For simplicity, assume that $\mathfrak{p} \nmid W_1$. Then, by lemma A.7,

$$\operatorname{ord}_{\mathfrak{p}}(\Delta) \leqslant c_{36} \log m + \operatorname{ord}_{\mathfrak{p}}\left(-\left(\frac{W_2}{W_1}\right)^n \frac{p_2(m)}{p_1(m)} - 1 \right). \tag{52}$$

We apply theorem B.3 with $n=3$, $d=c_{37}$, $p=c_{38}$, $\log A_1 = \log A_2 = c_{39}$, $\log A_3 = c_{40} \log m$ and $B = n \leqslant m$ to conclude that

$$\operatorname{ord}_{\mathfrak{p}}\left(-\left(\frac{W_2}{W_1}\right)^n \frac{p_2(m)}{p_1(m)} - 1 \right) \leqslant c_{41} (\log m)^3. \tag{53}$$

Combining (51), (52), (53) and $n+1 \geqslant m/h$, we obtain $m \leqslant c_{42} (\log m)^3$ which implies that $m \leqslant c_{43}$. □

Notes

In §3 of chapter C the theorem of Skolem–Mahler–Lech was stated. The result was proved by Skolem (1935) for rational recurrence sequences, by Mahler (1935*a*) for algebraic sequences, by Lech (1953) for sequences in a field of characteristic 0, and later once more for complex sequences by Mahler (1956). Robba (1978) derived an upper bound for the period length of the vanishing terms in an algebraic recurrence sequence with infinite 0-multiplicity (see also Mignotte (1978)).

Pólya (1921) proved that if $\{u_m\}_{m=0}^{\infty}$ is a non-degenerate rational recurrence sequence of order at least 2, then there exist infinitely many primes p with $p \mid u_m$ for some m. This result was rediscovered by Ward (1954, 1955*b*) for integer sequences of order 2 or 3, and by Laxton (1974) for integer sequences of arbitrary order (see van Leeuwen (1980)). It was proved by Berstel and Mignotte (1976) that for a given integer recurrence sequence the following two questions are decidable. (i) Is the 0-multiplicity finite? (ii) Is the set of prime numbers dividing at least one term of the sequence finite?

Mignotte (1974) and Loxton and van der Poorten (1977) explained how p-adic methods can be used to compute an upper bound for the 0-

multiplicity of a given non-degenerate integer recurrence sequence, and in certain cases to determine all vanishing terms. Loxton and van der Poorten further proved some results on the growth order of $|u_m|$ and $P(u_m)$. Far better, and in many respects the best possible, results can be obtained by the p-adic analogue of the Thue–Siegel–Roth–Schmidt method. Applications of these results of van der Poorten and Schlickewei (1982) and Evertse (1984*b*) to exponential equations were already mentioned in the notes of chapter 1. The fundamental result is the so-called Main Theorem on S-units of which a special case reads as follows (cf. van der Poorten and Schlickewei (1982), Evertse (1984*b*) and Laurent (1984)).

Let K be an algebraic number field. If G is a finitely generated subgroup of K^* then, for m fixed, the equation

$$X_1+X_2+\cdots+X_m=1$$

has only finitely many solutions $(X_1, X_2, \ldots, X_m)\in G^m$ such that no non-empty proper subsum $X_{i_1}+\cdots+X_{i_t}$ $(1\leqslant i_1<\cdots<i_t\leqslant m)$ vanishes.

The following results can be derived for a non-degenerate recurrence sequence of arbitrary order k.

(i) (see van der Poorten and Schlickewei (1982)). *Let Ω be the maximal absolute value of the roots of the non-degenerate algebraic recurrence sequence $\{u_m\}_{m=0}^{\infty}$. Then for every $\varepsilon>0$ there exists an m_0 such that*

$$|u_m|>\Omega^{m(1-\varepsilon)}$$

for $m>m_0$. Note that it is obvious that $|u_m|<\Omega^{m(1+\varepsilon)}$ for sufficiently large m.

(ii) (see Glass, Loxton and van der Poorten (1981, 1986)). *The total multiplicity of a non-degenerate recurrence sequence $\{u_m\}_{m=0}^{\infty}$ is finite,* that is, the number of solutions of the equation $u_m=u_n$ with $m>n$ is finite. It follows directly that for any complex a the a-multiplicity of $\{u_m\}$ is finite.

(iii) (see van der Poorten and Schlickewei (1982), Evertse (1984*b*)). *If $\{u_m\}_{m=0}^{\infty}$ is a non-degenerate recurrence sequence of rational integers with at least two distinct roots, then $P(u_m/(u_m, u_n))\to\infty$ if $m\to\infty, m>n, u_n\neq 0$.* This implies both (ii) and $P(u_m)\to\infty$ as $m\to\infty$. The extensions to algebraic recurrence sequences hold true too. The example $\{m^3a^m\}_{m=0}^{\infty}$ with $a\in\mathbb{Z}$, $a\geqslant 2$, where u_{a^l} is a power of a for every positive integer l, shows that the assertion does not hold if there is only one characteristic root.

Because of the ineffective nature of the Thue–Siegel–Roth–Schmidt method, the arguments do not, in general, permit computation of m_0 in (i), all solutions of $u_m = u_n$ in (ii) and a lower bound for $P(u_m)$ in (iii) for a given sequence $\{u_m\}_{m=0}^{\infty}$.

Effective results of type (i), (ii), (iii) have only been obtained under restrictions on r and s. First we state some results for r bounded and s free, then we consider the case when s is bounded. Suppose first that $r = 1$, hence there is only one root of maximal absolute value. Stewart (1982) proved that, if $u_m \neq f_1(m)\omega_1^m$ and $\varepsilon > 0$ then

$$P(u_m) > (1-\varepsilon)\log m, \quad Q(u_m) > m^{1-\varepsilon}$$

for $m > C_7$, where C_7 is a computable number depending only on ε and the sequence $\{u_m\}_{m=0}^{\infty}$. Weaker estimates for $P(u_m)$ were obtained by Sparlinskij (1980) and Kiss (1982). Mignotte (1979) investigated the equation $u_m = v_n$ where both $\{u_m\}_{m=0}^{\infty}$ and $\{v_m\}_{m=0}^{\infty}$ are integer recurrence sequences with exactly one root of maximal absolute value. Kiss (1986) considered the more general equation $s_1 u_m = s_2 v_n$ with $s_1, s_2 \in S$ and gave a lower bound for $|s_1 u_m - s_2 v_n|$. Mignotte (1974, 1975) was the first to investigate the case $r = 3$. He proved that if there are at most three roots of maximal absolute value and all these roots are simple, then

$$|u_m| \geqslant C_8 |\omega_1|^m m^{-C_9}$$

when $f_1(m)\omega_1^m + \cdots + f_r(m)\omega_r^m \neq 0$ and $m \geqslant C_{10}$. Here C_8, C_9, C_{10} are computable positive numbers depending only on the sequence. The results in Mignotte, Shorey and Tijdeman (1984), which are treated in this chapter, were the first ones in which the condition on three roots of maximal absolute value being simple was dropped. It was already noticed by Mignotte (1974) that certain cases of four simple roots with $\omega_1 = \overline{\omega_2}, \omega_3 = \overline{\omega_4}$ can be handled in the same way as three simple roots.

The arguments in the proof of theorem 4.6 can be used to prove the following result. If $r = s = 5$ and $\{u_m\}_{m=0}^{\infty}$ is a non-degenerate sequence of rational integers, then the equation $u_m = 0$ implies that m is bounded by a computable number depending only on the sequence $\{u_m\}_{m=0}^{\infty}$.

Now we turn to the study of multiplicity of recurrence sequences of order 2 and 3. Results on the multiplicity of integer recurrence sequences of order 2 have already been given in the notes of chapter 3. Suppose we have a non-degenerate integer sequence of order 3. If the three roots are distinct, non-zero and real, then the 0-multiplicity is at most 3. This was shown by Smiley (1956) and later also proved by Scott (1960) and Picon (1978). In general the 0-multiplicity of the sequence can be as large as 6 as is shown by the following example due to Berstel (1974) (cf. Loxton and van der Poorten (1977)): $u_0 = u_1 = 0$, $u_2 = 1$, $u_{m+3} = 2u_{m+2} - 4u_{m+1} + 4u_m$ for $m = 0, 1, 2, \ldots$

which satisfies $u_0 = u_1 = u_4 = u_6 = u_{13} = u_{52} = 0$. This example contradicts a conjecture of Ward. Kubota (1977*b*) claimed that he could prove that the 0-multiplicity of a non-degenerate ternary integer recurrence sequence is at most 6 but he has not substantiated his claim. Beukers (1982) proved that 7 is indeed an upper bound. Beukers and Tijdeman (1984) showed that the 0-multiplicity of a non-degenerate ternary rational recurrence sequence is at most 44.

A slightly more general problem is to find upper bounds for the multiplicity of non-degenerate binary algebraic recurrence sequences. Kubota (1977*b*) proved that if all terms of such a sequence belong to some number field K, then its multiplicity is bounded from above by a number depending only on the degree of K over $\mathbb{Q}$. An explicit bound can be found in Beukers and Tijdeman (1984). Presumably there is an absolute upper bound, but this has not been established yet. Beukers and Tijdeman further showed that if a binary complex recurrence sequence has multiplicity at least 4, then it is equivalent to an algebraic recurrence sequence. They used this result to prove that for any complex number z with $|z| \geqslant 2$ at most seven powers z^n $(n \in \mathbb{Z})$ are on a given line not passing through the origin. They applied the ineffective hypergeometric method. Tijdeman (1981) sketched how the effective method on linear forms in logarithms can be used instead, yielding the slightly worse bound of nine powers.

Lewis and Turk (1985) investigated the cardinality $r(a)$ of the set of pairs (m, n) with $u_m = au_n$ and $m > n$ where a is a number and $\{u_m\}_{m=0}^{\infty}$ a non-degenerate binary recurrence sequence in a field of characteristic 0. Thus $r(1)$ corresponds to total multiplicity. They call the sequence $\{u_m\}_{m=0}^{\infty}$ transcendental if it cannot be made algebraic by multiplying all the terms by some fixed constant. Lewis and Turk proved that if $r(a) \geqslant 3$ for a transcendental sequence, then $a = 1$ and $r(1) = 3$. Further, they proved that, given integers $m > n > p > q$, there are , up to multiplication by a constant, only finitely many non-degenerate binary recurrence sequences (necessarily algebraic) such that $u_m = u_n = u_p = u_q$.

CHAPTER 5

The Thue equation

Suppose that $f(X, Y)$ is a binary form with rational integer coefficients and with at least three pairwise non-proportional linear factors in its factorisation over $\mathbb{C}$. Let k be a non-zero rational integer. We consider the solutions of

$$f(x, y) = k \tag{1}$$

in rational integers x and y. Equation (1) is known as a *Thue equation*. Thue (1909) proved that equation (1) has only finitely many solutions in rational integers x and y. Thue's result was a direct consequence of his fundamental result on the approximations of algebraic numbers by rationals. His argument is ineffective, that is, it fails to provide a bound for the solutions x, y of (1). Baker (1968*b*), having established his fundamental inequality on linear forms in the logarithms of algebraic numbers, applied his work to give a proof of Thue's theorem which was effective. His estimates for the solutions of (1) were improved by several authors. Feldman (1971*a*) and Baker (1973), independently, proved the following theorem. (They made the assumption that f is irreducible, but it was not necessary to do so.)

Theorem 5.1. *If* x *and* y *are rational integers satisfying* (1), *then*

$$\max(|x|, |y|) \leqslant C_1 |k|^{C_2}$$

for some computable numbers C_1 *and* C_2 *depending only on* f.

The infinitely many solutions of the Pell equation $x^2 - dy^2 = 1$ show that theorem 5.1 is not true if f has only two distinct linear factors. In case y is composed of bounded primes, we have the following result.

Theorem 5.2. *Let* $A_1, A_2, A_3, B_1 \in \mathbb{Z}$ *such that* $B_1(A_2^2 - 4A_1A_3) \neq 0$. *Put* $B_2 = |B_1| + 1$. *Let* $x, y \in \mathbb{Z}$. *The equation*

$$A_1 x^2 + A_2 xy + A_3 y^2 = B_1$$

implies that

$$\max(|x|,|y|) \leqslant B_2^{C_3}$$

where C_3 is a computable number depending only on A_1, A_2, A_3 and $P(y)$.

Combining corollary 2.1 with theorem 5.1, Tijdeman (1975) derived the following result.

Theorem 5.3. *If $A \neq 0, B \neq 0, k \neq 0, n \geqslant 3, x > 1$ and $y \geqslant 0$ are rational integers satisfying*

$$Ax^n + By^n = k,$$

then

$$\max(x, y, n) \leqslant C_4$$

where C_4 is a computable number depending only on A, B and k.

We state notation for a generalisation of theorem 5.1. Let d and n be positive integers with $n \geqslant 3$. Let K be a finite extension of degree d over $\mathbb{Q}$. Denote by $\mathcal{O}_K$ the ring of integers of K. Let R be the regulator of K. Let $\alpha_1, \ldots, \alpha_n \in \mathcal{O}_K$ with $H(\alpha_i) \leqslant G$ for $1 \leqslant i \leqslant n$. Suppose that at least three of $\alpha_1, \ldots, \alpha_n$ are distinct. Let

$$g(X, Y) = (X - \alpha_1 Y) \cdots (X - \alpha_n Y)$$

be a binary form. Let $0 \neq \mu \in \mathcal{O}_K$ such that $\overline{|\mu|} \leqslant M$, where we assume $M \geqslant 2$. Siegel (1921) proved that equation (2) has only finitely many solutions in $x, y \in \mathcal{O}_K$. Baker (1969) (see also Baker and Coates (1970, p. 601)) gave an effective upper bound for these solutions. The following improvement is due to Győry and Papp (1978).

Theorem 5.4. *If $x, y \in \mathcal{O}_K$ satisfy*

$$g(x, y) = \mu, \tag{2}$$

then

$$\max(\overline{|x|}, \overline{|y|}) \leqslant M^{C_5}$$

where C_5 is a computable number depending only on d, R, n and G.

By combining theorem 5.4 with corollary A.7, we obtain the following generalisation of theorem 5.1 due to Győry and Papp (1978). Denote by $\mathcal{D}$ the discriminant of K.

Theorem 5.5. *Let $f(X, Y)$ be a binary form with coefficients in $\mathcal{O}_K$ and with at least three pairwise non-proportional linear factors in its factorisation over*

C. *If* $x, y \in \mathcal{O}_K$ *satisfy*

$$f(x, y) = \mu,$$

then

$$\max(|\overline{x}|, |\overline{y}|) \leqslant M^{C_6}$$

where C_6 *is a computable number depending only on* d, $\mathcal{D}$ *and* f.

Proofs

Proof of theorem 5.2. Let

$$A_1x^2 + A_2xy + A_3y^2 = B_1 \quad \text{for } x, y \in \mathbb{Z}. \tag{3}$$

We may assume that $A_1A_3xy \neq 0$; otherwise the assertion follows immediately. Further, there is no loss of generality in assuming that $(x, y) = 1$. Put $\mathscr{X} = \max(|x|, |y|)$. We may assume that $|y| \geqslant 2$ and therefore $\mathscr{X} \geqslant 2$. Further, put $P = P(y) \geqslant 2$. Multiplying both sides of (3) by A_1 and replacing x by A_1x, we may assume that $A_1 = 1$. Let α_1 and α_2 be the roots of $X^2 + A_2X + A_3$. By $A_2^2 - 4A_3 \neq 0$, the numbers α_1 and α_2 are distinct algebraic integers. For $i = 1, 2$, put

$$\beta_i = x - \alpha_i y.$$

Then $\beta_1 \neq \beta_2$. Put $L = \mathbb{Q}(\alpha_1)$ and denote by $\mathcal{O}_L$ the ring of algebraic integers of L. If L is a real quadratic field, let ε_1 be the fundamental unit of L. Further, put

$$\varepsilon = \begin{cases} \varepsilon_1 & \text{if } L \text{ is a real quadratic field,} \\ 1 & \text{otherwise.} \end{cases}$$

Denote by $c_1, c_2, \ldots, c_{11}$ computable positive numbers depending only on A_1, A_2, A_3 and P. It follows from (3) that

$$\beta_1 = \varepsilon^a \gamma_1 \tag{4}$$

and

$$\beta_2 = \varepsilon^{-a} \gamma_2 \tag{5}$$

where $a \in \mathbb{Z}$ such that $a = 0$ if $\varepsilon = 1$ and $\gamma_1, \gamma_2 \in \mathcal{O}_L$ satisfying

$$\max(|\overline{\gamma_1}|, |\overline{\gamma_2}|) \leqslant c_1 B_2. \tag{6}$$

Observe that $|\beta_i| \leqslant |x| + |\alpha_i||y| \leqslant c_2\mathscr{X}$ for $i = 1, 2$. Hence, by (4), (5), (6) and a Liouville-type argument,

$$|a| \leqslant c_3 \log(B_2\mathscr{X})$$

where $c_3 > 1$. Subtracting (5) from (4), we have

$$(\alpha_2 - \alpha_1)y = \varepsilon^a\gamma_1 - \varepsilon^{-a}\gamma_2. \tag{7}$$

For a rational prime $p \leqslant P$, let $\mathfrak{p}$ be a prime ideal in $\mathcal{O}_L$ dividing p. Then, by $(x, y)=1$, we have

$$\min(\text{ord}_{\mathfrak{p}}(\beta_1), \text{ord}_{\mathfrak{p}}(\beta_2)) \leqslant c_4.$$

For simplicity, assume that $\text{ord}_{\mathfrak{p}}(\beta_2) \leqslant c_4$. Then, by (7),

$$\text{ord}_p(y) \leqslant \text{ord}_{\mathfrak{p}}(y) \leqslant c_4 + \text{ord}_{\mathfrak{p}}\left(\varepsilon^{2a}\frac{\gamma_1}{\gamma_2} - 1\right).$$

We apply theorem B.4 with $n=2$, $d \leqslant 2$, $p \leqslant P$, $\log A = c_5 \log B_2$, $B' = 1$, $B = 2c_3 \log(B_2 \mathscr{X})$ and $\delta = (16Pc_3)^{-1}$ to conclude that

$$\text{ord}_p(y) \leqslant c_6 \log B_2 + (8P)^{-1} \log \mathscr{X}. \tag{8}$$

Now it follows from (N.1) and (8) that

$$\log |y| \leqslant \sum_{p \leqslant P} \text{ord}_p(y) \log p \leqslant 2P(c_6 \log B_2 + (8P)^{-1} \log \mathscr{X}).$$

Thus

$$\log |y| \leqslant c_7 \log B_2 + \tfrac{1}{4} \log \mathscr{X}. \tag{9}$$

If $\mathscr{X} \leqslant y^2$, then it follows from (9) that

$$\log |y| \leqslant 2c_7 \log B_2,$$

hence, by (3),

$$\log |x| \leqslant c_8 \log B_2.$$

Thus we may assume that $\mathscr{X} > y^2$. Consequently $|x| = \mathscr{X} > y^2$. Now it follows from (7) that

$$0 < \left|\varepsilon^{2a}\frac{\gamma_1}{\gamma_2} - 1\right| = \left|\frac{(\alpha_2 - \alpha_1)y}{x - \alpha_2 y}\right| < c_9 |x|^{-1/2}.$$

We apply theorem B.2 with $n=2$, $d \leqslant 2$, $\log A = c_5 \log B_2$, $B' = 1$, $B'' = 2c_3 \log(B_2 \mathscr{X}) = 2c_3 \log(B_2 |x|)$ and $\delta = (16c_3)^{-1}$ to conclude that

$$\left|\varepsilon^{2a}\frac{\gamma_1}{\gamma_2} - 1\right| \geqslant B_2^{-c_{10}} |x|^{-1/4}.$$

Therefore $|x| \leqslant B_2^{c_{11}}$, hence $|y| \leqslant |x| \leqslant B_2^{c_{11}}$. □

We now turn to the proof of theorem 5.4. Assume that K has s real conjugate fields and t pairs of non-real conjugate fields so that $d = s + 2t$. We shall signify the conjugates of any element θ of K by $\theta^{(1)}, \ldots, \theta^{(d)}$ with $\theta^{(1)}, \ldots, \theta^{(s)}$ real and $\theta^{(s+1)}, \ldots, \theta^{(s+t)}$ the complex conjugates of $\theta^{(s+t+1)}, \ldots, \theta^{(s+2t)}$ respectively. Set $r = s + t - 1$ and denote by $\eta_1, \ldots, \eta_r$ an independent system of units for K satisfying (A.45). The constants v, v_1, v_2 and v_3 are computable positive numbers depending only on d and R.

Let $x, y \in \mathcal{O}_K$, $xy \neq 0$. For $i = 1, 2, \ldots, n$ we write

$$\beta_i = x - \alpha_i y \tag{10}$$

and

$$m_i = |N(\beta_i)|. \tag{11}$$

For every i with $1 \leqslant i \leqslant n$, it follows from lemma A.15 that there exists an associate γ_i of β_i (with respect to $\eta_1, \ldots, \eta_r$) such that

$$|\log(m_i^{-1/d}|\gamma_i^{(j)}|)| \leqslant v \quad (1 \leqslant j \leqslant d). \tag{12}$$

For $i = 1, \ldots, n$, write

$$\gamma_i = \beta_i \eta_1^{b_{i,1}} \cdots \eta_r^{b_{i,r}} \tag{13}$$

and

$$H_i = \max_{1 \leqslant k \leqslant r} |b_{i,k}|.$$

Let l be the subscript for which

$$H_l = \max_{1 \leqslant i \leqslant n} H_i.$$

If $r = 0$, we may suppose $H_l = 0$. The proof of theorem 5.4 depends on the following result.

Lemma 5.1. *There exist an integer σ with $1 \leqslant \sigma \leqslant d$ and computable positive constants v_1, v_2 depending only on d and R such that*

$$\log(m_l^{-1/d}|\beta_l^{(\sigma)}|) \leqslant -v_1 H_l \quad \textit{if } H_l \geqslant v_2.$$

Proof. If $r = 0$, then $H_l = 0$ and the assertion follows. Thus we may assume $r > 0$. Consequently $d \geqslant 2$. By (13) with $i = l$, we have

$$\log\left|\frac{\gamma_l^{(k)}}{\beta_l^{(k)}}\right| = b_{l,1} \log|\eta_1^{(k)}| + \cdots + b_{l,r} \log|\eta_r^{(k)}| \quad (1 \leqslant k \leqslant r).$$

This system of r linear equations in $b_{l,j}$ with $1 \leqslant j \leqslant r$ has a non-zero determinant E with $|E| \geqslant R$. On solving $b_{l,j}$ in this system of equations we have

$$H_l \leqslant v_3 \left|\log\left|\frac{\gamma_l^{(J)}}{\beta_l^{(J)}}\right|\right| \tag{14}$$

where

$$\left|\log\left|\frac{\gamma_l^{(J)}}{\beta_l^{(J)}}\right|\right| = \max_{1 \leqslant k \leqslant r} \left|\log\left|\frac{\gamma_l^{(k)}}{\beta_l^{(k)}}\right|\right|.$$

By (14) and (12), we have

$$|\log(m_l^{-1/d}|\beta_l^{(J)}|)| \geqslant \left|\log\left|\frac{\beta_l^{(J)}}{\gamma_l^{(J)}}\right|\right| - |\log(m_l^{-1/d}|\gamma_l^{(J)}|)|$$

$$\geqslant v_3^{-1}H_l - v.$$

We may assume $H_l \geqslant 2vv_3$. Then

$$|\log(m_l^{-1/d}|\beta_l^{(J)}|)| \geqslant (2v_3)^{-1}H_l. \tag{15}$$

Re-writing (11) with $i=l$, we have

$$\sum_{\substack{j=1\\ j\neq J}}^{d} \log(m_l^{-1/d}|\beta_l^{(j)}|) = -\log(m_l^{-1/d}|\beta_l^{(J)}|). \tag{16}$$

Now the lemma follows immediately from (16) and (15). □

Proof of theorem 5.4. Denote by $c_{12}, c_{13}, \ldots, c_{19}$ computable positive numbers depending only on d, R, n and G.

Let $x, y \in \mathcal{O}_K$ satisfy (2). We may assume $xy \neq 0$; otherwise the theorem follows immediately. By taking norms on both sides in (2), we see from (10) and (11) that

$$m_1 \cdots m_n = |N(\mu)| \leqslant M^d. \tag{17}$$

Since at least three and so two of $\alpha_1, \ldots, \alpha_n$ are distinct, there exist indices λ and ν between 1 and n inclusively such that $\alpha_\lambda \neq \alpha_\nu$. Solving the equations (10) with $i=\lambda$ and $i=\nu$ for x and y, we have

$$x = \frac{\alpha_\lambda\beta_\nu - \alpha_\nu\beta_\lambda}{\alpha_\lambda - \alpha_\nu}, \quad y = \frac{\beta_\nu - \beta_\lambda}{\alpha_\lambda - \alpha_\nu}.$$

Thus we have

$$\max(\overline{|x|}, \overline{|y|}) \leqslant c_{12} \max(\overline{|\beta_\lambda|}, \overline{|\beta_\nu|}).$$

Consequently, in view of (13), (12) and (17), the proof of theorem 5.4 is complete if we show that $H_l \leqslant c_{13} \log M$.

We assume that $H_l \geqslant c_{14} \log M$ with c_{14} sufficiently large. Let c_{14} be so large that $H_l \geqslant v_2$ and thus the assertion of lemma 5.1 is valid. Hence there exists an integer σ with $1 \leqslant \sigma \leqslant d$ such that

$$|\beta_l^{(\sigma)}| \leqslant m_l^{1/d} \mathrm{e}^{-v_1 H_l}$$

which, together with (17), implies that

$$|\beta_l^{(\sigma)}| \leqslant M \mathrm{e}^{-v_1 H_l}.$$

Consequently, if c_{14} is sufficiently large,

$$|\beta_l^{(\sigma)}| \leqslant e^{-c_{15}H_l} \tag{18}$$

where $c_{15} = v_1/2$. Since μ is a non-zero algebraic integer, $|N(\mu)| \geqslant 1$. Hence, by a Liouville-type argument,

$$|\beta_1^{(\sigma)} \cdots \beta_n^{(\sigma)}| = |\mu^{(\sigma)}| \geqslant M^{-d+1}. \tag{19}$$

By (18), we have

$$\left|\sum_{\substack{i=1\\ \beta_i \neq \beta_l}}^{n} \beta_i^{(\sigma)}\right| > \left|\prod_{i=1}^{n} \beta_i^{(\sigma)}\right|. \tag{20}$$

Since at least three of $\alpha_1, \ldots, \alpha_n$ are distinct, observe that the product on the left-hand side of (20) is non-empty. Now it follows from (20) and (19) that there exists an integer m with $1 \leqslant m \leqslant n$ such that $\beta_m \neq \beta_l$ and

$$|\beta_m^{(\sigma)}| \geqslant M^{-d+1}. \tag{21}$$

Consequently, by (18) and (21), we have

$$\left|\frac{\beta_l^{(\sigma)}}{\beta_m^{(\sigma)}}\right| \leqslant M^{d-1}\, e^{-c_{15}H_l} \leqslant e^{-c_{15}H_l/2}, \tag{22}$$

if c_{14} is sufficiently large. Since at least three of $\alpha_1, \ldots, \alpha_n$ are distinct, we can find an integer p with $1 \leqslant p \leqslant n$ such that β_l, β_m and β_p are distinct. By permuting $1, 2, \ldots, n$, there is no loss of generality in assuming that $\beta_l = \beta_1$, $\beta_m = \beta_2$, $\beta_p = \beta_3$.

The proof of theorem 5.4 depends on the following identity, which can be verified by direct computation,

$$(\alpha_2^{(\sigma)} - \alpha_3^{(\sigma)})\beta_1^{(\sigma)} + (\alpha_3^{(\sigma)} - \alpha_1^{(\sigma)})\beta_2^{(\sigma)} + (\alpha_1^{(\sigma)} - \alpha_2^{(\sigma)})\beta_3^{(\sigma)} = 0.$$

For simplicity we omit the superscript (σ). It follows from this identity and (13) that

$$\left|\frac{\alpha_1 - \alpha_2}{\alpha_1 - \alpha_3}\eta_1^{u_1} \cdots \eta_r^{u_r}\frac{\gamma_3}{\gamma_2} - 1\right| = \left|\frac{\alpha_2 - \alpha_3}{\alpha_1 - \alpha_3}\right| \left|\frac{\beta_1}{\beta_2}\right|,$$

where $u_k = b_{2,k} - b_{3,k}$ for $1 \leqslant k \leqslant r$. By (22), we obtain

$$0 < \left|\frac{\alpha_1 - \alpha_2}{\alpha_1 - \alpha_3}\eta_1^{u_1} \cdots \eta_r^{u_r}\frac{\gamma_3}{\gamma_2} - 1\right| \leqslant e^{-c_{16}H_l}.$$

By (12) and (17) the height of γ_3/γ_2 does not exceed $M^{c_{17}}$. We apply theorem B.2 with $n = r + 2 \leqslant d + 2$, $\delta = \min(\frac{1}{4}, c_{16}/4n)$, $A' = c_{18}$, $A = M^{c_{17}}$, $B' = 1$ and

$B''=2H_l$. We obtain

$$\left|\frac{\alpha_1-\alpha_2}{\alpha_1-\alpha_3}\eta_1^{u_1}\cdots\eta_r^{u_r}\frac{\gamma_3}{\gamma_2}-1\right|\geqslant M^{-c_{19}}\,\mathrm{e}^{-c_{16}H_l/2}.$$

Thus $2^{-1}c_{16}H_l\leqslant c_{19}\log M$ which is not possible if $c_{14}>2c_{19}c_{16}^{-1}$. □

Proof of theorem 5.5. Let n be the degree of f, and H an upper bound for the heights of the coefficients of f. There is a rational integer a such that $0\leqslant a\leqslant n$ and $f(1,a)\neq 0$. Thus $g(X,Y):=f(X,aX+Y)$ is also a binary form with coefficients in $\mathcal{O}_K$ and with at least three pairwise non-proportional linear factors. Further, $a_0:=g(1,0)=f(1,a)\neq 0$ and the heights of the coefficients of g do not exceed c_{20} where c_{20}, and c_{21}, c_{22} below, are computable numbers depending only on d, n and H. It suffices to derive a bound of the form M^C for the solutions x, y in $\mathcal{O}_K$ of the equation $g(x,y)=\mu$. If $a_0\neq 1$, then we multiply both sides by a_0^{n-1} and replace a_0x by x and $a_0^{n-1}\mu$ by μ. We may therefore assume that the coefficient of X^n in g is 1, that the heights of the coefficients of g do not exceed c_{21} and that $\overline{|\mu|}\leqslant M^{c_{22}}$. Denote by

$$g(X,Y)=(X-\beta_1Y)\cdots(X-\beta_nY)$$

the factorisation of g over $\mathbb{C}$. Put $L=K(\beta_1,\ldots,\beta_n)$. Further, let G be the maximum of the heights of $\beta_1,\ldots,\beta_n$. By theorem 5.4, there exists a computable number c_{23} depending only on d_L, R_L, n, H and G such that

$$\max(\overline{|x|},\overline{|y|})\leqslant M^{c_{23}}.$$

By corollary A.7 applied to $g(X,1)$ the number c_{23} can be bounded from above by a computable number depending only on d, $\mathcal{D}$ and f (cf. the remark after corollary A.5).

Proof of theorem 5.1. This theorem is a special case of theorem 5.5. □

Proof of theorem 5.3. By corollary 2.1, n is bounded from above. For each possible $n\geqslant 3$ we can apply theorem 5.1, since the linear factors of the binary form Ax^n+By^n are all distinct. □

Notes

We present some explicit estimates for the magnitude of the solutions of Thue's equation (1). The first bounds, due to Baker (1968*b*), were further improved by Feldman (1971*a*), Sprindžuk (1970*a*, 1971*a*, 1972), Baker (1973), Stark (1973) and Győry and Papp (1978, 1983). In fact, the first two papers of Sprindžuk deal with Thue–Mahler equations and for such results we refer to the notes of chapter 7. Let $f(X, Y)$ be an irreducible binary form of degree $n \geqslant 3$ with rational integer coefficients whose absolute values are bounded from above by A ($\geqslant 2$). Let α be a root of $f(X, 1) = 0$ and $\tilde{D}$ and $\tilde{R}$ ($\geqslant 2$) upper bounds for the absolute value of the discriminant and for the regulator, respectively, of $\mathbb{Q}(\alpha)$. Let x, y and $k \neq 0$ be rational integers satisfying (1). Then there exist computable numbers C_7, C_8 and C_9 depending only on n such that

$$\log \max(|x|, |y|) < C_7 \tilde{R} (\log \tilde{R})^2 (\tilde{R} + \log A + \log |k|), \tag{23}$$

$$\log \max(|x|, |y|) < C_8 \tilde{D}^{1/2} (\log \tilde{D})^{2n} (\tilde{D}^{1/2} + \log A + \log |k|), \tag{24}$$

$$\log \max(|x|, |y|) < C_9 A^{[n(n-1)]/2} (\log A)^{2n} (A^{[n(n-1)]/2} + \log |k|). \tag{25}$$

This was proved by Győry and Papp (1983) with completely explicit constants C_7, C_8 and C_9. Inequality (25) is an improvement on an estimate of Baker (1968*b*), (23) is an improvement on Sprindžuk (1972) and (23)–(25) give slight improvements on the estimates of Stark (1973). For further history, see Sprindžuk (1982).

In some exceptional cases the hypergeometric method of Thue and Siegel leads to better upper bounds for the solutions of (1). This has been worked out for some cubic equations by Baker (1964*a*, *b*) and Faddeev (1966). Related results have been obtained by Osgood (1970*a*, *b*, 1971), Bombieri (1982), Bombieri and Mueller (1983, 1986), Mueller (1984) and Chudnovsky (1983*a*, *b*). It is possible to use similar, but ineffective, methods to obtain general upper bounds for the numbers of solutions of Thue equations. In special cases the bounds are quite good. See Lewis and Mahler (1960), Hyyrö (1964*b*), Mahler (1984), Evertse (1983*a*, *b*, 1986), Silverman (1982*a*, 1983*a*, *b*) and Evertse and Győry (1985) obtained upper bounds which are independent of the coefficients of f. A considerable improvement on results of Mahler and Evertse has been given by Bombieri and Schmidt (1986).

Many results mentioned in the preceding paragraph are not formulated in terms of diophantine equations, but in the equivalent way of irrationality measures of algebraic numbers. Let α be an algebraic number of degree $d \geqslant 2$. An immediate consequence of theorem 5.1 is the following result.

Let $d > 2$. There exist computable numbers $C_{10} > 0$ and κ with $0 < \kappa < d$ depending only on α such that for every rational number p/q with $q > 0$, we have

$$|\alpha - p/q| > C_{10} q^{-\kappa}. \tag{26}$$

This is the best-known general effective improvement of Liouville's inequality. The first effective general improvement is due to Baker (1968*b*).

For explicit values of the constants C_{10} and κ and extensions of the result to certain linear forms with algebraic coefficients in an arbitrary number of variables, see Győry (1980*h*). His results cover those of Kotov and Sprindžuk (1977) and Győry and Papp (1983). If we disregard the effective nature of inequality (26), the method of Thue–Siegel–Roth gives more. By improving the results of Thue (1909), Siegel (1921), Dyson (1947) and Gelfond (1952), Roth (1955) proved the following.

Given $\varepsilon>0$, there exists a constant $C_{11}>0$ depending only on α and ε such that

$$|\alpha-p/q|\geqslant C_{11}q^{-2-\varepsilon} \tag{27}$$

for every rational number p/q with $q>0$.

Inequality (27) is the best possible in the sense that $-2-\varepsilon$ cannot be replaced by $-2+\varepsilon$ in general. It is conjectured (see Lang, 1965, p. 184) that $q^{-2-\varepsilon}$ can be replaced by $q^{-2}(\log q)^{-1-\varepsilon}$ with $q\geqslant 2$. For generalisations to linear forms with algebraic coefficients see Schmidt (1970, 1971*a*, *b*, 1980*b*). If the constant C_{11} is allowed to depend on the greatest prime factor of pq, the theory of linear forms in logarithms gives a considerable improvement of inequality (27). As an immediate consequence of corollary B.1 we have the following result of Feldman (1968*a*, *b*).

Let $P\geqslant 2$. There exists a computable number C_{12} depending only on α and P such that for every rational number p/q with $q\geqslant 2$ and $P(pq)\leqslant P$,

$$|\alpha-p/q|\geqslant(\log q)^{-C_{12}}. \tag{28}$$

This represents a considerable improvement on an ineffective result of Ridout (1957). For an analogue of (28) with p/q replaced by the quotient of two elements of a non-degenerate binary recurrence, see theorem 3.5. Further, theorem 5.2 implies:

Let α be an algebraic number of degree 2. Then

$$|\alpha-p/q|\geqslant C_{13}q^{-2+\delta} \tag{29}$$

for every rational number p/q with $q>0$, where C_{13} and δ are computable positive numbers depending only on α and $P(q)$.

This inequality strengthens a result of Schinzel (1967).

Irrationality measures (26)–(29) yield upper bounds for the solutions of equation (1) in terms of k. The best general effective bound (26) leads to theorem 5.1. The better ineffective bound (27) enables us to replace k by a polynomial in x and y of degree less than $\deg(f)-2$, provided that f is irreducible. See Davenport and Roth (1955). Davenport and Lewis (unpublished) and later Schinzel (1969) proved that if $f\in\mathbb{Z}[X,Y]$ is an irreducible binary form of degree at least 3 and $g \in \mathbb{Z}[X, Y]$ has total degree less than the degree of f, then

$$f(x, y) = g(x, y) \tag{30}$$

has only finitely many solutions in rational integers x, y. The proof is ineffective. In corollary 7.1 we shall give an effective proof of the result on equation (30) for the case that g is also a binary form.

Let $\alpha \notin \mathbb{Q}$ be a real algebraic number. Denote by p_n/q_n the nth convergent in the simple continued fraction expansion of α. Mahler (1936) proved that $P(p_n q_n)$ tends to infinity with n. Further, Ridout (1957) proved that both $P(p_n)$ and $P(q_n)$ tend to infinity with n. However, these results are not effective. We see from (28) an effective version of Mahler's result. Further, Shorey (1976*b*) proved that

$$P(p_n q_n) \geqslant C_{14} \log\log q_n \quad (n \geqslant 3)$$

where $C_{14} > 0$ is a computable number depending only on α. For a quadratic irrational α, inequality (29) implies that $P(q_n) \to \infty$ effectively whenever $n \to \infty$. For further results in this direction, see Erdös and Mahler (1939) and Shorey (1983*b*).

Let H ($\geqslant 2$) be an upper bound for the heights of the coefficients of f, and n the degree of f. Győry (1981*b*) proved a general result which implies the explicit estimate

$$\log \max(\overline{|x|}, \overline{|y|}) \leqslant (5(d+1))^{50(d+2)} n^6 (h_K R_K)^7 \log(HM)$$

in theorem 5.5. Here $h_K R_K$ can be estimated by theorem A.3. See also Győry and Papp (1978, 1983).

There are many generalisations of results mentioned above. It follows from a theorem of Lang (1960, 1983) that if μ is a non-zero element of a finitely generated (but not necessarily algebraic) extension K of $\mathbb{Q}$ and $f(X, Y) \in K[X, Y]$ denotes a binary form having at least three distinct linear factors (over the algebraic closure of K), then $f(x, y) = \mu$ has only finitely many solutions in any subring of K finitely generated over $\mathbb{Z}$. An effective version of this result was proved by Győry (1983). For other generalisations, see the notes of chapters 6–8, Lang (1960, 1978, 1983), Schmidt (1971*a*, 1980*a*), Sprindžuk (1982) and Győry (1984*b*).

Let K be a finite extension of $\mathbb{Q}$. Let $f \in K[X_1, \ldots, X_m]$ be a form of degree n. We call f a *decomposable form* if there exist n linear forms $\mathscr{L}_i(X_1, \ldots, X_m) = \alpha_{i1} X_1 + \cdots + \alpha_{im} X_m$ ($1 \leqslant i \leqslant n$) with algebraic coefficients such that

$$f(X_1, \ldots, X_m) = \prod_{i=1}^{n} \mathscr{L}_i(X_1, \ldots, X_m).$$

Let $0 \neq \mu \in K$. In this case the equation

$$f(x_1, \dots, x_m) = \mu \quad \text{in } x_1, \dots, x_m \text{ in } \mathcal{O}_K \tag{31}$$

will be called a *decomposable form equation* over $\mathcal{O}_K$. Obviously all binary forms are decomposable, but forms of more than two variables need not be. Consequently, all Thue equations are decomposable form equations. Further important examples of decomposable forms are norm forms, discriminant forms and index forms. The corresponding equations of the form (31) are said to be norm form equations, discriminant form equations and index form equations over $\mathcal{O}_K$. These equations play an important role in algebraic number theory (see e.g. Győry, 1980*e*).

We call f a *norm form* over K if there exists a linear form $\mathscr{L}(X_1, \dots, X_m) = \alpha_1 X_1 + \dots + \alpha_m X_m$ with coefficients in a finite extension L of K such that

$$f(X_1, \dots, X_m) = \prod_{i=1}^{n} \mathscr{L}^{(i)}(X_1, \dots, X_m),$$

where $n = [L:K]$ and for $i = 1, \dots, n$, $\mathscr{L}^{(i)}(X_1, \dots, X_m) = \alpha_1^{(i)} X_1 + \dots + \alpha_m^{(i)} X_m$ and the $\alpha^{(i)}$ denote the images of $\alpha \in L$ under the K-embeddings of L. Then we write $N_{L/K}(\mathscr{L}(X_1, \dots, X_m))$ for f. We note that all irreducible binary forms over K are, up to a constant factor, norm forms over K. Consider the solutions $(x_1, \dots, x_m) \in \mathcal{O}_K^m$ of the norm form equation

$$N_{L/K}(\mathscr{L}(x_1, \dots, x_m)) = \mu, \tag{32}$$

where μ is a given non-zero element in K. We may assume without loss of generality that $\alpha_1 = 1$. If in particular $m = 2$ and the degree of α_2 is at least 3 over K, then (32) is a Thue equation over $\mathcal{O}_K$ and, by theorem 5.4, computable upper bounds for the heights of the solutions can be given. Several authors obtained ineffective finiteness results for certain special norm form equations of more than two variables. For references, see Skolem (1938) and Schmidt (1971*b*, 1980*b*). In case $K = \mathbb{Q}$, Schmidt (1971*a*, 1972) characterised all linear forms $\mathscr{L}$ for which (32) has only finitely many solutions $(x_1, \dots, x_m) \in \mathbb{Z}^m$ for any $\mu \in \mathbb{Q}, \mu \neq 0$. Schmidt (1972, 1973, 1980*b*) also obtained finiteness results for equations of the form $f(x_1, \dots, x_m) = g(x_1, \dots, x_m)$ over $\mathbb{Z}$, where g is a polynomial whose degree is small with respect to $\deg(f)$. For $K = \mathbb{Q}$, Győry and Pethö (1977, 1980) gave asymptotic estimates for the number of solutions of (32) with $\max_j |x_j| \leq \mathscr{X}$ in case (32) has infinitely many solutions.

When $f(X, Y)$ is irreducible, equation (1) can be considered as a norm form equation over $\mathbb{Z}$ in two variables. Sprindžuk (1974*a*, 1982) extended theorem 5.1 to the case of three variables in the norm form equation among which two are dominating. For certain special extensions L/K and certain numbers $\alpha_1, \ldots, \alpha_m$ of special types, Skolem (1937, 1938), Baker (1967) and Feldman (1970, 1971*b*, 1979) proved effective finiteness theorems on norm form equations in more than two unknowns. Furthermore, Győry and Lovász (1970) and Vojta (1983) gave effective versions of Schmidt's (1971*a*) finiteness criterion mentioned above in case $m=3$, $K=\mathbb{Q}$ and L is special. Győry and Lovász assumed L to be a totally imaginary quadratic extension of a totally real algebraic number field. Vojta assumed L to be a normal and complex extension of $\mathbb{Q}$. Győry, partly in collaboration with Papp, extended Baker's method described in this chapter to several classes of decomposable form equations in an arbitrary number of variables. Győry (1976) obtained general effective finiteness theorems for discriminant form equations and index form equations over $\mathbb{Z}$ which make it possible to solve any such equation. These theorems have many applications in algebraic number theory (see Győry (1976, 1980*e*)). Under the general hypothesis that in (32) α_{i+1} is of degree at least 3 over $K(\alpha_1, \ldots, \alpha_i)$ for $i=1, \ldots, m-1$, Győry and Papp (1978) derived explicit bounds for the solutions of (32), too. Similar results were proved by Kotov (1980*a*), Sprindžuk (1982) and Győry and Papp (1983). Recently Győry (1981*a*, *b*) and Kotov (1981, 1983), independently, obtained a further improvement. Gaál (1986) gave a common generalisation of results of Sprindžuk (1974*a*) and Győry and Papp (1983). We refer to the notes of chapter 7 for references of papers on decomposable form equations of Mahler type (see chapter 7, (38)). General effective results for decomposable form equations were given by Győry and Papp (1978) and Győry (1981*a*, *b*). Both the main result of Győry (1981*a*) and that of Győry (1981*b*) cover theorems 5.1, 5.4 and 5.5 and the above-mentioned general effective results on norm form, discriminant form and index form equations. An extension to the case that the ground ring is a finitely generated (but not necessarily algebraic) extension of $\mathbb{Z}$ is given in Győry (1983). For a good survey of these results, we refer to Győry (1980*e*, 1984*b*).

In the notes of chapter 1 we mentioned some results on unit equations over function fields. Corresponding theorems have been proved for Thue equations over function fields. Schmidt (1978, 1980*a*) gave upper bounds for the heights of both the integer and the so-called rational solutions of these equations. Mason (1981; see also 1984*a*) derived a better bound for the integer solutions and gave an algorithm to find all solutions. Mason (1986) generalised his results to norm form equations and Győry (1983) generalised Schmidt's result concerning integer solutions to decomposable form equations over function fields. Recently, Evertse (1986) gave an analogue to the function field case of the bound which he obtained in

Evertse (1984*a*) for the numbers of integer solutions of Thue equations over number fields. Further, Evertse and Győry (1985) generalised the result of Evertse (1984*a*) to a wide class of decomposable form equations in an arbitrary number of variables over finitely generated (but not necessarily algebraic) extension rings of $\mathbb{Z}$. A general finiteness criterion for decomposable form equations is given in Evertse and Győry (1986*b*).

CHAPTER 6

The superelliptic equation

In this chapter we prove that, under suitable conditions, the superelliptic equation (1) has only finitely many integral solutions. The resulting theorems 6.1 and 6.2 are applied to a system of two quadratic equations in corollary 6.1.

Denote by K a finite extension of $\mathbb{Q}$. Suppose that $\alpha_1, \ldots, \alpha_n$ are algebraic numbers in K. Write

$$f(X) = (X - \alpha_1) \cdots (X - \alpha_n).$$

For a rational integer $m \geqslant 2$ and a non-zero algebraic number b in K, we consider the superelliptic equation

$$f(x) = by^m \tag{1}$$

in algebraic integers $x \in K$ and $y \in K$. We shall apply a method of Siegel (1926) and theorem 5.5 to prove the following theorems of Baker (1969, 1975) on the integral solutions of (1).

Theorem 6.1. *Let $m \geqslant 3$. Suppose that $f(X)$ has at least two simple roots. If x and y are algebraic integers in K satisfying* (1), *then*

$$\max(\overline{|x|}, \overline{|y|}) \leqslant C_1$$

for some computable number C_1 depending only on b, m, f and K.

Theorem 6.2. *Suppose that $m = 2$ and $f(X)$ has at least three simple roots. Then all the solutions of* (1) *in algebraic integers $x \in K$, $y \in K$ satisfy*

$$\max(\overline{|x|}, \overline{|y|}) \leqslant C_2$$

where C_2 is a computable number depending only on b, f and K.

The ineffective versions of theorems 6.1 and 6.2 are consequences of a well-known theorem of Siegel (1929) which implies that all irreducible algebraic curves over any algebraic number field K on which there are infinitely many integer points in K must be of genus 0. Theorem 6.2 includes the elliptic equation

$$y^2 = Ax^3 + Bx^2 + Cx + D \quad \text{in } x \in \mathbb{Z},\ y \in \mathbb{Z}$$

with A, B, C, $D \in \mathbb{Z}$. For this equation, Baker (1968*c*) established the assertion of theorem 6.2 from his variant of theorem 5.1 by a method, due to Mordell (1922, 1923), involving the theory of reduction of binary quartic forms. For the case $K = \mathbb{Q}$, Baker (1969) proved theorems 6.1 and 6.2 with explicit bounds for the solutions. Furthermore, Baker and Coates (1970) established an effective version of Siegel's theorem for curves of genus 1.

Theorem 6.2 has the following consequences for systems of two quadratic equations.

Corollary 6.1. *Let $A, B, C, D, E, F, k, l \in \mathbb{Z}$ with $(B^2 - 4AC)(E^2 - 4DF)kl \neq 0$ be given. If the system*

$$\left.\begin{aligned} Ax^2 + Bxy + Cy^2 &= k \\ Dx^2 + Exz + Fz^2 &= l \end{aligned}\right\} \qquad (2)$$

in x, y, $z \in \mathbb{Z}$ has infinitely many solutions, then

$$\frac{Ck}{Fl} = \frac{B^2 - 4AC}{E^2 - 4DF}$$

and this ratio is the square of a rational number. If $CFkl$ is not a square, then there exists a computable upper bound for $\max(|x|, |y|, |z|)$.

The proofs of theorems 6.1 and 6.2 depend on the following lemma.

Lemma 6.1. *Let $m \geqslant 2$. Suppose that $b \neq 0$ and $\alpha_1, \ldots, \alpha_n$ are algebraic integers and that $f(X)$ has a simple root, say α. If x and y are algebraic integers in K satisfying (1), then there exist algebraic integers $\zeta \neq 0$, $\phi \neq 0$ and δ in K such that*

$$x - \alpha = (\zeta/\phi)\delta^m$$

and

$$\max(\overline{|\zeta|}, \overline{|\phi|}) \leqslant C_3$$

where C_3 is a computable number depending only on b, m, f and K.

Proofs

The constants $c_1, c_2, \ldots$ in the proofs of lemma 6.1, theorem 6.1 and theorem 6.2 will denote computable positive numbers depending only on b, m, f and K.

Proof of lemma 6.1. Let $\eta_1, \ldots, \eta_r$ be an independent system of units for K satisfying (A.45). By permuting the suffixes of $\alpha_1, \ldots, \alpha_n$, there is no loss of generality in assuming that $\alpha = \alpha_1$ is a simple root of f. Put

$$\Delta = [b] \prod_{j=2}^{n} [\alpha - \alpha_j].$$

Since α is a simple root of f, we find that Δ is a non-zero ideal of $\mathcal{O}_K$.

Suppose that x and y are in $\mathcal{O}_K$ satisfying (1). We may assume that $x \neq \alpha$; otherwise the lemma follows with $\zeta = 1$, $\phi = 1$ and $\delta = 0$. By (1), we have the following ideal equation in $\mathcal{O}_K$:

$$[x - \alpha_1] \cdots [x - \alpha_n] = [by^m].$$

From here, we obtain

$$[x - \alpha] = \mathfrak{a}\mathfrak{b}^m \tag{3}$$

where $\mathfrak{a}$ and $\mathfrak{b}$ are non-zero ideals in $\mathcal{O}_K$ and $\mathfrak{a}$ divides Δ^{m-1} in $\mathcal{O}_K$. By lemma A.11, there exist non-zero ideals $\mathfrak{a}_1$ and $\mathfrak{b}_1$ in $\mathcal{O}_K$ such that

$$\max(N(\mathfrak{a}_1), N(\mathfrak{b}_1)) \leqslant c_1$$

and the ideals

$$\mathfrak{a}\mathfrak{a}_1 = [\zeta_1], \quad \mathfrak{b}\mathfrak{b}_1 = [\delta_1]$$

are principal. Multiplying both sides of (3) by $\mathfrak{a}_1\mathfrak{b}_1^m$, we see that $\mathfrak{a}_1\mathfrak{b}_1^m$ is principal. Write

$$\mathfrak{a}_1\mathfrak{b}_1^m = [\phi_1].$$

Observe that

$$|N(\phi_1)| = N\mathfrak{a}_1(N\mathfrak{b}_1)^m \leqslant c_1^{m+1}.$$

Further

$$|N(\zeta_1)| = (N\mathfrak{a})(N\mathfrak{a}_1) \leqslant (N\Delta)^{m-1}c_1 = c_2,$$

since $\mathfrak{a}$ divides Δ^{m-1} and Δ is a non-zero ideal. Hence, by corollary A.6, we can find associates ζ_2 and ϕ_2 of ζ_1 and ϕ_1, respectively, such that

$$\max(\overline{|\zeta_2|}, \overline{|\phi_2|}) \leqslant c_3.$$

From the ideal equation (3) we get

$$\mathfrak{a}_1\mathfrak{b}_1^m[x - \alpha] = (\mathfrak{a}\mathfrak{a}_1)(\mathfrak{b}\mathfrak{b}_1)^m.$$

Thus we obtain

$$x - \alpha = \varepsilon(\zeta_2/\phi_2)\delta_1^m$$

where ε is a unit in $\mathcal{O}_K$. Now, by corollary A.5, we may write

$$\varepsilon = \varepsilon_1\varepsilon_2^m$$

where ε_1 and ε_2 are units in $\mathcal{O}_K$ and $\overline{|\varepsilon_1|} \leqslant c_4$. Hence

$$x-\alpha=\left(\frac{\varepsilon_1\zeta_2}{\phi_2}\right)(\varepsilon_2\delta_1)^m.$$

Set $\zeta=\varepsilon_1\zeta_2$, $\phi=\phi_2$ and $\delta=\varepsilon_2\delta_1$. Observe that

$$\max(\overline{|\zeta|}, \overline{|\phi|}) \leqslant c_5. \qquad \square$$

Proof of theorem 6.1. Let x and y be in $\mathcal{O}_K$ satisfying (1). By multiplying both sides of (1) by r^n where r is the product of the denominators of $b, \alpha_1, \ldots, \alpha_n$, and replacing rx by x, we may assume without loss of generality that $b, \alpha_1, \ldots, \alpha_n$ are algebraic integers. By permuting the suffixes of $\alpha_1, \ldots, \alpha_n$, we may assume that α_1 and α_2 are distinct simple roots of f. Then, by lemma 6.1, there exist elements ξ_1, ξ_2, ψ_1, ψ_2 and γ_1, γ_2 in $\mathcal{O}_K$ such that $\xi_1\xi_2\psi_1\psi_2 \neq 0$ and

$$x-\alpha_1=(\xi_1/\psi_1)\gamma_1^m, \tag{4}$$

$$x-\alpha_2=(\xi_2/\psi_2)\gamma_2^m \tag{5}$$

and

$$\max(\overline{|\xi_1|}, \overline{|\xi_2|}, \overline{|\psi_1|}, \overline{|\psi_2|}) \leqslant c_6. \tag{6}$$

Subtracting (5) from (4) and multiplying by $\psi_1\psi_2$, we obtain

$$\xi_1\psi_2\gamma_1^m - \xi_2\psi_1\gamma_2^m = (\alpha_2-\alpha_1)\psi_1\psi_2.$$

We apply theorem 5.5 with $f(X, Y)=\xi_1\psi_2X^m-\xi_2\psi_1Y^m$ and $\mu=(\alpha_2-\alpha_1)\psi_1\psi_2$. Since $m \geqslant 3$, $\xi_1\xi_2\psi_1\psi_2 \neq 0$ and $\alpha_1 \neq \alpha_2$, all assumptions are fulfilled. Hence, by (6),

$$\max(\overline{|\gamma_1|}, \overline{|\gamma_2|}) \leqslant c_7.$$

Now apply (5) and (6) to conclude that $\overline{|x|} \leqslant c_8$. Hence, by (1), $\overline{|y|} \leqslant c_9$. $\square$

Proof of theorem 6.2. By the same argument as in the proof of theorem 6.1, we may assume that $b, \alpha_1, \ldots, \alpha_n$ are algebraic integers. By permuting the suffixes of $\alpha_1, \ldots, \alpha_n$ we may assume that α_1, α_2 and α_3 are distinct simple roots of f.

Let x and y be in $\mathcal{O}_K$ satisfying (1) with $m=2$. It follows from lemma 6.1 that there exist elements $\xi_i \neq 0$, $\psi_i \neq 0$ and γ_i $(1 \leqslant i \leqslant 3)$ of $\mathcal{O}_K$ such that for $i=1, 2, 3$, we have

$$x-\alpha_i=(\xi_i/\psi_i)\gamma_i^2 \tag{7}$$

and

$$\max(\overline{|\xi_i|}, \overline{|\psi_i|}) \leqslant c_{10}. \tag{8}$$

From equations (7) we obtain

$$\sigma_1\gamma_1^2-\sigma_2\gamma_2^2=\alpha_2-\alpha_1,$$
$$\sigma_2\gamma_2^2-\sigma_3\gamma_3^2=\alpha_3-\alpha_2,$$
$$\sigma_3\gamma_3^2-\sigma_1\gamma_1^2=\alpha_1-\alpha_3,$$

where $\sigma_i=(\xi_i/\gamma_i)$ for $i=1,2,3$. Let $\sigma_1^{1/2}$, $\sigma_2^{1/2}$ and $\sigma_3^{1/2}$ be an arbitrary choice of the square roots. Put

$$L=K(\sigma_1^{1/2},\sigma_2^{1/2},\sigma_3^{1/2}).$$

Observe that $d_L\leqslant c_{11}$. Let $\eta'_1,\ldots,\eta'_r$ be an independent system of units for L satisfying (A.45). Put

$$\beta_3=\sigma_1^{1/2}\gamma_1-\sigma_2^{1/2}\gamma_2,$$
$$\beta_1=\sigma_2^{1/2}\gamma_2-\sigma_3^{1/2}\gamma_3,$$
$$\beta_2=\sigma_3^{1/2}\gamma_3-\sigma_1^{1/2}\gamma_1$$

and $A=\psi_1\psi_2\psi_3$. By (8), $\overline{|A|}\leqslant c_{10}^3$ and consequently $|A|\geqslant c_{12}$. Hence, by (8) and lemma A.2, $A\sigma_i^{1/2}$ $(i=1,2,3)$ is an algebraic integer of height at most c_{13}. Since also $d_L\leqslant c_{11}$, lemma A.16 gives $|\mathscr{D}_L|\leqslant c_{14}$. Observe that $A\beta_1$, $A\beta_2$ and $A\beta_3$ are non-zero algebraic integers in L, for α_1, α_2 and α_3 are distinct. Further,

$$\max_{1\leqslant i\leqslant 3}|N_L(A\beta_i)|\leqslant c_{15}.$$

Consequently, by lemmas A.15 and A.16, there exist associates β'_1, β'_2 and β'_3 of $A\beta_1$, $A\beta_2$ and $A\beta_3$, respectively, such that $\max_{1\leqslant i\leqslant 3}\overline{|\beta'_i|}\leqslant c_{16}$. Hence we can write

$$A\beta_i=\nu_i\varepsilon_i^3\quad(i=1,2,3) \tag{9}$$

where ε_1, ε_2, ε_3 are units in $\mathcal{O}_L$ and ν_1, ν_2, ν_3 are non-zero elements of $\mathcal{O}_L$ satisfying

$$\max_{1\leqslant i\leqslant 3}\overline{|\nu_i|}\leqslant c_{17}. \tag{10}$$

Observe that

$$A\beta_1+A\beta_2+A\beta_3=0.$$

Thus, by (9),

$$\nu_1\left(\frac{\varepsilon_1}{\varepsilon_3}\right)^3+\nu_2\left(\frac{\varepsilon_2}{\varepsilon_3}\right)^3=-\nu_3.$$

We apply theorem 5.5 with $f(X, Y)=v_1X^3+v_2Y^3$ and $\mu=-v_3$. By $d_L \leqslant c_{11}$, $\mathscr{D}_L \leqslant c_{14}$, $v_1v_2v_3 \neq 0$ and (10), we obtain

$$\max\left(\overline{\left|\frac{\varepsilon_1}{\varepsilon_3}\right|}, \overline{\left|\frac{\varepsilon_2}{\varepsilon_3}\right|}\right) \leqslant c_{18}. \tag{11}$$

It follows from (10) that $|v_i| \geqslant c_{19}$ for $i=1,2,3$. We may fix any choice of the sign of $\sigma_2^{1/2}$. Then we can select the sign of $\sigma_1^{1/2}$ such that $|\beta_3| \leqslant c_{20}$. Now it follows from (9) with $i=3$ that $|\varepsilon_3| \leqslant c_{21}$, since $|A| \leqslant c_{10}^3$ and $|v_3| \geqslant c_{19}$. Further, by (11) and $|\varepsilon_3| \leqslant c_{21}$, we find $|\varepsilon_1| \leqslant c_{22}$. We obtain from (9) with $i=1$ that $|\beta_1| \leqslant c_{23}$, since $|A| \geqslant c_{12}$ and $|v_1| \leqslant c_{17}$. The inequality $|\beta_1| \leqslant c_{23}$ holds for either choice of the sign of $\sigma_3^{1/2}$. Consequently, by (8), we find that $\max(|\gamma_2|, |\gamma_3|) \leqslant c_{24}$. Now, by (7) and (8), we conclude that $|x| \leqslant c_{25}$. We argue similarly from the equations conjugate to (1). We obtain $\overline{|x|} \leqslant c_{26}$. Hence, by (1), we obtain $\overline{|y|} \leqslant c_{27}$. □

Proof of corollary 6.1. Suppose that the system (2) has infinitely many solutions $x, y, z \in \mathbb{Z}$. Then each of the equations has an infinity of solutions. Hence both B^2-4AC and E^2-4DF are positive and non-square. Consequently, $CF \neq 0$. Furthermore,

$$(Bx+2Cy)^2-(B^2-4AC)x^2=4Ck,$$

$$(Ex+2Fz)^2-(E^2-4DF)x^2=4Fl$$

and therefore

$$((B^2-4AC)x^2+4Ck)((E^2-4DF)x^2+4Fl)=((Bx+2Cy)(Ex+2Fz))^2. \tag{12}$$

By theorem 6.2 the polynomial at the left-hand side has at most two simple zeros. Since $(B^2-4AC)(E^2-4DF)(4Ck)(4Fl) \neq 0$, we have

$$\frac{4Ck}{B^2-4AC}=\frac{4Fl}{E^2-4DF}.$$

Hence, by (12), $CkFl$ and therefore Ck/Fl is the square of a rational number.

If Ck/Fl is not a square, then we may apply theorem 6.2 to (12). Since $f(X)=((B^2-4AC)X^2+4Ck)((E^2-4DF)X^2+4Fl)$ now has four simple roots, the second assertion follows. □

Notes

Mordell (1914) established the connection between the equation $y^2=Ax^3+Bx^2+Cx+D$ $(A, B, C, D \in \mathbb{Z})$ and equations of the type $f(x, y)=$

1, where f denotes an irreducible binary form with rational integer coefficients of degree 3 or 4. By using the theorem of Thue, Mordell (1922, 1923) proved that the equation $Ey^2=Ax^3+Bx^2+Cx+D$ ($A,B,C,D,E\in\mathbb{Z}$, $E\neq 0$) has only finitely many solutions if the polynomial Ax^3+Bx^2+Cx+D has three distinct roots. In particular the equation $y^2=x^3+k$ ($k\neq 0$) has only finitely many rational integer solutions x, y. In a letter to Mordell, Siegel (1926) showed how Mordell's result can be extended to the ineffective analogues of theorems 6.1 and 6.2 by using a result of Siegel (1921). Subsequently, Siegel (1929) classified all irreducible algebraic curves defined over any algebraic number field K on which there are infinitely many points in $\mathcal{O}_K$. In particular these curves must be of genus zero and have at most two infinite valuations. Lang (1960) generalised Siegel's result by showing that one may take any finitely generated (but not necessarily algebraic) extension of $\mathbb{Q}$ in place of K and any finitely generated subring of this extension in place of $\mathcal{O}_K$. LeVeque (1964) proved that equation (1) has only finitely many integer points in $\mathcal{O}_K$ unless m divides the multiplicities of all but one root of f or, if m is even, the multiplicities of all but two roots of f are divisible by m and the remaining two by $m/2$. Faltings (1983) proved the remarkable result that any irreducible algebraic curve defined over K which is of genus greater than or equal to 2 contains at most finitely many points in K^2. This result was conjectured by Mordell. Hurwitz (1917) proved that the assertion does not hold for all curves of genus 1. The proofs in the above-mentioned papers are ineffective.

Baker (1968*b*) gave the first effective solution of the Mordell equation

$$y^2=x^3+k \tag{13}$$

where k is a non-zero integer. Namely, if x and y are rational integer solutions of (13), then

$$\log\max(|x|,|y|)\leqslant 10^{10}|k|^{10000}. \tag{14}$$

Stark (1973) applied his version of estimate (24) of chapter 5 to improve the estimate (14) as follows: Given $\varepsilon>0$, there is a computable number C_4 depending only on ε such that, if x and y are rational integer solutions of (13), then

$$\log\max(|x|,|y|)\leqslant C_4|k|^{1+\varepsilon}.$$

Sprindžuk (1982, p. 149) replaced the right-hand side of (14) by $C_5|k|\times(\log|k|+1)^6$, where C_5 is a computable absolute constant. Baker (1968*c*) gave an estimate similar to (14) for the integral solutions of the equation $y^2=Ax^3+Bx^2+Cx+D$.

Let $f(X)$ be a polynomial with rational integer coefficients. Denote by n,

$\mathscr{D}$ and H the degree, the discriminant and the height of f, respectively. Baker (1969) gave upper bounds for $\max(|x|, |y|)$, where x, y is a rational integer solution of $f(x) = y^m$, namely

$$\exp \exp\{(5m)^{10} n^{10n^3} H^{n^2}\} \tag{15}$$

if $m \geqslant 3$ and f has at least two simple zeros, and

$$\exp \exp \exp\{n^{10n^3} H^{n^2}\} \tag{16}$$

if $m = 2$ and f has at least three simple zeros. These bounds were improved by Sprindžuk (1973*a*, 1976, 1977, 1982). He proved that if $m = 2$ and f is monic with $n \geqslant 3$ simple zeros and $A \neq 0$ a rational integer, then the rational integer solutions x, y of $Ay^2 = f(x)$ satisfy

$$\log \max(|x|, |y|) \leqslant C_6 |A|^{12+\varepsilon} |\mathscr{D}|^{24(n+2)+\varepsilon} (\log H)^{1+\varepsilon} \tag{17}$$

where $\varepsilon > 0$ and C_6 is a computable number depending only on n and ε (cf. Sprindžuk, 1982, p. 164). The bounds in case $m \geqslant 3$ can be found in Sprindžuk (1977; 1982, p. 182). A more explicit version of (17) was given by Turk (1986, theorem 2). We remark that we could have applied theorem 1.4 in place of theorem 5.5 in the proof of theorem 6.2. Further, we note that it follows from the proofs that, in theorems 6.1 and 6.2 and lemma 6.1, the dependence on K can be refined to dependence on the degree and the discriminant of K only.

Brindza (1984*a*) gave an effective proof of LeVeque's result (cf. theorem 8.3). Baker and Coates (1970) proved that if $f \in \mathbb{Z}[X, Y]$ is an absolutely irreducible polynomial of degree n and height H such that the associated curve $f(X, Y) = 0$ has genus 1, then all rational integer solutions of $f(x, y) = 0$ satisfy

$$\max(|x|, |y|) < \exp \exp \exp\{(2H)^{10^{n^{10}}}\}.$$

Baker and Coates observed that there is no difficulty in dealing in a similar way with curves of genus 0 when there are at least three infinite valuations, but an effective extension to curves of genus greater than 1 remains an important quest.

The bounds (16) and (17) can be used to derive lower bounds for the greatest square-free part $Q^*(x)$ of $f(x)$ where $f \in \mathbb{Z}[X]$ has at least three simple zeros and x is a rational integer. Sprindžuk (1982, p. 164) derived from (17) that

$$Q^*(f(x)) > C_7 (\log |x|)^{(1-\varepsilon)/3(n-1)(n-2)}$$

where n is the degree of f and $C_7 > 0$ is a computable number depending only on f and ε. Sprindžuk (1977; 1982, p. 193) and Turk (1982, 1986) gave similar bounds for the greatest m-free part of $f(x)$ where m is any rational integer with $m \geqslant 3$. Sprindžuk (1976, 1977) used these bounds to derive lower bounds for the discriminants and class numbers of certain algebraic number fields.

Turk (1984) gave an explicit upper bound for the solutions x, y of (2) in the special case $B=E=0$ and under the assumption that $CFkl$ is not a square.

For certain special equations all solutions have been determined by using estimates for linear forms in logarithms. Ellison *et al.* (1972) determined all the rational integer solutions x, y of the equation $y^2=x^3-28$. Boyd and Kisilevsky (1972) solved the equation $y^2=x^3-x+1$. The first effective result of determining all the solutions by linear forms in logarithms was given by Baker and Davenport (1969) in relation to a problem of Diophantos. They solved the system of equations

$$\left.\begin{aligned} 3x^2-y^2&=2 \\ 8x^2-z^2&=7 \end{aligned}\right\} \tag{18}$$

(cf. corollary 6.1). The proof also depends on a useful lemma on inhomogeneous diophantine approximation (cf. Grinstead, 1978). Elementary solutions for (18) were derived by Kanagasabapathy and Ponnudurai (1975) and by Sansone (1976). Similar systems of quadratic equations were solved by Jones (1976, 1978), Veluppillai (1980) and Mohanty and Ramasamy (1984). Theorem 6.2 has been applied to tight designs by Bannai (1979). For a similar application of a diophantine equation, see Bremner (1979).

Estimates like (14)–(17) can also be used to derive lower bounds for $|y^2-x^3|$ and, more generally, for $|by^m-f(x)|$, provided that these expressions are non-zero. For another type of lower bound for $|y^2-x^3|$, see Nair (1978). For general information on the equation $y^2=x^3+k$ we refer to Hall Jr (1971), London and Finkelstein (1973) and Danilov (1982).

It follows from theorem 6.2 that equation

$$(x+1)(x+2)=(y+1)(y+2)(y+3) \tag{19}$$

has only finitely many solutions in non-negative integers x, y. For this, observe that equation (19) can be re-written as $(2x+3)^2=4(y+1)(y+2)\times(y+3)+1$. Mordell (1963) proved that the only solutions of (19) in non-negative integers x, y are given by $x=1$, $y=0$ and $x=13$, $y=4$. Further, MacLeod and Barrodale (1970) conjectured that the product of two consecutive positive integers is never equal to the product of $l \geqslant 4$ consecutive positive integers. They proved the conjecture for $l=4$ and $l=8$. Boyd and Kisilevsky used their result on $y^2=x^3-x+1$ to determine all solutions of the equation $(x+1)(x+2)(x+3)=(y+1)(y+2)(y+3)(y+4)$ in rational integers x, y. This equation has some relation with sporadic simple groups.

For positive integers a, b, x, y and k with $b>a$ and $x-y\geqslant k$, Shorey (1984*c*) proved that the equation

$$a(x+1)\cdots(x+k)=b(y+1)\cdots(y+k) \tag{20}$$

implies that either $k\leqslant C_8$ or $k=[\alpha+1]$ where $\alpha=(\log(b/a))(\log(x/y))^{-1}$ and C_8 is a computable number depending only on a and b. Further, he applied corollary B.1 to show that equation (20) implies that $\max(x, y, k)$ is bounded by a computable number depending only on a, b and the greatest prime factor of xy. For given a, b and $k\geqslant 3$ such that $a(X+1)\cdots(X+k)-b(Y+1)\cdots(Y+k)$ is irreducible over rationals, it follows from a theorem of Schinzel (1969), stated below, that equation (20) has only finitely many solutions in integers x, y. Cohn (1971) proved that equation (20) with $a=1$, $b=2$, $k=4$, has only one solution, $x=4$, $y=3$, and Ponnudurai (1975) showed that equation (20) with $a=1$, $b=3$, $k=4$ has only two solutions in positive integers, namely $x=2$, $y=1$ and $x=6$, $y=4$.

The first general result concerning the integer solutions of a diophantine equation in two unknowns is due to Runge (1887).

Let $F(X,Y)=\sum_{i=0}^{r}\sum_{j=0}^{r}a_{ij}X^iY^j$ be an irreducible polynomial with integer coefficients. Suppose that the equation $F(x, y)=0$ has infinitely many rational integer solutions. Then there exist integers m, n such that (i) $a_{m0}\neq 0$, $a_{0n}\neq 0$, (ii) $a_{ij}=0$ *if* $ni+mj>mn$, (iii) $\sum_{ni+mj=mn}a_{ij}X^iY^j$ *is a constant multiple of a power of an irreducible polynomial.*

Runge's method of proof is effective. Hilliker and Straus (1983) proved that if f does not satisfy (i)–(iii), then each rational integer solution x, y of $F(x, y)=0$ satisfies

$$\max(|x|,|y|)<(8rH)^{r^{(2r^3)}}$$

where H is the height of F. Hilliker (1982) showed how to find all solutions of some diophantine equations of this kind in practice. An immediate consequence of Runge's theorem is that if $F(x, y)=0$ has infinitely many solutions, then the highest homogeneous part of F is a constant multiple of a power of an irreducible polynomial. By combining this assertion with the theorem of Siegel (1929), Schinzel (1969) showed, ineffectively, that this irreducible polynomial is either a linear form or an indefinite quadratic form. In fact the result was proved in greater generality. It implies the result of Davenport and Lewis on equation (30) of chapter 5. Skolem (1929) deduced from Runge's result that, if $a_{00}=0$, then $F(x, y)=0$ has only finitely many solutions x, y for which (x, y) is bounded. For more results in this direction we refer to Skolem (1938).

Kleiman (1976) used theorem 6.2 to derive conditions on F such that computable upper bounds can be given for all rational integer solutions x, y of $F(x, y)=0$.

We shall consider the solutions of superelliptic equations in rational numbers whose denominators are composed of primes from some fixed finite set in chapter 8 and of the corresponding equations over function fields in the notes of chapter 8.

Lang (1960) also proved the analogue of Siegel's (1929) result for function fields of characteristic 0. In particular his result applies to superelliptic equations over function fields. Upper bounds for the solutions of such equations were given by Schmidt (1978), Mason (1983, 1984*a*) and Mason and Brindza (1986). The papers of Mason and Brindza also provide efficient algorithms for determining all solutions.

CHAPTER 7

The Thue–Mahler equation

Let $f(X,Y)$ be a binary form of degree n with rational integer coefficients and with at least three pairwise non-proportional linear factors in its factorisation over $\mathbb{C}$. Mahler (1933*a*) generalised the theorem of Thue (1909) by proving that $P(f(x,y)) \to \infty$ whenever $\max(|x|,|y|)$, with $x, y \in \mathbb{Z}$ and $(x,y)=1$, tends to infinity. Mahler proved this result by way of his p-adic analogues of the methods of Thue (1909) and Siegel (1921) on the approximations of algebraic numbers by rationals and by algebraic numbers. Thus Mahler's result is not effective. Coates (1970*a*), having established a p-adic analogue of an inequality of Baker on linear forms in logarithms, proved an effective version of Mahler's result. This result has been improved as follows.

Theorem 7.1. *For all rational integers x, y with $(x,y)=1$ and $f(x,y)\neq 0$, we have*

$$P(f(x,y)) \geq C_1 \log\log \mathscr{X} \tag{1}$$

where $\mathscr{X}=\max(|x|,|y|,3)$ and $C_1>0$ is a computable number depending only on f.

For irreducible forms, Coates (1970*a*) proved (1) with the right-hand side replaced by $C_2(\log\log \mathscr{X})^{1/4}$ where C_2 is an explicitly given positive constant depending only on f. Sprindžuk (1971*c*) proved (1) for all irreducible forms of degree greater than or equal to 5 and for so-called non-exceptional forms of degree 4. Shorey *et al.* (1977) proved theorem 7.1 in its presented form. Note that the assertion of theorem 7.1 does not hold for forms with at most two non-proportional linear factors in their factorisations over $\mathbb{C}$.

Theorem 7.1 implies a bound for the solutions of equation (2), the so-called *Thue–Mahler equation* (*over* $\mathbb{Z}$). Here we present an upper bound which is the best known in terms of P and s.

Theorem 7.2. *Let k and s be rational integers with $k \neq 0$ and $s > 0$. Let $p_1, \dots, p_s$ be primes with $p_1 < p_2 < \cdots < p_s =: P$. All solutions of the equation*

$$f(x, y) = kp_1^{z_1} \cdots p_s^{z_s} \quad \text{in } x, y, z_1, \dots, z_s \in \mathbb{Z} \tag{2}$$

with $(x, y) = 1$ and $z_1 \geqslant 0, \dots, z_s \geqslant 0$, satisfy

$$\max(|x|, |y|, e^{\max_j z_j}) \leqslant \exp\{((C_3 s \log P)^s P)^{C_4}\}$$

where C_3 and C_4 are computable numbers such that C_3 depends only on f and k and C_4 only on n.

For the equation $f(x, y) = k$ one may take $s = 1$, $p_1 = 2$, $z_1 = 0$, but theorem 5.1 provides a more precise bound for this equation. Coates (1970*a*) proved theorem 7.2 for irreducible forms with another explicit upper bound. The presented upper bound is due to Györy (1980*c*) who gave explicit values of C_3 and C_4.

Let $P \geqslant 3$ and denote by S the set of all rational integers composed of primes not exceeding P. Theorem 7.2 implies that the equation $f(x, y) = z$ in rational integers x, y, z with $(x, y) = 1$ and $z \in S$ has only finitely many solutions. In the following result f may be multiplied by any non-zero rational integer and z by a binary form in x and y.

Theorem 7.3. *Let $f(X, Y)$ and $g(X, Y)$ be binary forms with rational integer coefficients. Suppose f has at least three pairwise non-proportional linear factors in its factorisation over $\mathbb{C}$ which do not divide g over $\mathbb{C}$. Then all solutions of the equation*

$$wf(x, y) = zg(x, y) \quad \text{in rational integers } w, x, y, z \tag{3}$$

with $wf(x, y) \neq 0$, $z \in S$, $(w, z) = (x, y) = 1$ satisfy $\max(|w|, |x|, |y|, |z|) \leqslant C_5$ where C_5 is a computable number depending only on P, f and g.

The supposition of theorem 7.3 certainly holds if f is irreducible of degree at least 3 and the degree of g is less than the degree of f.

The restriction $(x, y) = 1$ does not occur in the following application of theorem 7.3.

Corollary 7.1. *Under the conditions of theorem 7.3, suppose $\deg(f) > \deg(g)$. Then all solutions of the equation*

$$f(x, y) = g(x, y) \quad \text{in rational integers } x, y$$

with $f(x, y) \neq 0$ are such that $\max(|x|, |y|)$ is bounded by a computable number depending only on f and g.

Combining theorem 2.4 with theorem 7.3, we obtain the following result.

Theorem 7.4. *Let $A\neq 0$, $B\neq 0$, C and D be rational integers. Then the equation*

$$w(Ax^m+By^m)=z(Cx^n+Dy^n) \tag{4}$$

has only finitely many solutions in integers w, x, y, z, m, n with $w\neq 0$, $z\in S$, $(w, z)=1$, $|xy|>1$, $(x, y)=1$, $0\leqslant n<m$, $m>2$, $wAx^m\neq zCx^n$ and $Ax^m+By^m\neq 0$ provided that

$m=4$, $n=2$ is excluded if CX^2+DY^2 divides AX^4+BY^4 over $\mathbb{Q}$,

$m=3$, $n=1$ is excluded if $CX+DY$ divides AX^3+BY^3 over $\mathbb{Q}$, and

$m=3$, $n=2$ is excluded if CX^2+DY^2 and AX^3+BY^3

have a common linear factor over $\mathbb{Q}$.

Further, $\max(|w|, |x|, |y|, |z|, m, n)$ is bounded by a computable number depending only on A, B, C, D and P.

The following result of Shorey (1982, 1984*a*) is a consequence of theorem 7.4.

Corollary 7.2. *Let $A\neq 0$, $B\neq 0$, C and D be rational integers. Then the equation*

$$Ax^m+By^m=Cx^n+Dy^n \tag{5}$$

has only finitely many solutions in rational integers x, y, m, n with $|x|\neq|y|$, $0\leqslant n<m$, $m>2$, $Ax^m\neq Cx^n$ and $Ax^m+By^m\neq 0$ provided that $m=4$, $n=2$ is excluded if CX^2+DY^2 divides AX^4+BY^4 over $\mathbb{Q}$. Further, $\max(|x|, |y|, m, n)$ is bounded by a computable number depending only on A, B, C and D.

The conditions are necessary. For example, the equation $x^4-4y^4=x^2+2y^2$ has infinitely many solutions x, y.

Another consequence of theorem 7.4 is the following result of van der Poorten (1977*b*).

Corollary 7.3. *For every pair A, B of non-zero rational integers,*

$$P(Ax^m+By^m)\to\infty, \quad \textit{effectively},$$

as $\max(x, y, m)$ tends to infinity through positive integers $x>1, y, m$ with $(x, y)=1$ and $m>2$.

We state notation for another generalisation of theorem 7.1. Let K be a finite extension of degree d over $\mathbb{Q}$ with discriminant $\mathcal{D}$. Denote by $\mathcal{O}_K$ the ring of integers of K. For $\alpha, \beta\in\mathcal{O}_K$, denote by $[\alpha]$ the ideal generated by α in

$\mathcal{O}_K$, by (α, β) the greatest common divisor of the ideals $[\alpha]$ and $[\beta]$ and by $N((\alpha, \beta))$ the norm of the ideal (α, β) with respect to the field K. Let $f(X, Y)$ be a binary form of degree n with coefficients from $\mathcal{O}_K$. Suppose that f has at least three pairwise non-proportional linear factors in its factorisation over $\mathbb{C}$. Let N_0 be a positive integer.

Theorem 7.5. *Suppose that x and y are in $\mathcal{O}_K$ such that $f(x, y) \neq 0$ and*

$$N((x, y)) \leqslant N_0. \tag{6}$$

Then

$$P(N(f(x, y))) \geqslant C_6 \log\log \mathscr{X} \tag{7}$$

where $\mathscr{X} = \max(|N(x)|, |N(y)|, 3)$ and $C_6 > 0$ is a computable number depending only on K, f and N_0.

An ineffective proof that $P(N(f(x, y))) \to \infty$ as $\mathscr{X} \to \infty$ is due to Parry (1950). By generalising the method of Sprindžuk (1972), Kotov (1975) proved theorem 7.5 for all irreducible forms $f \in \mathcal{O}_K[X, Y]$ of degree $\geqslant 5$. Theorem 7.5 in its full generality is due to Györy (1979*b*). Explicit lower bounds for $P(N(f(x, y)))$ were given by Coates (1970*a*) in case $K = \mathbb{Q}$ and f irreducible, and by Györy (1979*b*) in case $\mathscr{X}$ exceeds a certain bound.

The next result is a more explicit version of theorem 7.5. Let $t \geqslant 1$. Let $\{\mathfrak{p}_1, \ldots, \mathfrak{p}_t\}$ be a finite set of prime ideals of $\mathcal{O}_K$. Denote by $\mathscr{S}$ the set of all non-zero elements α of $\mathcal{O}_K$ such that $[\alpha]$ has no prime ideal divisors other than $\mathfrak{p}_1, \ldots, \mathfrak{p}_t$. Let P be the maximal rational prime which is divisible by at least one of these prime ideals. Observe that $\mathscr{S}$ contains all the units of $\mathcal{O}_K$.

Theorem 7.6. *For every solution of*

$$f(x, y) = z \quad \text{in } x, y \in \mathcal{O}_K, \ z \in \mathscr{S} \tag{8}$$

with $N((x, y)) \leqslant N_0$, there exists a unit ε in $\mathcal{O}_K$ such that

$$\max(\overline{|\varepsilon x|}, \overline{|\varepsilon y|}) \leqslant \exp\{((C_7 t \log P)^t P)^{C_8}\} \tag{9}$$

where C_7 and C_8 are computable numbers such that C_7 depends only on $d, \mathscr{D}, f$ and N_0, and C_8 only on d and n.

The following corollary of theorem 7.6 may be compared with theorems 5.5 and 7.2.

Corollary 7.4. *Let $\mu \in \mathcal{O}_K$, $\mu \neq 0$. Let $\{\pi_1, \ldots, \pi_s\}$ be a set of non-zero non-units in $\mathcal{O}_K$ where $s \geqslant 1$. All solutions of*

$$f(x, y) = \mu \pi_1^{z_1} \cdots \pi_s^{z_s} \tag{10}$$

in non-negative rational integers $z_1, \ldots, z_s$ *and* $x, y \in \mathcal{O}_K$ *with* $N((x, y)) \leqslant N_0$ *satisfy*

$$\max(\overline{|x|}, \overline{|y|}, z_1, \ldots, z_s) \leqslant C_9$$

where C_9 *is a computable number depending only on* $K, f, \mu, \pi_1, \ldots, \pi_s$ *and* N_0.

Equations (8) and (10) are called *Thue–Mahler equations* (*over* $\mathcal{O}_K$). Kotov (1975) and Kotov and Sprindžuk (1977) proved results like theorem 7.6 and corollary 7.4 for binary forms f of degree greater than or equal to 5 which are irreducible over K. Theorem 7.6 and corollary 7.4 are due to Győry (1980*c*, 1981*a*) who provided explicit constants $C_7 - C_9$.

Proofs

We first prove theorem 7.6 for a special class of binary forms. Denote by h and R the class number and regulator of K, respectively.

Lemma 7.1. *Let* $\alpha_1, \ldots, \alpha_n$ *be elements of* $\mathcal{O}_K$ *such that at least three of them are distinct. Let* $\max_i \overline{|\alpha_i|} \leqslant A$. *Put*

$$g(X, Y) = (X - \alpha_1 Y) \cdots (X - \alpha_n Y).$$

For every solution of

$$g(x, y) = z \quad \text{in } x, y \in \mathcal{O}_K,\ z \in \mathcal{S} \tag{11}$$

with $N((x, y)) \leqslant N_0$, *there exists a unit* ε *in* $\mathcal{O}_K$ *such that*

$$\max(\overline{|\varepsilon x|}, \overline{|\varepsilon y|}) \leqslant \exp\{((C_{10} t \log P)^t P)^{C_{11}}\} \tag{12}$$

where C_{10} *and* C_{11} *are computable numbers such that* C_{10} *depends only on* d, h, R, N_0, n *and* A, *and* C_{11} *only on* d.

Proof. We shall denote by $c_1, \ldots, c_6$ computable positive numbers depending only on d, h, R, N_0, n and A, and by k_1 a computable positive number depending only on d. Suppose x, y, z is a solution of (11) as specified in the lemma. Put $\beta_i = x - \alpha_i y$ for $i = 1, \ldots, n$. Then (11) implies $\beta_i \in \mathcal{S}$ for $i = 1, \ldots, n$. We may assume without loss of generality that $\alpha_1, \alpha_2, \alpha_3$ are distinct. We have

$$(\alpha_2 - \alpha_3)\beta_1 + (\alpha_3 - \alpha_1)\beta_2 + (\alpha_1 - \alpha_2)\beta_3 = 0 \tag{13}$$

and $\max(\overline{|\alpha_2 - \alpha_3|}, \overline{|\alpha_3 - \alpha_1|}, \overline{|\alpha_1 - \alpha_2|}) \leqslant 2A$. By theorem 1.4 applied to (13), there exist $\sigma \in \mathcal{S}$ and $\rho_1, \rho_2, \rho_3 \in \mathcal{S}$ such that $\beta_i = \sigma \rho_i$ $(i = 1, 2, 3)$ and

$$\max_{i=1,2,3} \overline{|\rho_i|} \leqslant \exp\{((c_1 t \log P)^t P)^{k_1}\} =: T.$$

From the system of equations

$$x-\alpha_i y=\sigma\rho_i \quad (i=1,2),$$

we obtain

$$x=\sigma\phi/\kappa, \quad y=\sigma\psi/\kappa \tag{14}$$

for suitable non-zero elements ϕ, ψ, κ in $\mathcal{O}_K$ satisfying

$$\max(\overline{|\phi|}, \overline{|\psi|})\leqslant c_2 T, \quad \overline{|\kappa|}\leqslant c_3. \tag{15}$$

Since σ divides κx and κy in $\mathcal{O}_K$, the ideal $[\sigma]$ divides the ideal $(\kappa x, \kappa y)=[\kappa](x, y)$. Consequently we obtain, by taking norms with respect to K,

$$|N(\sigma)|\leqslant|N(\kappa)|N((x, y))\leqslant c_4.$$

By corollary A.6, there exists a unit ε in K such that

$$\overline{|\varepsilon\sigma|}\leqslant c_5. \tag{16}$$

Since εx, $\varepsilon y\in\mathcal{O}_K$, we have, by using (16), (14), (15) and a Liouville-type argument for κ,

$$\max(\overline{|\varepsilon x|}, \overline{|\varepsilon y|})\leqslant c_6 T. \qquad \square$$

Proof of theorem 7.6. By $c_7, \ldots, c_{16}$ we shall denote computable positive numbers depending only on d, $\mathscr{D}$, f and N_0, and by $k_2, \ldots, k_7$ computable positive numbers depending only on d and n. Without loss of generality, we may assume that $f(1,0)\neq 0$. Indeed, there is a rational integer a with $0\leqslant a\leqslant n$ such that $f(1, a)\neq 0$. Since the ideals (x, y) and $(x, ax+y)$ are equal, and hence $N((x, ax+y))\leqslant N_0$ for $x, y\in\mathcal{O}_K$, it suffices to prove the assertion for $f(X, aX+Y)$ in place of $f(X, Y)$.

Let $\alpha_1, \ldots, \alpha_n$ be the zeros of $f(X, 1)$ in $\mathbb{C}$ and put $L=K(\alpha_1, \ldots, \alpha_n)$. Note that the degree of L is at most k_2. By corollary A.7, the heights of the numbers α_i as well as the class number and the regulator of L are less than c_7. Put $a_0=f(1,0)$. Let x, y, z be a solution of (8) with the properties as specified in the theorem. Then, putting $x'=a_0 x$, $y'=y$, $\alpha_i'=a_0\alpha_i$ for $i=1, \ldots, n$ and $z'=a_0^{n-1}z$, we have $\alpha_i'\in\mathcal{O}_L$ $(i=1, \ldots, n)$ and

$$(x'-\alpha_1' y')\cdots(x'-\alpha_n' y')=z'. \tag{17}$$

We are going to apply lemma 7.1 in the field L to this equation. Observe that $N_K((x', y'))\leqslant c_8$ and $[L:K]\leqslant k_2$, hence $N_L((x', y'))\leqslant c_9$. Further, since $[L:K]\leqslant k_2$, the number of distinct prime ideal divisors in $\mathcal{O}_L$ of z' is at most $k_2 t+c_{10}$ and $P(N(z'))\leqslant P+c_{11}$ where we choose $c_{11}\geqslant 1$. Hence, by applying lemma 7.1 with $k_2 t+c_{10}$ in place of t and $P+c_{11}$ in place of P, there exists a unit ε' in $\mathcal{O}_L$ such that

$$\max(\overline{|\varepsilon' x'|}, \overline{|\varepsilon' y'|}) \leqslant \exp\{((c_{11} t \log P)^{t+c_{10}P})^{k_3}\}.$$

Observe, by distinguishing cases $c_{10} \leqslant t$, $t < c_{10} \leqslant \log P/\log\log P$ and $c_{10} > t$, $c_{10} > \log P/\log\log P$, that

$$(c_{11} t \log P)^{c_{10}} \leqslant (c_{11} t \log P)^t + (c_{11} c_{10})^{c_{10}P} + (c_{11} c_{10})^{3c_{10}}.$$

Hence

$$\max(\overline{|\varepsilon' x'|}, \overline{|\varepsilon' y'|}) \leqslant \exp\{((c_{12} t \log P)^t P)^{k_4}\} =: T_1. \tag{18}$$

This implies

$$\max(|N_L(x)|, |N_L(y)|) = \max(|N_L(\varepsilon' x)|, |N_L(\varepsilon' y)|) \leqslant T_1^{k_2}.$$

Now, considering norms with respect to K over $\mathbb{Q}$, we have

$$\max(|N(x)|, |N(y)|) \leqslant T_1^{k_2}.$$

By corollary A.4 and lemma A.15, there exist units ε_1 and ε_2 in $\mathcal{O}_K$ such that

$$\max(\overline{|\varepsilon_1 x|}, \overline{|\varepsilon_2 y|}) \leqslant c_{13} T_1^{k_2}. \tag{19}$$

From (18) we obtain, by a Liouville-type argument,

$$\overline{|x/y|} \leqslant \overline{|a_0^{-1}|} \,.\, \overline{|\varepsilon' x'/\varepsilon' y'|} \leqslant c_{14} T_1^{k_5}.$$

Hence

$$\overline{|\varepsilon_2/\varepsilon_1|} \leqslant \overline{|\varepsilon_2 y/\varepsilon_1 x|} \,.\, \overline{|x/y|} \leqslant c_{15} T_1^{k_6}. \tag{20}$$

From (19) and (20), we deduce

$$\overline{|\varepsilon_2 x|} \leqslant \overline{|\varepsilon_1 x|} \,.\, \overline{|\varepsilon_2/\varepsilon_1|} \leqslant c_{16} T_1^{k_7}. \tag{21}$$

Combining (19) and (21), we complete the proof. □

Proof of corollary 7.4. By $c_{17}, \ldots, c_{23}$ we shall denote computable numbers depending only on $K, f, \mu, \pi_1, \ldots, \pi_s$ and N_0. Let $x, y, z_1, \ldots, z_s$ be a solution of (10) as specified in the corollary. Denote by $\mathfrak{p}_1, \ldots, \mathfrak{p}_t$ all the prime ideals which divide $[\mu\pi_1 \cdots \pi_s]$. Since $t \leqslant c_{17}$ and $P(N(\mathfrak{p}_1 \cdots \mathfrak{p}_t)) \leqslant c_{18}$, there exists by theorem 7.6, a unit ε in $\mathcal{O}_K$ such that

$$\max\{\overline{|\varepsilon x|}, \overline{|\varepsilon y|}\} \leqslant c_{19}. \tag{22}$$

From (10), we have

$$f(\varepsilon x, \varepsilon y) = \varepsilon^n \mu \pi_1^{z_1} \cdots \pi_s^{z_s}. \tag{23}$$

Since $\pi_1, \ldots, \pi_s$ are non-zero non-units, this gives, by (23) and (22),

$$2^{z_j} \leqslant |N(\pi_j)|^{z_j} \leqslant |N(f(\varepsilon x, \varepsilon y))| \leqslant c_{20} \quad (1 \leqslant j \leqslant s). \tag{24}$$

Hence, by (23) and (22), $\overline{|\varepsilon|} \leqslant \overline{|\varepsilon^n|} \leqslant c_{21} \overline{|f(\varepsilon x, \varepsilon y)|} \leqslant c_{22}$. From (22), we obtain

$$\max(\overline{|x|}, \overline{|y|}) \leqslant c_{23}. \tag{25}$$

The combination of (24) and (25) proves the corollary. □

Proof of theorem 7.2. By c_{24}, c_{25} and k_8 we shall denote computable positive numbers which depend on the same parameters as C_3, C_3 and C_4, respectively. Let $x, y, z_1, \ldots, z_s$ be a solution as specified in the theorem. By theorem 7.6 with $K = \mathbb{Q}$, we have

$$\max(|x|, |y|) \leqslant \exp\{((c_{24} s \log P)^s P)^{k_8}\}. \tag{26}$$

Hence, by (2), for $j = 1, \ldots, s$,

$$2^{z_j} \leqslant p_j^{z_j} \leqslant |f(x, y)| \leqslant \exp\{((c_{25} s \log P)^s P)^{k_8}\}. \tag{27}$$

The combination of (26) and (27) proves the theorem. □

Proof of theorem 7.5. We shall denote by c_{26}, c_{27}, c_{28} computable positive numbers depending only on K, f and N_0. Put $f(x, y) = z$. Let $\mathfrak{p}_1, \ldots, \mathfrak{p}_t$ be all the prime ideal divisors of $[z]$. For $t = 0$, put $P = 2$ and, for $t > 0$, let P be the maximal rational prime divisible by at least one of these prime ideals. Then $P = \max(P(N(f(x, y))), 2)$. By theorem 7.6, there exists a unit ε in $\mathcal{O}_K$ such that

$$\max(\overline{|\varepsilon x|}, \overline{|\varepsilon y|}) \leqslant \exp\{((c_{26}(t+1) \log P)^{t+1} P)^{c_{27}}\} =: T_2. \tag{28}$$

Since there are at most d prime ideals which divide a given rational prime, we have, by (N.1),

$$t \leqslant d\pi(P) \leqslant 2\, dP/\log P.$$

Hence, by (28),

$$\mathscr{X} \leqslant 3 \max(|N(x)|, |N(y)|) \leqslant 3 T_2^d \leqslant \exp\{\exp(c_{28} P)\}.$$ □

Proof of theorem 7.1. Immediate consequence of theorem 7.5. □

Proof of theorem 7.3. By $c_{29}, \ldots, c_{32}$ we shall denote computable positive numbers depending only on P, f and g. There is a rational integer a with $0 \leqslant a \leqslant \deg(f) + \deg(g)$ such that $f(1, a) \neq 0$ and $g(1, a) \neq 0$. We may therefore assume that $f(1, 0) \neq 0$ and $g(1, 0) \neq 0$. Let $f(X, Y) = a_0(X - \alpha_1 Y) \cdots$

$(X-\alpha_m Y)$, $g(X,Y)=b_0(X-\beta_1 Y)\cdots(X-\beta_n Y)$ and $K=\mathbb{Q}(\alpha_1,\ldots,\alpha_m,\beta_1,\ldots,\beta_n)$. We may further assume that α_1,α_2 and α_3 are distinct and that none of these numbers is contained in the set $\{\beta_1,\ldots,\beta_n\}$.

Let $wf(x,y)=zg(x,y)$ with w,x,y,z as specified in the theorem. Recall that $a_0\alpha_i\in\mathcal{O}_K$ for $i=1,\ldots,m$ and $b_0\beta_j\in\mathcal{O}_K$ for $j=1,\ldots,n$. Let $\mathfrak{p}$ be a prime ideal in $\mathcal{O}_K$ which divides

$$[a_0b_0x-a_0b_0\alpha_1 y][a_0b_0x-a_0b_0\alpha_2 y][a_0b_0x-a_0b_0\alpha_3 y].$$

Then, by

$$wa_0\prod_{i=1}^{m}(a_0b_0x-a_0b_0\alpha_i y)=za_0^{m-n}b_0^{m-n+1}\prod_{j=1}^{n}(a_0b_0x-a_0b_0\beta_j y),$$

either $\mathfrak{p}\mid[za_0b_0]$ or $\mathfrak{p}\mid[a_0b_0x-a_0b_0\beta_j y]$ for some j. In the former case $N(\mathfrak{p})\leqslant c_{29}$; in the latter case $\mathfrak{p}\mid(a_0b_0x-a_0b_0\alpha_i y, a_0b_0x-a_0b_0\beta_j y)$ for some i,j with $\alpha_i\neq\beta_j$, hence $\mathfrak{p}\mid[a_0b_0][\alpha_i-\beta_j]$ in view of $(x,y)=1$, and therefore $N(\mathfrak{p})\leqslant c_{30}$. Thus

$$P\left(N\left(\prod_{i=1}^{3}(a_0b_0x-a_0b_0\alpha_i y)\right)\right)\leqslant\max(c_{29},c_{30}).$$

By applying theorem 7.5 to the polynomial $f(X,Y)=(X-a_0b_0\alpha_1 Y)\times(X-a_0b_0\alpha_2 Y)(X-a_0b_0\alpha_3 Y)$ we obtain $\max(|x|,|y|)\leqslant\max(|a_0b_0x|,|y|)\leqslant\mathscr{X}\leqslant c_{31}$ which, together with equation (3), $(w,z)=1$ and $wf(x,y)\neq 0$, implies that $\max(|w|,|z|)\leqslant c_{32}$. □

Proof of corollary 7.1. Let x,y be rational integers such that $f(x,y)=g(x,y)\neq 0$. Let $m=\deg(f)$, $n=\deg(g)$. Put $(x,y)=d$, $x_1=x/d$, $y_1=y/d$. Then $m>n$ and

$$d^{m-n}f(x_1,y_1)=g(x_1,y_1).$$

By theorem 7.3 applied with $w=d^{m-n}$, $x=x_1$, $y=y_1$, $z=1$, we obtain that $\max(d^{m-n},|x_1|,|y_1|)\leqslant C_5$. Hence $\max(|x|,|y|)\leqslant d\max(|x_1|,|y_1|)\leqslant C_5^2$. □

Proof of theorem 7.4. Suppose (4) holds for values of w, x, y, z, m, n as specified in the theorem. By $(x,y)=1$ and $|xy|>1$ we have $|x|\neq|y|$. If $Ax^mDy^n=By^mCx^n$, then, by (4), $wAx^m=zCx^n$ and this case is excluded. It therefore follows from theorem 2.4 that m is bounded by a computable number c_{33} depending only on A,B,C,D and P. Thus we may assume that m and n are fixed.

Define binary forms $f(X,Y)=AX^m+BY^m$, $g(X,Y)=CX^n+DY^n$. Since the zeros of $f(X,1)$ are the mth roots of unity all multiplied by some fixed constant and, if C is non-zero, $g(X, 1)$ has the corresponding property with respect to nth roots of unity, the number of common linear factors of $f(X, Y)$ and $g(X, Y)$ is at most (m, n). From the conditions of the theorem, we see that f has at least three non-proportional linear factors in its factorisation over $\mathbb{C}$ none of which is a divisor of g over $\mathbb{C}$. It follows from theorem 7.3 that, for all solutions as specified in theorem 7.4, $\max(|w|, |x|, |y|, |z|)$ is bounded by a computable number depending only on A, B, C, D, m and n, hence only on A, B, C, D and P. □

Proof of corollary 7.2. Suppose (5) holds for values of x, y, m, n as specified in the corollary. Observe that $x \neq 0$ and $y \neq 0$. Put

$$x_1 = \frac{x}{(x, y)}, \quad y_1 = \frac{y}{(x, y)}.$$

Then $(x_1, y_1) = 1$ and equation (5) can be written as

$$(x, y)^{m-n}(Ax_1^m + By_1^m) = Cx_1^n + Dy_1^n. \tag{29}$$

Apply theorem 7.4 to equation (29) with $w = (x, y)^{m-n}$ and $z = 1$. Since x_1, y_1, m, n satisfy all the conditions of the theorem if $m > 3$, the only remaining case is $m = 3$, and $AX^3 + BY^3$ has a common linear factor over $\mathbb{Q}$ with either $CX + DY$ or $CX^2 + DY^2$.

Suppose $m = 3, n = 1$ and $AX^3 + BY^3$ is divisible by $CX + DY$. Put $AX^3 + BY^3 = (CX + DY)(A_1X^2 + B_1XY + C_1Y^2)$ with A_1, B_1, $C_1 \in \mathbb{Q}$. Observe that both zeros of $A_1X^2 + B_1XY + C_1Y^2$ are non-real. By (5) and $Ax^m + By^m \neq 0$, we have $Cx + Dy \neq 0$ and

$$A_1x^2 + B_1xy + C_1y^2 = 1.$$

Since the discriminant of the quadratic form is negative, there exists a computable upper bound for $\max(|x|, |y|)$ in this case.

Suppose $m = 3, n = 2$. By a similar reasoning, we now obtain an equation

$$A_2x^2 + B_2xy + C_2y^2 = A_3x + B_3y$$

with A_2, B_2, C_2, A_3, $B_3 \in \mathbb{Q}$ and $B_2^2 - 4A_2C_2 < 0$. Hence there exists a computable upper bound for $\max(|x|, |y|)$ in this case too. □

Proof of corollary 7.3. Apply theorem 7.4 with $w = 1$, $C = 1$, $D = n = 0$. □

Notes

Let $g(X)$ be a polynomial with rational integer coefficients. If g is quadratic with distinct roots, then

$$P(g(x)) \to \infty \quad \text{as } |x| \to \infty,\ x \in \mathbb{Z}. \tag{30}$$

This was proved by Pólya (1918) who used Thue's method. Siegel (1921) improved Thue's approximation theorem. From this, he derived (30) for all polynomials g with at least two distinct roots.

The above results are ineffective. Now we describe effective results on the greatest prime factor of a polynomial at integer points. Størmer (1897) used his method on Pellian equations to prove that $P(x(x-1)) \to \infty$ as $|x| \to \infty$. Chowla (1935), Mahler (1935*b*) and Nagell (1937, 1955) used similar ideas to prove that $P(g(x)) \geqslant \log\log|x|$ for certain special polynomials of the form $g(X) = aX^2 + b$ and $g(X) = aX^3 + b$ with $a, b \in \mathbb{Z}$. Schinzel (1967) applied Gelfond's method on linear forms in logarithms of algebraic numbers to prove that there exists a computable number $C_{12} > 0$ depending only on g such that

$$P(g(x)) \geqslant C_{12} \log\log|x| \quad (x \in \mathbb{Z},\ |x| > 3) \tag{31}$$

for all quadratic polynomials g with distinct roots (cf. Langevin, 1976*b*). Keates (1969) used an estimate of Baker (1968*c*) to prove (31) for all polynomials g of degree 3 with distinct roots. Kotov (1973*a*) proved (31) for all irreducible polynomials g of degree at least 2. Theorem 7.1 applied to the binary form $f(X, Y) = Y^{n+1} g(X/Y)$, where n is the degree of g, implies that (31) is valid for all polynomials g with at least two distinct roots. Shorey and Tijdeman (1976*b*) gave a simple proof of this result without using p-adic methods. In fact they proved more, namely the following. Let $A > 0$. Suppose g has at least two distinct roots. There exists a computable number $C_{13} > 0$ depending only on A and g such that if

$$P(g(x)) \leqslant \exp((\log_2 x)^A) \quad (x \in \mathbb{Z},\ x > 3)$$

then

$$\omega(g(x)) \geqslant C_{13} \log_2 x / \log_3 x.$$

Inequality (31) follows by applying (N.1).

A particularly interesting polynomial is $g(X) = (X+1)\cdots(X+k)$ in which case $P(g(x))$ is the greatest prime factor of a block of consecutive integers. The first non-trivial estimate in this direction is due to Sylvester (1892) who proved that

$$P((x+1)\cdots(x+k)) > k$$

for all pairs $x, k \in \mathbb{Z}$ with $x \geqslant k > 0$. This theorem of Sylvester was rediscovered by Schur (1929). It follows from Hanson's (1973) inequality

$$\prod_{p^l \leqslant n} p < 3^n \quad \text{(cf. Rosser and Schoenfeld, 1962, p. 77)} \tag{32}$$

that $P((x+1)\cdots(x+k)) \geqslant 1.5\,k-1$ for $x \geqslant k > 0$. The best results in this direction obtained by elementary methods and the method of Størmer (1897) can be found in Langevin (1976*a*, 1978). Improvements of the results of Erdös (1934), Ramachandra (1970, 1971) and Tijdeman (1972) by Ramachandra and Shorey (1973), Jutila (1974) and Shorey (1974*b*) show that

$$P((x+1)\cdots(x+k)) \geqslant \begin{cases} C_{14}k\exp\left(\dfrac{C_{15}(\log k)^3}{(\log x)^2}\right) & \text{if } k^{3/2} < x \leqslant \exp\left(\dfrac{(\log k)^{3/2}}{\log_2 k}\right), \\ C_{16}k\log k\dfrac{\log_2 k}{\log_3 k} & \text{if } x > \exp\left(\dfrac{(\log k)^{3/2}}{\log_2 k}\right), \end{cases} \tag{33}$$

where C_{14}, C_{15}, C_{16} are computable positive absolute constants. The proof of (33) depends on estimates for exponential sums and for linear forms in logarithms of algebraic numbers. Langevin (1975*a*, *b*) proved that, for any $\varepsilon > 0$,

$$P((x+1)\cdots(x+k)) \geqslant (1-\varepsilon)k\log_2 k \quad \text{if } x \geqslant C_{17}$$

where C_{17} is a computable number depending only on k and ε. See also Erdös and Shorey (1976), Langevin (1975*c*, 1976*b*) and Stewart (1984) in this connection.

More generally, it is possible to derive lower bounds for $P(g(x+1)g(x+2)\cdots g(x+k))$ where $g \in \mathbb{Z}[X]$. For short intervals the method for estimating linear forms in logarithms of algebraic numbers yields better results. Shorey and Tijdeman (1976*b*) used their result mentioned above to prove the following. Let $B > 0$. Suppose $g \in \mathbb{Z}[X]$ has at least two distinct roots. Then for any positive integers $x(>3)$ and k with $k \leqslant \exp((\log_2 x)^B)$ there is a computable number $C_{18} > 0$ depending only on B and g such that

$$P\left(\prod_{i=1}^{k} g(x+i)\right) \geqslant C_{18}k\frac{\log_2 x}{\log_3 x}(\log k + \log_3 x). \tag{34}$$

See also Langevin (1975*b*, theorème 2; 1975*c*, 1976*b*, 1981) and Turk (1980*b*, theorem 2). Lower bounds for $P(\prod_{n \leqslant x} g(n))$ have been obtained by different methods. For an account of these results, see Hooley (1976, Ch. 1).

Let $k \in \mathbb{Z}$, $k > 3$. It follows from (33) that, if $n_1, n_2, \ldots$ is the increasing sequence of all positive integers with $P(n_i) > k$, then

$$n_{i+1} - n_i \leqslant C_{19}k\log_3 k/(\log k\log_2 k) \quad \text{for all } i$$

where C_{19} is a computable absolute constant. Erdös (1955) has conjectured that $\sup_i(n_{i+1}-n_i) \sim (\log k)^2$ as $k \to \infty$. Improving on the results of

Ramachandra (1973) and Shorey (1976*a*), Ramachandra, Shorey and Tijdeman (1976) proved that there is a computable absolute constant C_{20} such that if $x \geqslant \exp(C_{20}(\log k)^2)$ then the number of integers i with $1 \leqslant i \leqslant k$ and $P(x+i) \leqslant k$ is at most $\pi(k)$. Hence there exists a computable absolute constant C_{21} such that

$$\omega((x+1)\cdots(x+k)) \geqslant k$$

for $k \leqslant \exp(C_{21}(\log x)^{1/2})$. Extensions of these results, also to the corresponding case of polynomial values, were given by Turk (1978; 1979, Ch. 4; 1980*a*, *b*). A general theorem covering many of the above-mentioned results was given by Langevin (1978, 1981). Since $P((x+1)\cdots(x+k))$ is the greatest prime factor of the binomial coefficient

$$\binom{x+k}{k}$$

and $\omega((x+1)\cdots(x+k))$ differs from

$$\omega\left(\binom{x+k}{k}\right)$$

by at most $\pi(k)$, the above results imply lower bounds for

$$P\left(\binom{n}{k}\right) \quad \text{and} \quad \omega\left(\binom{n}{k}\right).$$

Various bounds can be found in Langevin (1979).

A related problem is the conjecture of Grimm (1969) that if p and $p+k$ are consecutive prime numbers, then there exist distinct prime numbers $p_1, \ldots, p_{k-1}$ with $p_i \mid p+i$ for $i = 1, \ldots, k-1$. Grimm also made the weaker conjecture $\omega((p+1)\cdots(p+k)) \geqslant k$. Erdös and Selfridge (1971) pointed out that the validity of these conjectures implies

$$\overline{\lim_{n\to\infty}} \frac{p_{n+1}-p_n}{p_n^{1/2}} = 0$$

where p_n denotes the nth prime. Let $G(x)$ be the largest integer such that there exist distinct prime numbers $p_1, \ldots, p_{G(x)}$ with $p_i \mid x+i$ for $i = 1, \ldots, G(x)$. Improving upon earlier results of Grimm (1969), Erdös and Selfridge (1971), Ramachandra (1973) and Cijsouw and Tijdeman (1973), Ramachandra, Shorey and Tijdeman (1975) showed that

$$G(x) > C_{22}(\log x/\log_2 x)^3$$

where $C_{22} > 0$ is a computable absolute constant. The corresponding problem for arithmetical progressions has been treated by Langevin (1978).

In the notes of chapter 6 we referred to bounds of Sprindžuk and Turk for the m-free part of $g(x)$ where $g \in \mathbb{Z}[X]$. These lower bounds can be combined with (32) to derive (31). Recall that the m-free part of an integer n is the smallest positive integer a such that $|n| = ay^m$ for some $y \in \mathbb{Z}_+$. Turk (1982) proved that, if $g \in \mathbb{Z}[X]$ has at least two distinct roots and $g_m(x)$ denotes the m-free part of $g(x)$, then

$$P(g_m(x)) \geqslant C_{23} \log_2 x \quad (m \geqslant 2, |x| > 3) \tag{35}$$

where $C_{23} > 0$ is a computable number depending only on g, hence independent of m. Obviously (35) implies (31).

The first bounds for the solutions of Thue–Mahler equations over $\mathbb{Z}$ were given by Coates (1969, 1970*a*), Sprindžuk (1968, 1969, 1970*a*, 1971*b*, *c*, 1973*b*) and Vinogradov and Sprindžuk (1968). One may consider Thue equations in S-integers as well. A rational number is called an *S-integer* if it is the quotient of a rational integer and an element of S. Every solution of the equation

$$f(x, y) = k \quad \text{in } S\text{-integers } x, y \tag{36}$$

induces a solution of (2). Hence theorem 7.2 implies an upper bound for the solutions of (36). For the best-known explicit bound see Győry (1981*b*). Conversely, equation (2) can be reduced to a finite number of equations of type (36) in S-integers.

The method described in this chapter may be used to solve a given Thue–Mahler equation in practice. An example of historical interest is the equation $x^2 + 7 = 2^n$ in rational integers n, x. Ramanujan (1913) conjectured that the only solutions are $(n, x) = (3, 1), (4, 3), (5, 5), (7, 11)$ and $(15, 181)$. This was confirmed by Nagell (1948, 1961). Other papers in which this or related equations are solved by arguments from elementary and algebraic number theory are Skolem, Chowla and Lewis (1959), Browkin and Schinzel (1956, 1960). Chowla, Dunton and Lewis (1960), Lewis (1961), Mordell (1962), Cohen (1978), Inkeri (1979), Bremner *et al.* (1983) and Tzanakis (1983, 1984). For solving these equations, Mignotte (1984) developed a method using recurrence sequences which turns out to be efficient in practice. See, further, Apéry (1960*a*, *b*) and the survey paper by Hasse (1966). If the hypergeometric method can be applied, it provides more general results; see Beukers (1979, 1981) and Tzanakis and Wolfskill (1986). All equations referred to up to now in this paragraph are of the form (2) with y constant. Algebraic methods may not suffice to solve an equation (2) with both x and y variable, but the method described in this chapter may be used. In this way Agrawal *et al.* (1980) solved the equation $x^3 - x^2y + xy^2 + y^3 = \pm 11^n$ in rational integers x, y, n. Some of the above-mentioned results have been applied to other fields: The results of Nagell and Bremner *et al.* have applications in the theory of perfect codes. Agrawal *et al.* used their result to determine all rational elliptic curves of conductor 11.

Theorem 7.3 can be extended in the following way. The upper bound C_5 can be made to depend only on P and the rational prime factors and non-constant irreducible factors of f and g. A similar result can be proved for equations (3) with $wf(x, y) \neq 0$, $w, z \in S$ and $(w, z) = (x, y) = 1$ provided that f and g are relatively prime binary forms such that fg has at least three pairwise non-proportional linear factors in its factorisation over $\mathbb{C}$. It is also possible to give quantitative bounds, thereby generalising theorems 7.1 and 7.2. For these results, see Evertse *et al.* (1986). The proofs of theorems 7.3 and 7.4 can be considered as an elaboration of the ideas mentioned in the note added in proof in Shorey (1984*a*).

Skolem (1945*b*) dealt with some special Thue–Mahler equations over algebraic number fields. The ineffective versions of theorems 7.5 and 7.6 were obtained by Parry (1950). Bounds for the numbers of solutions of Thue–Mahler equations over $\mathbb{Z}$ were given by Mahler (1933*b*) and Lewis and Mahler (1960) and over $\mathcal{O}_K$ by Evertse (1983*b*, Ch. 6; 1984*a*), Silverman (1983*b*) and Evertse and Győry (1985). Kotov (1973*b*, 1975), Sprindžuk (1973*b*, 1974*b*), Sprindžuk and Kotov (1973, 1976) and Kotov and Sprindžuk (1977) gave effective proofs of results like theorems 7.5 and 7.6 but only valid for binary forms f which are divisible by an irreducible form of degree greater than or equal to 5 or an irreducible non-exceptional binary form of degree 4. See also Sprindžuk (1980, 1982). Győry obtained several versions of theorems 7.5, 7.6 and corollary 7.4 as consequences of more general theorems concerning decomposable form equations, see Győry (1979*b*, 1980*c*, *d*, *g*, 1981*a*, 1984*b*).

Theorem 7.6 and corollary 7.4 may also be formulated in terms of $\mathcal{S}$-integers. A number $\alpha \in K$ is called an *$\mathcal{S}$-integer* if $\mathrm{ord}_{\mathfrak{p}}(\alpha) \geqslant 0$ for all prime ideals $\mathfrak{p}$ in $\mathcal{O}_K$ with $\mathfrak{p} \notin \{\mathfrak{p}_1, \ldots, \mathfrak{p}_t\}$. Thus α is an $\mathcal{S}$-integer if and only if it is the quotient of an element of $\mathcal{O}_K$ and an element of $\mathcal{S}$. Denote the set of $\mathcal{S}$-integers by $\mathcal{O}_{\mathcal{S}}$. Every solution of the equation

$$f(x, y) = 1 \quad \text{in } x, y \in \mathcal{O}_{\mathcal{S}} \tag{37}$$

induces a solution of (8). Hence theorem 7.6 implies an upper bound for the solutions of (37). For explicit bounds for the solutions of (37) see Győry (1981*b*, 1983, 1984*b*).

In the notes of chapter 5 the connection between bounds for the solutions of Thue equations and inequalities on the approximations of algebraic numbers by rationals was indicated. Similarly, bounds for the solutions of Thue–Mahler equations can be transferred to inequalities on the p-adic approximations of algebraic numbers. The first result in this direction is due to Coates (1969). Sprindžuk (1970*b*, 1971*a*, *d*), Sprindžuk and Kotov (1976) and Kotov and Sprindžuk (1977) derived such inequalities for the approximations of algebraic numbers of degree at least 5. Györy (1980*h*) extended these approximation results to the case of all algebraic numbers of degree at least 3 as well as to a wide class of linear forms with algebraic coefficients in an arbitrary number of variables.

Let K be a finite extension of $\mathbb{Q}$ with ring of integers $\mathcal{O}_K$. Let $f \in \mathcal{O}_K[X_1, \ldots, X_m]$ be a decomposable form and $\mu \in \mathcal{O}_K$, $\mu \neq 0$. Let $\{\pi_1, \ldots, \pi_s\}$ be a finite set of non-zero non-units in $\mathcal{O}_K$ and let $N_0 \geqslant 1$ be a rational integer. As a generalisation of equation (31) of chapter 5, consider the *decomposable form equation of Mahler type*:

$$f(x_1, \ldots, x_m) = \mu \pi_1^{z_1} \cdots \pi_s^{z_s} \tag{38}$$

in non-negative rational integers $z_1, \ldots, z_s$ and $x_1, \ldots, x_m \in \mathcal{O}_K$ with $N((x_1, \ldots, x_m)) \leqslant N_0$.

The Thue–Mahler equations (2) and (10) are special cases of (38). Further examples are the norm form equations, discriminant form equations and index form equations of Mahler type. These equations play an important role in algebraic number theory (see e.g. Györy, 1980*e*, 1984*b*,*c*, and Evertse and Györy, 1985).

Let $\mathcal{L}(X_1, \ldots, X_m) = \alpha_1 X_1 + \cdots + \alpha_m X_m$ be a linear form with non-zero algebraic integer coefficients in a finite extension L of K and consider the *norm form equation of Mahler type*:

$$N_{L/K}(\mathcal{L}(x_1, \ldots, x_m)) = \mu \pi_1^{z_1} \cdots \pi_s^{z_s} \tag{39}$$

in non-negative rational integers $z_1, \ldots, z_s$ and $x_1, \ldots, x_m \in \mathcal{O}_K$ with $N((x_1, \ldots, x_m)) \leqslant N_0$.

We may assume without loss of generality that $\alpha_1 = 1$. If in particular $m = 2$ and the degree of α_2 is at least three over K, then (39) is a Thue–Mahler equation over $\mathcal{O}_K$ and, by corollary 7.4, there are only finitely many solutions which can all be bounded by a computable number depending only on $K, L, \alpha_2, \ldots, \alpha_m, \mu, s, \pi_1, \ldots, \pi_s$ and N_0. In case $K = \mathbb{Q}$, Schlickewei (1977*a*) extended Schmidt's (1971*a*, 1972) general ineffective finiteness theorems on norm form equations to norm form equations of Mahler type. For certain further ineffective extensions to the case of ground rings that are finitely generated (but not necessarily algebraic) over $\mathbb{Z}$, see Laurent (1984) and Evertse and Györy (1985, 1986*b*). Schlickewei (1977*b*, *c*) also established finiteness results for equations of the form

$$f(x_1, \ldots, x_m) = g(x_1, \ldots, x_m) \pi_1^{z_1} \cdots \pi_s^{z_s}$$

over $\mathbb{Z}$ where f is an appropriate decomposable form and g is a polynomial whose degree is small relative to deg(f). For $K=\mathbb{Q}$, Pethö (1982*b*) extended the results of Győry and Pethö (1977, 1980) on the distribution of the solutions of norm form equations to equations of the form (39).

Győry and Papp (1977) extended Győry's (1976) general effective finiteness results on discriminant form equations and index form equations to such equations of Mahler type. Independently, Trelina (1977*b*) obtained an extension of Győry's (1976) result on index form equations to index form equations of Mahler type. For certain special extensions L/K and certain numbers $\alpha_1, \ldots, \alpha_m$ of special types, Matveev (1979, 1980, 1981) obtained effective finiteness theorems for (39). The first general effective finiteness results for (39) were established by Győry (1979*b*, 1980*c*). Later, Győry (1980*d*, *g*) and Kotov (1980*b*) independently derived explicit bounds for the solutions of (39), under the general hypothesis that in (39) α_{i+1} is of degree at least 3 over $K(\alpha_1, \ldots, \alpha_i)$ for $i=1, \ldots, m-1$. Further, Győry (1981*a*) and Kotov (1981) independently established a further improvement. General effective finiteness results for decomposable form equations of Mahler type were given by Győry (1979*b*, 1980*c*, 1981*a*). The main result of Győry (1981*a*) implies theorems 7.2, 7.6 and corollary 7.4 and the above-mentioned general effective results on norm form, discriminant form and index form equations of Mahler type. In several of the above-mentioned papers, the authors deduced effective lower bounds for the greatest prime factors of the norms of decomposable forms at algebraic integer points. Further, Győry (1981*b*) deduced explicit bounds for the S-integral solutions of decomposable form equations, and, in particular, of norm form, discriminant form and index form equations. Extensions of the results of Győry (1981*a*) on decomposable form equations of Mahler type to the case that the ground ring is a finitely generated (but not necessarily algebraic) extension of $\mathbb{Z}$ are given in Győry (1984*b*). For a survey of the effective results mentioned above, we refer to Győry (1980*e*, 1984*b*).

Generalising earlier results of Evertse (1983*b*, 1984*a*) on Thue–Mahler equations, Evertse and Győry (1985) derived explicit bounds for the numbers of solutions of decomposable form equations of Mahler type in an arbitrary number of variables over finitely generated (but not necessarily algebraic) extension rings of $\mathbb{Z}$. Their bounds are independent of the coefficients of the decomposable forms involved, but their method is ineffective. Warkentin (1984*a*, *b*) extended results of Schmidt (1972) and Schlickewei (1977*b*) to norm form equations over a rational function field.

CHAPTER 8——

The generalised superelliptic equation

Denote by K a finite extension of $\mathbb{Q}$ and by $\mathcal{O}_K$ the ring of integers of K. Let $\{\mathfrak{p}_1, \ldots, \mathfrak{p}_t\}$ be a finite set of prime ideals of $\mathcal{O}_K$. Denote by $\mathcal{S}$ the set of all non-zero elements of $\mathcal{O}_K$ that are composed of $\mathfrak{p}_1, \ldots, \mathfrak{p}_t$. Let $\alpha_1, \ldots, \alpha_n$ be distinct elements of K. Write

$$f(X, Z) = (X - \alpha_1 Z)^{r_1} \cdots (X - \alpha_n Z)^{r_n}$$

where $r_1, \ldots, r_n$ are positive rational integers. For given rational integers $m \geqslant 2, \tau \geqslant 0$ and for a given non-zero algebraic number b in K, we consider the superelliptic equation

$$f(x, z) = by^m \tag{1}$$

in $x \in \mathcal{O}_K$, $z \in \mathcal{S}$ and $y \in \mathcal{O}_K$ satisfying

$$\max_{1 \leqslant i \leqslant t} \min(\mathrm{ord}_{\mathfrak{p}_i}(x), \mathrm{ord}_{\mathfrak{p}_i}(z)) \leqslant \tau. \tag{2}$$

The above notation will be used throughout the chapter without further reference.

We shall apply theorem 7.6 to generalise theorem 6.1 as follows.

Theorem 8.1. *Let $m \geqslant 3$. Suppose that $f(X, 1)$ has at least two distinct simple roots. Let $x \in \mathcal{O}_K$, $z \in \mathcal{S}$ and $y \in \mathcal{O}_K$ satisfy (2) and (1). There exist a unit $\varepsilon_1 \in \mathcal{O}_K$ and a computable number C_1 depending only on b, m, τ, f, K and $\mathcal{S}$ such that*

$$\max(\overline{|\varepsilon_1 x|}, \overline{|\varepsilon_1 z|}) \leqslant C_1.$$

For $m = 2$, we shall apply theorems 7.6 and 6.2 to generalise theorem 6.2 as follows.

Theorem 8.2. *Suppose that $f(X, 1)$ has at least three distinct simple roots. Let $x\in\mathcal{O}_K$, $z\in\mathcal{S}$ and $y\in\mathcal{O}_K$ satisfy* (2) *and* (1) *with $m=2$. There exist a unit ε_2 in $\mathcal{O}_K$ and a computable number C_2 depending only on b, τ, f, K and $\mathcal{S}$ such that*

$$\max(\overline{|\varepsilon_2 x|}, \overline{|\varepsilon_2 z|}) \leqslant C_2.$$

We shall apply theorems 8.1 and 8.2 to prove the following result which is an effective version of a theorem of LeVeque (1964).

Theorem 8.3. *Let $m\geqslant 2$ and $n\geqslant 2$. Put*

$$q_i = \frac{m}{(m, r_i)} \quad (i=1,\ldots,n).$$

Suppose that $(q_1,\ldots,q_n)$ is not a permutation of either of the n-tuples $(q, 1, 1, \ldots, 1)$ and $(2, 2, 1, 1, \ldots, 1)$. Let $x\in\mathcal{O}_K$, $z\in\mathcal{S}$ and $y\in\mathcal{O}_K$ satisfy (2) *and* (1). *There exists a unit ε_3 in $\mathcal{O}_K$ and a computable number C_3 depending only on b, m, τ, f, K and $\mathcal{S}$ such that*

$$\max(\overline{|\varepsilon_3 x|}, \overline{|\varepsilon_3 z|}) \leqslant C_3.$$

Clearly theorem 8.3 includes theorems 8.1 and 8.2. Theorem 8.3 was proved by Brindza (1984*a*); in fact, Brindza gave a quantitative version of theorem 8.3. An immediate consequence of theorem 8.1 is the following result which is an effective version of a theorem of Mahler (1953).

Corollary 8.1. *Let A and B be non-zero rational integers. Let $m\geqslant 2$ and $n\geqslant 2$ with $mn\geqslant 6$ be rational integers. Then*

$$P(Ax^m + By^n) \to \infty, \quad \textit{effectively},$$

as $\max(|x|, |y|)$ *tends to infinity through non-zero rational integers x, y with* $(x, y)=1$.

A quantitative version of corollary 8.1 follows from a result of Kotov (1976).

Proofs

The proofs of theorems 8.1, 8.2 and 8.3 depend on the following lemma.

Lemma 8.1. *Let $m\geqslant 2$. Suppose that $b\neq 0$ and $\alpha_1,\ldots,\alpha_n$ are in $\mathcal{O}_K$. Let α be a root of $f(X, 1)$ of order r. Put*

$$q = m/(m, r).$$

Let $x\in\mathcal{O}_K$, $y\in\mathcal{O}_K$ and $z\in\mathcal{S}$. Then there exist $\zeta\neq 0$, $\phi\neq 0$ and δ in $\mathcal{O}_K$ such that *equation* (1) *implies that*

$$x-\alpha z=(\zeta/\phi)\delta^q$$

where

$$\max(\overline{|\zeta|},\overline{|\phi|})\leqslant C_4$$

for some computable number C_4 depending only on b, m, f, K and $\mathscr{S}$.

Proof. By permuting the suffixes of $\alpha_1,\ldots,\alpha_n$, there is no loss of generality in assuming that $\alpha=\alpha_1$ and $r=r_1$. Put

$$\Delta=[b]\prod_{j=2}^{n}[\alpha-\alpha_j]\prod_{v=1}^{t}\mathfrak{p}_v.$$

Since $\alpha_1,\ldots,\alpha_n$ are distinct, we find that Δ is a non-zero ideal in $\mathcal{O}_K$.

Let $x\in\mathcal{O}_K$, $y\in\mathcal{O}_K$ and $z\in\mathscr{S}$ satisfy (1). We may assume that $x\neq\alpha z$; otherwise the theorem follows with $\zeta=1$, $\phi=1$ and $\delta=0$. By (1), we have the ideal equation in $\mathcal{O}_K$

$$[x-\alpha_1 z]^{r_1}\cdots[x-\alpha_n z]^{r_n}=[by^m].$$

Hence

$$[x-\alpha z]^r=\mathfrak{a}\mathfrak{b}^m \tag{3}$$

for some ideals $\mathfrak{a},\mathfrak{b}$ in $\mathcal{O}_K$ where $\mathfrak{a}$ is composed exclusively of prime ideal factors of Δ and $(\mathfrak{b},\Delta)=[1]$. If $\mathfrak{p}$ is a prime ideal in $\mathcal{O}_K$ and $\mathfrak{p}^u$ is the highest power of $\mathfrak{p}$ which divides $\mathfrak{b}^m$, then clearly $m\mid u$ and, by (3), $r\mid u$. This implies that the least common multiple of m and r divides u. Hence $qr\mid u$. Consequently, it follows from (3) that

$$[x-\alpha z]=\mathfrak{a}_1\mathfrak{b}_1^q$$

where $\mathfrak{a}_1$ and $\mathfrak{b}_1$ are ideals in $\mathcal{O}_K$ and $\mathfrak{a}_1$ divides Δ^{q-1} in $\mathcal{O}_K$. Now proceed as in the proof of lemma 6.1 to complete the proof of lemma 8.1. □

We denote by $c_1,c_2,\ldots$ computable positive numbers depending only on b, m, τ, f, K and $\mathscr{S}$.

Proof of theorem 8.1. As in the beginning of the proof of theorem 6.1, we may assume without loss of generality that $b,\alpha_1,\ldots,\alpha_n$ are algebraic integers. By permuting the suffixes of $\alpha_1,\ldots,\alpha_n$, we may further assume that α_1 and α_2 are simple roots of $f(X,1)$. Then, by lemma 8.1, there exist ξ_1,ξ_2, ψ_1,ψ_2 and γ_1,γ_2 in $\mathcal{O}_K$ such that $\xi_1\xi_2\psi_1\psi_2\neq 0$ and

$$x-\alpha_1 z=(\xi_1/\psi_1)\gamma_1^m, \tag{4}$$

$$x-\alpha_2 z=(\xi_2/\psi_2)\gamma_2^m \tag{5}$$

and

$$\max(\overline{|\xi_1|}, \overline{|\xi_2|}, \overline{|\psi_1|}, \overline{|\psi_2|}) \leqslant c_1. \tag{6}$$

If $\mathfrak{p}$ is a prime ideal in $\mathcal{O}_K$ and $\mathfrak{p}^u (u \geqslant 1)$ divides both $[x-\alpha_1 z]$ and $[x-\alpha_2 z]$, then $\mathfrak{p}^u \mid [\alpha_2-\alpha_1][x]$ and $\mathfrak{p}^u \mid [\alpha_2-\alpha_1][z]$. Since $z \in \mathcal{S}$, it follows that $N\mathfrak{p} \leqslant c_2$. Further, by (2), $u \leqslant c_3$. Consequently, by (4), (5) and (6), we find that

$$N((\gamma_1, \gamma_2)) \leqslant c_4. \tag{7}$$

Subtracting (5) from (4) and multiplying by $\psi_1\psi_2$, we obtain

$$\xi_1\psi_2\gamma_1^m - \xi_2\psi_1\gamma_2^m = (\alpha_2-\alpha_1)\psi_1\psi_2 z.$$

We apply theorem 7.6 to the binary form $\xi_1\psi_2 X^m - \xi_2\psi_1 Y^m$ and with γ_1, γ_2 and $(\alpha_2-\alpha_1)\psi_1\psi_2 z$ in place of x, y, z, respectively. Observe that $P = P(N((\alpha_2-\alpha_1)\psi_1\psi_2 z)) \leqslant c_5$. Therefore the number t of prime ideals in $\mathcal{O}_K$ dividing $(\alpha_2-\alpha_1)\psi_1\psi_2 z$ is bounded from above by some number c_6. Since $m \geqslant 3, \xi_1\xi_2\psi_1\psi_2 \neq 0, \alpha_1 \neq \alpha_2, z \neq 0$ and (7) holds, all assumptions are fulfilled with $N_0 = c_4$. Hence, by theorem 7.6, there exists a unit ε in $\mathcal{O}_K$ such that

$$\max(\overline{|\varepsilon\gamma_1|}, \overline{|\varepsilon\gamma_2|}) \leqslant c_7.$$

Now, by solving for x and y in the equations (4) and (5), we obtain that $\max(\overline{|\varepsilon^m x|}, \overline{|\varepsilon^m z|})$ is bounded by a computable number depending only on b, m, τ, f, K and $\mathcal{S}$. □

Proof of theorem 8.2. As in the proof of theorem 8.1, we may assume that b, $\alpha_1, \ldots, \alpha_n$ are algebraic integers and that α_1, α_2 and α_3 are simple roots of $f(X, 1)$. It follows from lemma 8.1 that there exist $\xi_i \neq 0$, $\psi_i \neq 0$ and γ_i in $\mathcal{O}_K$ such that, for $i = 1, 2, 3$, we have

$$x - \alpha_i z = (\xi_i/\psi_i)\gamma_i^2 \tag{8}$$

and

$$\max(\overline{|\xi_i|}, \overline{|\psi_i|}) \leqslant c_8. \tag{9}$$

From equations (8), we obtain

$$\sigma_1\gamma_1^2 - \sigma_2\gamma_2^2 = (\alpha_2-\alpha_1)z,$$

$$\sigma_2\gamma_2^2 - \sigma_3\gamma_3^2 = (\alpha_3-\alpha_2)z,$$

$$\sigma_3\gamma_3^2 - \sigma_1\gamma_1^2 = (\alpha_1-\alpha_3)z$$

where $\sigma_i = \xi_i/\psi_i$ for $i = 1, 2, 3$. Let $\sigma_1^{1/2}, \sigma_2^{1/2}$ and $\sigma_3^{1/2}$ be arbitrary, but fixed, choices of the square roots. Put

$$L = K(\sigma_1^{1/2}, \sigma_2^{1/2}, \sigma_3^{1/2}).$$

Observe that L is a finite extension of K of degree at most 8 and therefore $d_L \leqslant c_9$. Let $\eta'_1, \ldots, \eta'_r$ be an independent system of units for L satisfying (A.45). Put

$$\beta_3 = \sigma_1^{1/2}\gamma_1 - \sigma_2^{1/2}\gamma_2,$$
$$\beta_1 = \sigma_2^{1/2}\gamma_2 - \sigma_3^{1/2}\gamma_3,$$
$$\beta_2 = \sigma_3^{1/2}\gamma_3 - \sigma_1^{1/2}\gamma_1.$$

Put $A = \psi_1\psi_2\psi_3$. Notice that A is a non-zero element of $\mathcal{O}_K$ satisfying $\overline{|A|} \leqslant c_8^3$, hence $|A| \geqslant c_{10}$. By (9) and lemma A.2, $A\sigma_i^{1/2}$ is an element of $\mathcal{O}_L$ of height at most c_{11}. Since $d_L \leqslant c_9$, we infer from lemma A.16 that $\max(h_L, R_L, |\mathcal{D}_L|) \leqslant c_{12}$. Further observe that $A\beta_1$, $A\beta_2$ and $A\beta_3$ are non-zero elements of $\mathcal{O}_L$. Since $z \in \mathcal{S}$, we see that

$$P(N_L(A\beta_i)) \leqslant c_{13} \quad (i = 1, 2, 3).$$

Consequently we see from lemma A.12, corollary A.6 and corollary A.5 that

$$A\beta_i = f_i g_i^3 l \quad (i = 1, 2, 3) \tag{10}$$

where $f_1, f_2, f_3, g_1, g_2, g_3, l$ are non-zero elements of $\mathcal{O}_L$ satisfying

$$\max_{1 \leqslant i \leqslant 3} \overline{|f_i|} \leqslant c_{14}, \tag{11}$$

$$\max_{1 \leqslant i \leqslant 3} P(N_L(g_i l)) \leqslant c_{15} \tag{12}$$

and the ideals $[g_1]$, $[g_2]$ and $[g_3]$ are relatively prime. Observe that

$$A\beta_1 + A\beta_2 + A\beta_3 = 0.$$

By (10),

$$f_1 g_1^3 + f_2 g_2^3 = -f_3 g_3^3. \tag{13}$$

Since there are only finitely many possibilities for f_1, f_2 and f_3 in view of $d_L \leqslant c_9$ and (11), we may assume that f_1, f_2 and f_3 are fixed. We are going to apply theorem 7.6 to the binary form $f_1 X^3 + f_2 Y^3$ with $x = g_1$, $y = g_2$ and $z = -f_3 g_3^3$. Recall that $d_L \leqslant c_9$ and $|\mathcal{D}_L| \leqslant c_{12}$. Since $[g_1]$, $[g_2]$ and $[g_3]$ are relatively prime, we have, by (13) and (11), $N((g_1, g_2)) \leqslant c_{16}$. Furthermore, by (11) and (12), $P(N_L(f_3 g_3^3)) \leqslant c_{17}$, whence the number of prime ideal factors in $\mathcal{O}_L$ dividing $f_3 g_3^3$ does not exceed c_{18}. We conclude, by theorem 7.6, that there exists a unit ε_4 in $\mathcal{O}_L$ such that

$$\max(\overline{|\varepsilon_4 g_1|}, \overline{|\varepsilon_4 g_2|}) \leqslant c_{19}.$$

This, together with (10) and (11), implies that

$$\max\left(\left|\overline{\varepsilon_5 \frac{A\beta_1}{l}}\right|, \left|\overline{\varepsilon_5 \frac{A\beta_2}{l}}\right|\right) \leqslant c_{20}$$

where $\varepsilon_5 = \varepsilon_4^3$. Thus $\varepsilon_5 A\beta_1 = \lambda_1 l$ and $\varepsilon_5 A\beta_2 = \lambda_2 l$ for some $\lambda_1, \lambda_2 \in \mathcal{O}_L$ with $\max(\overline{|\lambda_1|}, \overline{|\lambda_2|}) \leqslant c_{20}$.

Recall that, in the definitions of β_1, β_2 and β_3, we have taken an arbitrary but fixed choice of square roots for σ_1, σ_2 and σ_3. Similarly we can find l', λ_3, $\lambda_4 \in \mathcal{O}_L$ with $\max(P(N_L(l')), \overline{|\lambda_3|}, \overline{|\lambda_4|}) \leqslant c_{21}$ and a unit $\varepsilon_6 \in \mathcal{O}_L$ such that

$$\varepsilon_6 A\tilde{\beta}_1 = \lambda_3 l', \quad \varepsilon_6 A\beta_2 = \lambda_4 l'$$

where

$$\tilde{\beta}_1 = -\sigma_2^{1/2}\gamma_2 - \sigma_3^{1/2}\gamma_3.$$

Consequently

$$-2A\sigma_3^{1/2}\gamma_3 = A(\beta_1 + \tilde{\beta}_1) = \varepsilon_5^{-1}\lambda_1 l + \varepsilon_6^{-1}\lambda_3 l', \tag{14}$$

$$2A\sigma_2^{1/2}\gamma_2 = A(\beta_1 - \tilde{\beta}_1) = \varepsilon_5^{-1}\lambda_1 l - \varepsilon_6^{-1}\lambda_3 l' \tag{15}$$

and

$$\varepsilon_6 \lambda_2 l = \varepsilon_5 \lambda_4 l'. \tag{16}$$

If $\mathfrak{p}$ is a prime ideal in $\mathcal{O}_K$ and $\mathfrak{p}^u$ divides both $[x - \alpha_2 z]$ and $[x - \alpha_3 z]$, then $\mathfrak{p}^u \mid [\alpha_3 - \alpha_2][x]$ and $\mathfrak{p}^u \mid [\alpha_3 - \alpha_2][z]$. Hence, by (2), $u \leqslant c_{22}$. Since $d_L \leqslant c_9$, we obtain from (8) that

$$\min(\mathrm{ord}_{\mathfrak{p}}(\gamma_2), \mathrm{ord}_{\mathfrak{p}}(\gamma_3)) \leqslant c_{23}$$

for every prime ideal $\mathfrak{p}$ in $\mathcal{O}_L$. Consequently, the equations (14) and (15) imply that

$$\min(\mathrm{ord}_{\mathfrak{p}}(l), \mathrm{ord}_{\mathfrak{p}}(l')) \leqslant c_{24}$$

for every prime ideal $\mathfrak{p}$ in $\mathcal{O}_L$. Now it follows from (16) and $P(N_L(ll')) \leqslant c_{15} + c_{21}$ that $\max(|N_L(l)|, |N_L(l')|) \leqslant c_{25}$ which implies that

$$\max(|N_L(\beta_1)|, |N_L(\tilde{\beta}_1)|) \leqslant c_{26}.$$

Since $\beta_1 \tilde{\beta}_1 = (\alpha_3 - \alpha_2)z$, we find that $|N_K(z)| \leqslant c_{26}^2$. By corollary A.6, there exists a unit $\varepsilon_7 \in \mathcal{O}_K$ such that $\overline{|\varepsilon_7 z|} \leqslant c_{27}$. Hence $\varepsilon_7 z$ can be assumed to be fixed. Write $N = r_1 + r_2 + \cdots + r_n$. By corollaries A.4 and A.5, we may further write $\varepsilon_7^N = \varepsilon_8 \varepsilon_9^2$ where $\overline{|\varepsilon_8|} \leqslant c_{28}$. Multiplying both sides of (1) by ε_7^N, we obtain

$$(\varepsilon_7 x - \alpha_1 \varepsilon_7 z)^{r_1} \cdots (\varepsilon_7 x - \alpha_n \varepsilon_7 z)^{r_n} = \varepsilon_7^N f(x, z) = b\varepsilon_8(\varepsilon_9 y)^2.$$

Now apply theorem 6.2 to conclude that $\overline{|\varepsilon_7 x|} \leqslant c_{29}$. □

Proof of theorem 8.3. Since $(q_1, \ldots, q_n)$ is not a permutation of either of the n-tuples $(q, 1, \ldots, 1)$ and $(2, 2, 1, \ldots, 1)$, we have the following two possibilities.

(a) There exist distinct i, j with $i, j \in \{1, \ldots, n\}$ such that $q_i \geqslant 3$ and $q_j \geqslant 2$.
(b) There exist distinct i, j, k with $i, j, k \in \{1, \ldots, n\}$ such that $q_i = q_j = q_k = 2$.

Case (a). By permuting the suffixes of $\alpha_1, \ldots, \alpha_n$, we may assume that $q_1 \geqslant 3$ and $q_2 \geqslant 2$. It follows from lemma 8.1 that there exist $\xi_1, \xi_2, \psi_1, \psi_2$ and γ_1, γ_2 in $\mathcal{O}_K$ such that $\xi_1 \xi_2 \psi_1 \psi_2 \neq 0$ and

$$x - \alpha_1 z = (\xi_1/\psi_1)\gamma_1^{q_1}, \tag{17}$$

$$x - \alpha_2 z = (\xi_2/\psi_2)\gamma_2^{q_2}, \tag{18}$$

$$\max(\overline{|\xi_1|}, \overline{|\xi_2|}, \overline{|\psi_1|}, \overline{|\psi_2|}) \leqslant c_{30}. \tag{19}$$

Subtracting (17) from (18) and multiplying by $\psi_1 \psi_2$, we obtain

$$\xi_2 \psi_1 \gamma_2^{q_2} - \xi_1 \psi_2 \gamma_1^{q_1} = (\alpha_1 - \alpha_2)\psi_1 \psi_2 z. \tag{20}$$

Since $z \in \mathcal{S}$, we may write

$$z = z_1 z_2^{q_2} \tag{21}$$

where $z_1, z_2 \in \mathcal{S}$ and $\overline{|z_1|} \leqslant c_{31}$. Hence, by (20),

$$\xi_2 \psi_1 \gamma_2^{q_2} - (\alpha_1 - \alpha_2)\psi_1 \psi_2 z_1 z_2^{q_2} = \xi_1 \psi_2 \gamma_1^{q_1}. \tag{22}$$

We are going to apply theorem 8.1 to the binary form $\xi_2 \psi_1 X^{q_2} - (\alpha_1 - \alpha_2)\psi_1 \psi_2 z_1 Y^{q_2}$ and with $b = \xi_1 \psi_2$ and $m = q_1$. By (22), (21), (17), (18), (19) and (2), we have

$$\max_{1 \leqslant i \leqslant t} \min(\mathrm{ord}_{\mathfrak{p}_i}(\gamma_2), \mathrm{ord}_{\mathfrak{p}_i}(z_2)) \leqslant c_{32}.$$

Hence, by theorem 8.1, there exists a unit ε_{10} in $\mathcal{O}_K$ such that

$$\max(\overline{|\varepsilon_{10}\gamma_2|}, \overline{|\varepsilon_{10} z_2|}) \leqslant c_{33}.$$

This, together with (21) and (18), implies that $\overline{|\varepsilon_{10}^{q_2} z|} \leqslant c_{34}$ and $\overline{|\varepsilon_{10}^{q_2} x|} \leqslant c_{35}$.

Case (b). By permuting the indices of $\alpha_1, \ldots, \alpha_n$, we may assume that $q_1 = q_2 = q_3 = 2$. By lemma 8.1, the relations (8) and (9) are valid. Consequently

$$(x - \alpha_1 z)(x - \alpha_2 z)(x - \alpha_3 z) = \frac{\xi_1 \xi_2 \xi_3}{\psi_1 \psi_2 \psi_3}(\gamma_1 \gamma_2 \gamma_3)^2.$$

Now apply theorem 8.2 to conclude the assertion of theorem 8.3. □

Proof of corollary 8.1. There is no loss of generality in assuming that $m \geqslant 3$ and $n \geqslant 2$. Suppose that x and y with $(x, y) = 1$ are non-zero integers. Since $(x, y) = 1$, we have

$$Ax^m + By^n \neq 0$$

whenever $\max(|x|, |y|) \geqslant c$ for some computable number c depending only on A and B. We assume that $\max(|x|, |y|) \geqslant c$. Then write

$$Ax^m + By^n = \pm p_1^{a_1} \cdots p_s^{a_s}$$

where $p_1 < p_2 < \cdots < p_s =: P$ are rational prime numbers and $a_1, \ldots, a_s$ are non-negative rational integers. We can write

$$\pm p_1^{a_1} \cdots p_s^{a_s} = wz^n$$

where w and z are non-zero rational integers and $|w| \leqslant P^{sn}$. Thus

$$Ax^m = -By^n + wz^n.$$

Since $(x, y) = 1$, we have $(y, z) \leqslant |A|$. Now we apply theorem 8.1 to conclude that $|y|$, $|z|$ and hence $|x|$ are bounded by a computable number depending only on A, B, m, n and P. □

Notes

Denote by $\mathbb{Z}_S$ the set of rational numbers with denominators in S, the so-called S-integers. Mahler (1934*a*) used his p-adic analogue of Thue's theorem to prove that if for $F \in \mathbb{Z}[X, Y]$ the equation $F(X, Y) = 0$ represents an (irreducible) curve of genus 1, then there are only finitely many $x, y \in \mathbb{Z}_S$ with $F(x, y) = 0$. Mahler's result is ineffective. Kotov (1977, 1979) used the (p-adic) method on linear forms in logarithms to derive an effective analogue of Mahler's result. For absolutely irreducible polynomials F of degree at least 3, Kotov computed an upper bound for the absolute values of the numerators of $x, y \in \mathbb{Z}_S$ with $F(x, y) = 0$. Kotov and Trelina (1979) improved upon this bound by showing that in this case

$$\max(H(x), H(y)) \leqslant \exp(\exp(C_4 P))$$

where C_4 is an almost explicitly given function of the degree and the height of F. In case the curve has complex multiplication, Bertrand (1978) derived an essentially better upper bound $\exp(P^{C_5})$, but with an ineffective constant C_5.

The full p-adic analogue of Siegel's (1929) theorem was established by Lang (1960), and independently by LeVeque (1961): Let K be a finite extension of $\mathbb{Q}$. (Lang assumed K to be any finitely generated (not necessarily algebraic) extension of $\mathbb{Q}$ and considered the solutions in an arbitrary subring of K which is finitely generated over $\mathbb{Z}$.) Let $F \in K[X, Y]$ be such that the equation $F(X, Y)=0$ represents an (irreducible) curve of genus $g \geqslant 1$. Denote the $\mathscr{S}$-integers by $\mathcal{O}_{\mathscr{S}}$. Then there are only finitely many $x, y \in \mathcal{O}_{\mathscr{S}}$ with $F(x, y)=0$. As in chapter 6, the general effective analogue has not been proved yet, but effective results are available for the special class of superelliptic equations $y^m=G(x)$ where $m \in \mathbb{Z}$, $m \geqslant 2$ and $G \in \mathbb{Z}[X]$. Trelina (1978) generalised results of Sprindžuk on the integral solutions of superelliptic equations to the $\mathscr{S}$-integral solutions of these equations. Brindza (1984*a*) gave also a p-adic extension of his effective proof of LeVeque's (1964) theorem. Essentially Brindza's result is equivalent to theorem 8.3, but he specified the bound too. He proved that, under the conditions of theorem 8.3, the solutions $x, y \in \mathcal{O}_{\mathscr{S}}$ of $f(x, 1)=y^m$ satisfy

$$\max\{H(x), H(y)\} < \exp(\exp(C_6 P^2 (t+1)^3))$$

where $P=\max_{1 \leqslant j \leqslant t} P(N_K(\mathfrak{p}_j))$ and C_6 is a computable number depending only on K, f and m. We could have applied theorem 1.4 instead of theorem 7.6 in the proof of theorem 8.2. We recall that in theorems 8.1–8.3 the dependence on K can be refined to dependence on the degree and the discriminant of K only.

It is a consequence of Mahler's (1934*a*) paper that $P(Ax^3+By^2) \to \infty$ as $\max(|x|, |y|)$ tends to infinity through non-zero rational integers x, y with $(x, y)=1$. An effective proof of this result for $A=1, B=-1$ was obtained by Coates (1970*b*), who also computed an effective lower bound for $P(x^3-y^2)$, namely: if $x, y \in \mathbb{Z}$ with $(x, y)=1$ and $x^3 \neq y^2$ then

$$P(x^3-y^2) \geqslant 10^{-3}(\log\log \mathscr{X})^{1/4}$$

where $\mathscr{X}=\max(|x|, |y|)$. Kotov (1979) and Kotov and Trelina (1979) extended Coates' result to arbitrary elliptic curves and improved on the lower bound in such a way that, as a particular case, it follows that

$$P(x^3-y^2) \geqslant C_7 \log\log \mathscr{X}$$

where $C_7>0$ is a computable absolute constant. For an entirely algebraic approach to results on $P(Ax^3+By^2)$, see Herzberg (1975).

The ineffective version of corollary 8.1 is due to Mahler (1953). The effective corollary 8.1 can be derived from Coates' (1970*a*) result on the Thue–Mahler equation. Kotov (1976) generalised corollary 8.1 to algebraic number fields. He derived explicit bounds which, under the conditions of corollary 8.1, yield

$$P(Ax^m+By^n) \geqslant C_8(\log_2 \mathscr{X} \log_3 \mathscr{X})^{1/2}$$

where $C_8>0$ is a computable constant depending only on A, B, m and n.

CHAPTER 9

Perfect powers in binary recurrence sequences

In this chapter, we show that there are only finitely many perfect powers of the form u_1+u_2 where u_1 and u_2 are relatively prime rational integers composed of a given finite set of primes. We consider the analogous question in a number field. As applications, we prove that there are only finitely many perfect powers in a non-degenerate binary recurrence of algebraic numbers and that, under suitable restrictions on rational integers A_1, A_2, A_3,

$$P(A_1x^{2t}+A_2x^ty+A_3y^2)\to\infty$$

uniformly in positive integers $t>1$, $x>1$ and y with $(x^t, y)=1$ whenever $\max(t, x, y)$ tends to infinity.

Let $P\geqslant 3$ and $a\geqslant 1$. Denote by S the set of all rational integers composed of primes not exceeding P. Then we have

Theorem 9.1. *Let u, u_1 and u_2 with $(u_1, u_2)\leqslant a$ be members of S. Let $q\geqslant 2$ and y with $|y|>1$ be rational integers. If*

$$u_1+u_2=uy^q, \tag{1}$$

then $P(q)$ is bounded by a computable number depending only on P and a.

Combining theorem 9.1 with theorem 7.2, we obtain

Theorem 9.2. *Under the conditions of theorem* 9.1, *equation* (1) *implies that*

$$\max(|u|, |u_1|, |u_2|, q)\leqslant C_1$$

for some computable number C_1 depending only on P and a.

Now we state notation for a generalisation which includes theorem 9.2. Denote by K a finite extension of $\mathbb{Q}$, by $\mathcal{O}_K$ the ring of integers of K and by R

the regulator of K. Put $[K:\mathbb{Q}]=d$. Let $v \geqslant 1$ and $s \geqslant 0$. Let $\{\pi_1, \ldots, \pi_s\}$ be a set of non-zero non-units of $\mathcal{O}_K$. Denote by $\mathcal{S}_0 = \mathcal{S}_{K,v}(\pi_1, \ldots, \pi_s)$ the set of all non-zero elements of $\mathcal{O}_K$ of the form

$$\mu \pi_1^{v_1} \cdots \pi_s^{v_s}$$

where $v_1, \ldots, v_s$ are non-negative rational integers and $\mu \in \mathcal{O}_K$ with $\overline{|\mu|} \leqslant v$. If $s=0$, then $\mathcal{S}_0$ is the set of all $\mu \in \mathcal{O}_K$ with $\mu \neq 0$ and $\overline{|\mu|} \leqslant v$. We consider the equation

$$\varepsilon_1 \gamma_1 + \varepsilon_2 \gamma_2 = \gamma y^q \tag{2}$$

in $\varepsilon_1, \varepsilon_2, y \in \mathcal{O}_K, \gamma_1, \gamma_2, \gamma \in \mathcal{S}_0, q \in \mathbb{Z}_+$ with $\varepsilon_1, \varepsilon_2$ units and $y \neq 0$. We shall use the above notation throughout the chapter without any further reference.

Theorem 9.1 is contained in the following result.

Theorem 9.3. *Let y be a non-zero non-unit in $\mathcal{O}_K$. Let $\tau \geqslant 0$. Suppose*

$$\min(\text{ord}_{\wp}(\gamma_1), \text{ord}_{\wp}(\gamma_2)) \leqslant \tau \tag{3}$$

for every prime ideal $\wp$ in $\mathcal{O}_K$. Then equation (2) implies that $P(q)$ is bounded by a computable number depending only on τ, K and $\mathcal{S}_0$.

It is impossible to prove the assertion of theorem 9.3 if y is a unit. However, if (2) is satisfied with a unit $y \in \mathcal{O}_K$ and (3) is valid, then we may apply theorem 1.3 to obtain

$$\max(\overline{|\gamma|}, \overline{|\gamma_1|}, \overline{|\gamma_2|}) \leqslant C_2$$

and

$$\max(\overline{|\varepsilon_1^{-1} y^q|}, \overline{|\varepsilon_2^{-1} y^q|}, \overline{|\varepsilon_1^{-1} \varepsilon_2|}) \leqslant C_2$$

where C_2 is a computable number depending only on τ, K and $\mathcal{S}_0$ (and not on q).

For a given $q \geqslant 2$, we apply theorem 7.6 to equation (2) to derive the following result.

Theorem 9.4. *Let $q \geqslant 2$. For $\tau \geqslant 0$, assume that (3) is valid. Then there exists a computable number C_3 depending only on q, τ, K and $\mathcal{S}_0$ such that equation (2) implies that*

$$\max(\overline{|\gamma|}, \overline{|\gamma_1|}, \overline{|\gamma_2|}) \leqslant C_3 \tag{4}$$

and

$$\max(\overline{|\varepsilon_1^{-1} y^q|}, \overline{|\varepsilon_2^{-1} y^q|}, \overline{|\varepsilon_1^{-1} \varepsilon_2|}) \leqslant C_3. \tag{5}$$

Thus, under the assumptions of theorem 9.4, it follows from (4) and (5) that

$$\max(\overline{|\varepsilon_1^{-1}\gamma y^q|}, \overline{|\varepsilon_1^{-1}\varepsilon_1\gamma_1|}, \overline{|\varepsilon_1^{-1}\varepsilon_2\gamma_2|}) \leqslant C_3. \tag{6}$$

The assertion of theorem 9.2 may be obtained by combining theorems 9.3 and 9.4.

We apply theorems 9.3, 9.4 and 1.3 to prove the following result.

Theorem 9.5. *Let $0 \leqslant m \in \mathbb{Z}$. Let α, β be non-zero elements of $\mathcal{O}_K$ such that α/β is not a root of unity and $(\alpha, \beta) = [1]$. Let $\tau_1 \geqslant 0$. Suppose*

$$\min(\operatorname{ord}_\wp(\gamma_1\alpha^m), \operatorname{ord}_\wp(\gamma_2\beta^m)) \leqslant \tau_1$$

for every prime ideal $\wp$ in $\mathcal{O}_K$. Assume

$$\gamma_1\alpha^m + \gamma_2\beta^m = \varepsilon\gamma y^q \tag{7}$$

with $0 \neq y \in \mathcal{O}_K$, a unit $\varepsilon \in \mathcal{O}_K$ and $2 \leqslant q \in \mathbb{Z}$. Then

$$\max(\overline{|\gamma|}, \overline{|\gamma_1|}, \overline{|\gamma_2|}, m) \leqslant C_4 \tag{8}$$

for some computable number C_4 depending only on τ_1, α, β, K and $\mathcal{S}_0$.

An immediate consequence of theorem 9.5 is the following result.

Corollary 9.1. *Suppose that the assumptions of theorem 9.5 are satisfied. Further, assume that $\varepsilon = 1$ and y is not a root of unity. Then*

$$\max(\overline{|\gamma|}, \overline{|\gamma_1|}, \overline{|\gamma_2|}, \overline{|y|}, q, m) \leqslant C_5$$

where C_5 is a computable number depending only on τ_1, α, β, K and $\mathcal{S}_0$.

Further, we apply theorem 9.5 to prove the following result on perfect powers in a non-degenerate binary recurrence.

Theorem 9.6. *Let $\{u_m\}_{m=0}^{\infty}$ be a simple non-degenerate algebraic recurrence of order 2 and let $q \geqslant 2$. If*

$$u_m = \gamma y^q \tag{9}$$

with $0 \neq y \in K$, then m is bounded by a computable number depending only on K, $\mathcal{S}_0$ and $\{u_m\}_{m=0}^{\infty}$.

Theorem 9.6 admits the following consequence.

Corollary 9.2. *Let $\{u_m\}_{m=0}^{\infty}$ be a simple non-degenerate algebraic recurrence of order 2. Let $0 \neq \delta \in K$ and $q \geqslant 2$. If*

$$u_m = \delta y^q \tag{10}$$

where $0 \neq y \in K$ *is not a root of unity, then*

$$\max(\lceil y \rceil, q, m) \leqslant C_6$$

for some computable number C_6 *depending only on* δ, K *and the sequence* $\{u_m\}_{m=0}^{\infty}$.

Some special cases of this corollary can be found in the literature. Shorey and Tijdeman (1976*a*) proved that there are only finitely many perfect powers in the Lucas sequence given by $u_0=0$, $u_1=1$ and

$$u_{m+2}=(x+1)u_{m+1}-xu_m \quad (m=0, 1, \ldots)$$

where $x>1$ is a fixed integer. Here $u_m=(x^m-1)/(x-1)$. Thus there are only finitely many perfect powers among the integers whose digits in the x-adic expansion are all equal to 1. We shall return to this problem in chapter 12. For a non-degenerate recurrence sequence $\{u_m\}_{m=0}^{\infty}$ of order 2 induced by a (rational) integral recurrence, it has been proved, independently, by Pethö (1982*a*) and Shorey and Stewart (1983) that equation (10) with $0 \neq \delta \in \mathbb{Z}$, $y \in \mathbb{Z}$, $|y|>1$ and $q \geqslant 2$ implies that $\max(|y|, q, m)$ is bounded by a computable number depending only on δ and the sequence $\{u_m\}_{m=0}^{\infty}$; in fact, Pethö proved that $\max(|y|, q, m)$ is bounded by a computable number depending only on $P(\delta)$ and $\{u_m\}_{m=0}^{\infty}$ provided that the companion polynomial to $\{u_m\}_{m=0}^{\infty}$ has relatively prime integral coefficients.

Shorey and Stewart (1983) applied their result to prove the following.

Theorem 9.7. *Let* $A_1, A_2, A_3, B \in \mathbb{Z}$ *with* $A_1A_2B \neq 0$ *and* $A_2^2-4A_1A_3 \neq 0$. *Let* x, y, t *with* $|x|>1$ *and* $t>1$ *be rational integers satisfying*

$$A_1x^{2t}+A_2x^ty+A_3y^2=B. \tag{11}$$

Then

$$\max(|x|, |y|, t) \leqslant C_7$$

where C_7 *is a computable number depending only on* A_1, A_2, A_3 *and* B.

Let $A_1, A_2, A_3, B \in \mathbb{Z}$ with $B \neq 0$ and $A_2^2-4A_1A_3$ positive and not a square. It is well known that if the equation

$$A_1x^2+A_2xy+A_3y^2=B \tag{12}$$

has one solution in integers x and y, then it has infinitely many solutions in integers x and y. Theorem 9.7 states that among these solutions there are only finitely many in which x is a perfect power. Theorem 9.7 is contained in the following result.

Theorem 9.8. *Let* $a_1 \geqslant 1$. *For* $P \geqslant 3$, *denote by* S *the set of all integers* *composed of primes not exceeding P. Let $A_1, A_2, A_3 \in \mathbb{Z}$ with $A_1A_3 \neq 0$ and $A_2^2 - 4A_1A_3 \neq 0$. Suppose x, y, t with $|x| > 1$, $y \neq 0$ and $t > 1$ are rational integers satisfying $(x^t, y) \leqslant a_1$ and*

$$A_1x^{2t} + A_2x^ty + A_3y^2 \in S. \tag{13}$$

There exists a computable number C_8 depending only on a_1, A_1, A_2, A_3 and the set S such that

$$\max(|x|, |y|, t) \leqslant C_8.$$

Proofs

Proof of theorem 9.1. Suppose that equation (1) holds. Denote by $k_1, \ldots, k_6$ computable positive numbers depending only on P and a. There is no loss of generality in assuming that q is prime. Further, we may assume that $q > k_1$ with k_1 sufficiently large. Put

$$U = \max(|u_1|, |u_2|, e).$$

Write

$$u = \pm \prod_{p \leqslant P} p^{a_p}, \quad u_1 = \pm \prod_{p \leqslant P} p^{b_p}, \quad u_2 = \pm \prod_{p \leqslant P} p^{c_p}$$

where a_p, b_p and c_p are non-negative integers. Observe that

$$\max_{p \leqslant P} \max(b_p, c_p) \leqslant 2 \log U.$$

It follows from (1) that

$$\max_{p \leqslant P} a_p \leqslant 2 \log|u| \leqslant 2 \log(2U).$$

Let p_1 be a prime dividing u_1 such that $p_1^{\mathrm{ord}_{p_1}(u_1)} > a$. Then equation (1) implies that

$$\mathrm{ord}_{p_1}(u_1) = \mathrm{ord}_{p_1}(uy^q - u_2) \leqslant \mathrm{ord}_{p_1}(uu_2^{-1}y^q - 1) + k_2,$$

since $(u_1, u_2) \leqslant a$. In theorem B.4, set $n = \pi(P) + 2$, $d = 1$, $p = p_1 \leqslant P$, $\delta = 1/2$, $A' = P$, $A = 3|y|$, $B = 2\log(2U)$ and $B' = b_n = q$. Let $k_1 > P$. Then, since $p_1 \leqslant P < k_1 < q$ and q is prime, we have $q \not\equiv 0 (\mathrm{mod}\ p_1)$. Now we apply theorem B.4 to conclude that

$$\mathrm{ord}_{p_1}(uu_2^{-1}y^q - 1) \leqslant k_3\left(\log q \log|y| + \frac{\log U}{q}\right).$$

Thus

$$\mathrm{ord}_{p_1}(u_1) \leqslant k_4\left(\log q \log|y| + \frac{\log U}{q}\right).$$

Consequently

$$\log|u_1| = \sum_{p|u_1} \operatorname{ord}_p(u_1)\log p \leqslant k_5\left(\log q \log|y| + \frac{\log U}{q}\right).$$

Similarly

$$\log|u_2| \leqslant k_5\left(\log q \log|y| + \frac{\log U}{q}\right).$$

Let $k_1 > 2k_5$. Then, since $k_5 q^{-1} < k_5 k_1^{-1} < 2^{-1}$, the above inequalities imply that

$$\log U \leqslant 2k_5 \log q \log|y|. \tag{14}$$

Further, it follows from (1) that

$$q \log|y| \leqslant \log(2U)$$

which, together with (14), implies that $q \leqslant k_6$. □

Proof of theorem 9.2. Suppose that equation (1) is valid. There is no loss of generality in assuming that q is prime. By theorem 9.1, we see that q is bounded by a computable number depending only on P and a. Write

$$u = u_3 u_4^q \quad \text{and} \quad u_1 = u_5 u_6^q$$

where

$$\max(|u_3|, |u_5|) \leqslant P^{(q-1)\pi(P)}.$$

Then, by multiplying both sides of (1) by $u_5^q u_6$, we obtain

$$u_5 u_6((u_5 u_6)^q - u_3 u_5^{q-1}(u_4 y)^q) = -u_2 u_5^q u_6.$$

Note that $-u_2 u_5^q u_6 \in S$. Since $(u_1, u_2) \leqslant a$, equation (1) implies $(u_1, yu) \leqslant a$ which gives $b := (u_5 u_6, u_4 y) \leqslant a$. Divide both sides by b^q. Now apply theorem 7.2 with

$$f(X, Y) = bX(X^q - u_3 u_5^{q-1} Y^q), \quad x = \frac{u_5 u_6}{b}, \quad y = \frac{u_4 y}{b}$$

to complete the proof of theorem 9.2. □

Now we turn to the proof of theorem 9.3. Suppose that (2) is fulfilled for some non-zero non-unit y in $\mathcal{O}_K$. First we make some simplifications. We may assume $q > 1$. Further, since every power of y is a non-zero non-unit in $\mathcal{O}_K$, there is no loss of generality in assuming that q is prime.

Let $\eta_1, \ldots, \eta_r$ be an independent system of units for K satisfying (A.45). By corollary A.5, we may write

$$\varepsilon_1 = \rho_1 \eta_1^{a_1} \cdots \eta_r^{a_r} \quad \text{and} \quad \varepsilon_2 = \rho_2 \eta_1^{b_1} \cdots \eta_r^{b_r} \tag{15}$$

where $a_1, \ldots, a_r, b_1, \ldots, b_r \in \mathbb{Z}$ and $\rho_1, \rho_2 \in \mathcal{O}_K$ with $\rho_1\rho_2 \neq 0$ and

$$\max(\overline{|\rho_1|}, \overline{|\rho_2|}) \leqslant c \tag{16}$$

for some computable number c depending only on K. For $1 \leqslant i \leqslant r$, write

$$b_i = qb_{i,1} + b_{i,2}, \quad 0 \leqslant b_{i,2} < q$$

and

$$\varepsilon_3 = \eta_1^{b_{1,1}} \cdots \eta_r^{b_{r,1}}, \quad \varepsilon_4 = \eta_1^{b_{1,2}} \cdots \eta_r^{b_{r,2}}.$$

Thus $\varepsilon_2 = \rho_2\varepsilon_4\varepsilon_3^q$. On dividing both the sides of (2) by ε_3^q and observing that $y\varepsilon_3^{-1}$ is a non-zero non-unit in $\mathcal{O}_K$, we may assume that

$$0 \leqslant b_i < q \quad (1 \leqslant i \leqslant r). \tag{17}$$

Let σ_0 be an embedding of K such that $|\sigma_0(y)| = \overline{|y|}$. By taking images under σ_0 on both sides of (2), we may assume $|y| = \overline{|y|}$. Finally, by corollary A.2, observe that there are only finitely many $\rho_1, \rho_2 \in \mathcal{O}_K$ satisfying (16). By lemma A.2, their heights are bounded and we may therefore assume that ρ_1 and ρ_2 are fixed.

Thus, for the proof of theorem 9.3, it suffices to prove the following result.

Theorem 9.3′. *Let c be a computable number depending only on K. Let $0 \neq \rho_1 \in \mathcal{O}_K, 0 \neq \rho_2 \in \mathcal{O}_K$ satisfy (16). Let ε_1 and ε_2 be given by (15). Assume that $0 \leqslant b_i < q$ for $1 \leqslant i \leqslant r$. Suppose that y is a non-zero non-unit in $\mathcal{O}_K$ satisfying $|y| = \overline{|y|}$. Assume that equation (2) is satisfied. For $\tau \geqslant 0$, suppose that inequality (3) is valid. Then q is bounded by a computable number depending only on τ, K and $\mathcal{S}_0$.*

Put

$$W_1 = \max(|a_1|, \ldots, |a_r|, b_1, \ldots, b_r).$$

We may write

$$\gamma_1 = \mu_1\pi_1^{f_1} \cdots \pi_s^{f_s} \quad \text{and} \quad \gamma_2 = \mu_2\pi_1^{g_1} \cdots \pi_s^{g_s} \tag{18}$$

where $f_1, \ldots, f_s, g_1, \ldots, g_s$ are non-negative rational integers and $\mu_1, \mu_2 \in \mathcal{O}_K$ with $\mu_1\mu_2 \neq 0$ and

$$\max(\overline{|\mu_1|}, \overline{|\mu_2|}) \leqslant v. \tag{19}$$

Set

$$W_2 = \max(f_1, \ldots, f_s, g_1, \ldots, g_s)$$

and

$$W = \max(W_1, W_2, e).$$

Finally we write

$$\gamma = \mu_3\pi_1^{h_1} \cdots \pi_s^{h_s}$$

where $h_1, \ldots, h_s$ are non-negative rational integers and $0 \neq \mu_3 \in \mathcal{O}_K$ with $\overline{|\mu_3|} \leq v$. Denote by $c_1, c_2, \ldots, c_{29}$ computable positive numbers depending only on τ, K and $\mathcal{S}_0$. We assume that c_1 is sufficiently large.

We split the proof of theorem 9.3′ into four lemmas. We apply theorem B.3 to obtain the following estimate for integers $h_1, \ldots, h_s$.

Lemma 9.1. *Suppose that the assumptions of theorem 9.3′ are satisfied. Then*

$$h_i \leq c_2(\log W)^2 \quad (1 \leq i \leq s). \tag{20}$$

Proof. Let $1 \leq i \leq s$ and $h_i > 0$. Since π_i is not a unit, there exists a prime ideal $\wp$ in $\mathcal{O}_K$ dividing π_i. In view of (3), either $\mathrm{ord}_\wp(\gamma_1) \leq \tau$ or $\mathrm{ord}_\wp(\gamma_2) \leq \tau$. If $\mathrm{ord}_\wp(\gamma_1) \leq \tau$, then equation (2) implies that

$$h_i \leq \mathrm{ord}_\wp(\gamma) \leq \mathrm{ord}_\wp(\varepsilon_1\gamma_1 + \varepsilon_2\gamma_2) \leq \mathrm{ord}_\wp(-\varepsilon_1^{-1}\varepsilon_2\gamma_1^{-1}\gamma_2 - 1) + \tau.$$

We apply theorem B.3 with $n = r + s + 3$, $p \leq c_3$, $A_1 = A_2 = \cdots = A_n = c_4$ and $B = W$ to conclude that

$$\mathrm{ord}_\wp(-\varepsilon_1^{-1}\varepsilon_2\gamma_1^{-1}\gamma_2 - 1) \leq c_5(\log W)^2.$$

Thus

$$h_i \leq c_6(\log W)^2$$

whenever $\mathrm{ord}_\wp(\gamma_1) \leq \tau$. This inequality follows similarly when $\mathrm{ord}_\wp(\gamma_2) \leq \tau$. □

Further, we apply theorem B.4 and lemma 9.1 to obtain

Lemma 9.2. *Suppose that the assumptions of theorem 9.3′ are satisfied and* $q > c_1$. *Then*

$$W_2 \leq c_7(\log q \log |y| + Wq^{-2}). \tag{21}$$

Proof. Let $1 \leq i \leq s$ and $g_i > 0$. Since π_i is not a unit, we can find a prime ideal $\wp$ in $\mathcal{O}_K$ dividing π_i. Let p be the rational prime divisible by $\wp$. Then $p \leq c_8$. Let $c_1 > c_8$. Hence $q \not\equiv 0 \pmod p$, since q is prime. By (2) and (3),

$$g_i \leq \mathrm{ord}_\wp(\varepsilon_2\gamma_2) = \mathrm{ord}_\wp(\gamma y^q - \varepsilon_1\gamma_1) \leq \mathrm{ord}_\wp(\varepsilon_1^{-1}\gamma_1^{-1}\gamma y^q - 1),$$

if $\mathrm{ord}_\wp(\gamma_1) \leq \tau$. We apply theorem B.4 with $n = r + s + 3$, $p \leq c_8$, $\delta = 1/q$, $A' = c_9$, $A = c_{10}\overline{|y|}$, $b_n = q$, $B' = q$ and, by lemma 9.1, $B = \max(W, h_1, \ldots, h_s) \leq c_{11}W$ to conclude that

$$g_i \leq c_{12}(\log q \log |y| + Wq^{-2}).$$

By (3), this inequality is also valid if $\mathrm{ord}_\wp(\gamma_1) > \tau$. Here we have used the inequality $|y| = \overline{|y|} > 1 + c_{13}$ which holds by (A.13). It follows that

$$g_i \leqslant c_{14}(\log q \log |y| + Wq^{-2}) \quad (1 \leqslant i \leqslant s).$$

and, similarly,

$$f_i \leqslant c_{14}(\log q \log |y| + Wq^{-2}) \quad (1 \leqslant i \leqslant s).$$

Hence (21) follows. □

Now we apply lemma 9.1 and lemma 9.2 to obtain

Lemma 9.3. *Suppose that the assumptions of theorem 9.3′ are satisfied and $q > c_1$. Then*

$$W \leqslant c_{15} q \log |y|. \tag{22}$$

Proof. Let r_1 and $2r_2$ be the number of conjugate fields of K which are real and non-real, respectively. For $\alpha \in K$, we signify by $\alpha^{(1)}, \ldots, \alpha^{(r_1)}$ the real conjugates of α and by $\alpha^{(r_1+1)}, \ldots, \alpha^{(r_1+2r_2)}$ the conjugates of α satisfying $\overline{\alpha^{(j)}} = \alpha^{(j+r_2)}$ for $r_1 + 1 \leqslant j \leqslant r_1 + r_2$.

If $W = W_2$, then (22) follows from (21). Thus we may assume that $W = \max(W_1, \mathrm{e})$. Since $\max(b_1, \ldots, b_r, \mathrm{e}) < q$, we may assume that

$$W = \max(|a_1|, \ldots, |a_r|).$$

If $r = 0$, we may put $W = 0$. Thus we may assume $r > 0$. Re-writing equation (2), we have

$$\varepsilon_1 = \gamma_1^{-1}(\gamma y^q - \varepsilon_2 \gamma_2).$$

Thus, for $j = 1, \ldots, d$,

$$|\log |\varepsilon_1^{(j)}|| \leqslant |\log |\gamma_1^{(j)}|| + |\log |(\gamma y^q - \varepsilon_2 \gamma_2)^{(j)}||. \tag{23}$$

By (15), (16) and (17),

$$|\log |\varepsilon_2^{(j)}|| \leqslant c_{16} q \quad (1 \leqslant j \leqslant d). \tag{24}$$

By lemma 9.1,

$$|\log |\gamma^{(j)}|| \leqslant c_{17}(\log W)^2 \tag{25}$$

and, by lemma 9.2,

$$\max(|\log |\gamma_1^{(j)}||, |\log |\gamma_2^{(j)}||) \leqslant c_{18}(\log q \log |y| + Wq^{-2}) \tag{26}$$

for $j = 1, \ldots, d$. Hence it follows from (15), (23), (24), (25), (26) and a Liouville-type argument that

$$|a_1 \log |\eta_1^{(j)}| + \cdots + a_r \log |\eta_r^{(j)}|| \leqslant c_{19}(q \log |y| + (\log W)^2 + Wq^{-2})$$

for $j=1,\ldots,r$. This is a system of linear inequalities with a non-zero determinant E with $|E|\geqslant R$. Consequently

$$W=\max(|a_1|,\ldots,|a_r|)\leqslant c_{20}(q\log|y|+(\log W)^2+Wq^{-2})$$

which implies (22) if $q>c_1$ with c_1 sufficiently large. □

Lemma 9.4. *Suppose that the assumptions of theorem* 9.3′ *are satisfied and* $q>c_1$. *Then*

$$|y|\leqslant c_{21}. \tag{27}$$

Proof. By (2),

$$|\varepsilon_2\gamma_2|=|\gamma y^q-\varepsilon_1\gamma_1|=|\gamma y^q||\varepsilon_1\gamma_1\gamma^{-1}y^{-q}-1|.$$

We apply theorem B.2 with $n=r+s+3$, $\delta=1/q$, $A'=c_{22}$, $A=c_{23}|y|$, $B'=q$ and, by (20) and (22), $B''\leqslant c_{24}q\log|y|$ to obtain

$$|\varepsilon_1\gamma_1\gamma^{-1}y^{-q}-1|\geqslant\exp(-c_{25}\log q\log|y|).$$

Thus

$$|\varepsilon_2\gamma_2|\geqslant|\gamma y^q|\exp(-c_{25}\log q\log|y|)$$

which, together with (25) and (22), gives

$$|\varepsilon_2\gamma_2|\geqslant|y^q|\exp(-c_{26}(\log q)^2\log|y|).$$

On the other hand, it follows from (24), (26) and (22) that

$$|\varepsilon_2\gamma_2|\leqslant\exp(c_{27}(q+\log q\log|y|)).$$

Hence

$$q\log|y|\leqslant c_{27}q+(c_{26}+c_{27})(\log q)^2\log|y|$$

which implies that

$$q\log|y|\leqslant 2c_{27}q$$

if c_1 is sufficiently large. Hence $\log|y|\leqslant 2c_{27}$. □

Proof of theorem 9.3′. We may assume that $q>c_1$ with c_1 sufficiently large so that (27) is valid. Then, since y is not a unit, we apply (27) and theorem 1.3 to equation (2) to obtain

$$\left|\overline{\gamma y^q}\right|\leqslant c_{28}.$$

Consequently, since y is not a unit, we have

$$2^q\leqslant|N(y)|^q\leqslant|N(\gamma y^q)|\leqslant c_{28}^d$$

which implies $q\leqslant c_{29}$. □

Proof of theorem 9.4. Suppose that equation (2) with $q \geqslant 2$ is satisfied. Denote by $c_{30}, \ldots, c_{38}$ computable positive numbers depending only on q, τ, K and $\mathscr{S}_0$. We may write

$$\gamma = \phi_1 \phi_2^q, \quad \gamma_1 = \phi_3 \phi_4^q, \quad \varepsilon_1 = \varepsilon_5 \varepsilon_6^q \tag{28}$$

where $\phi_1, \phi_2, \phi_3, \phi_4$ are elements of $\mathscr{S}_0$ and $\varepsilon_5, \varepsilon_6 \in \mathcal{O}_K$ are units such that

$$\max(\overline{|\phi_1|}, \overline{|\phi_3|}, \overline{|\varepsilon_5|}) \leqslant c_{30}. \tag{29}$$

Multiplying both sides of (2) by $(\varepsilon_5 \phi_3)^{q-1}$, we have

$$(\varepsilon_5 \varepsilon_6 \phi_3 \phi_4)^q - \varepsilon_5^{q-1} \phi_1 \phi_3^{q-1} (\phi_2 y)^q = -\varepsilon_2 \varepsilon_5^{q-1} \phi_3^{q-1} \gamma_2.$$

Set

$$f(X, Y) = X(X^q - \varepsilon_5^{q-1} \phi_1 \phi_3^{q-1} Y^q).$$

Note that, by (29), the function f belongs to a finite and computable set of polynomials. Observe that there is a constant c_{31} such that

$$P(N(f(\varepsilon_5 \varepsilon_6 \phi_3 \phi_4, \phi_2 y))) = P(N(\varepsilon_2 \varepsilon_5^{q-1} \phi_3^{q-1} \gamma_2)) \leqslant c_{31}$$

and that, by (28) and (3),

$$N((\varepsilon_5 \varepsilon_6 \phi_3 \phi_4, \phi_2 y)) \leqslant N((\gamma_1, \gamma y)) \leqslant N((\varepsilon_1 \gamma_1, \varepsilon_2 \gamma_2)) \leqslant c_{31}.$$

We apply theorem 7.6 to conclude that there exists a unit ε in $\mathcal{O}_K$ such that

$$\max(\overline{|\varepsilon \varepsilon_5 \varepsilon_6 \phi_3 \phi_4|}, \overline{|\varepsilon \phi_2 y|}) \leqslant c_{32}$$

which, together with (28) and (29), implies that

$$\max(\overline{|\varepsilon' \varepsilon_1 \gamma_1|}, \overline{|\varepsilon' \gamma y^q|}) \leqslant c_{33} \tag{30}$$

with $\varepsilon' = \varepsilon^q$. Now equation (2) and (30) imply that

$$\overline{|\varepsilon' \varepsilon_2 \gamma_2|} \leqslant c_{34}. \tag{31}$$

Since $\gamma, \gamma_1, \gamma_2 \in \mathscr{S}_0$, it follows from (30) and (31) that

$$\max(\overline{|\gamma|}, \overline{|\gamma_1|}, \overline{|\gamma_2|}) \leqslant c_{35}$$

which, together with (30) and (31), implies that

$$\max(\overline{|\varepsilon' y^q|}, \overline{|\varepsilon' \varepsilon_1|}, \overline{|\varepsilon' \varepsilon_2|}) \leqslant c_{36}.$$

Hence, by a Liouville-type argument,

$$\max(\overline{|\varepsilon_1^{-1} y^q|}, \overline{|\varepsilon_2^{-1} y^q|}, \overline{|\varepsilon_1^{-1} \varepsilon_2|}) \leqslant c_{37}. \qquad \square$$

Now we turn to the proof of theorem 9.5. It depends on the following application of theorem 1.3 to equation (7).

Lemma 9.5. *Suppose that the assumptions of theorem 9.5 are satisfied. Further, assume that y is a unit. Then*

$$\max(\overline{|\gamma|}, \overline{|\gamma_1|}, \overline{|\gamma_2|}, m) \leqslant c_{38}$$

where c_{38} is a computable number depending only on τ_1, α, β, K and $\mathscr{S}_0$.

Proof. Let γ_1 and γ_2 be given by (18) where μ_1, μ_2 satisfy (19). Observe that, by lemma A.2, the pair μ_1, μ_2 belongs to a computable finite set. Thus it suffices to prove lemma 9.5 with c_{38} depending only on $\tau_1, \alpha, \beta, K, \mathscr{S}_0, \mu_1$ and μ_2. Put

$$\mu_3 = \begin{cases} 1 & \text{if } \mu_1 \text{ is a non-unit,} \\ \mu_1 & \text{if } \mu_1 \text{ is a unit.} \end{cases}$$

Thus $\mu_3 \in \mathcal{O}_K$ is a unit. Denote by $c_{39}, \ldots, c_{46}$ computable positive numbers depending only on $\tau_1, \alpha, \beta, K, \mathscr{S}_0, \mu_1$ and μ_2. We split the proof of lemma 9.5 into two cases.

Case 1: At least one of α and β is a non-unit. By permuting α and β, it involves no loss of generality to assume that α is a non-unit. Note that $P(N(\alpha\beta\gamma\gamma_1\gamma_2)) \leqslant c_{39}$. After dividing both sides of (7) by μ_3, we apply theorem 1.3 with $\delta_1 = \mu_3^{-1}\varepsilon\gamma y^q$, $\delta_2 = -\mu_3^{-1}\gamma_2\beta^m$ and $\delta_3 = -\mu_3^{-1}\gamma_1\alpha^m$ to conclude that

$$\max(\overline{|\delta_1|}, \overline{|\delta_2|}, \overline{|\delta_3|}) \leqslant c_{40}.$$

Therefore, since $\gamma, \gamma_1, \gamma_2 \in \mathscr{S}_0$ and μ_3^{-1}, ε, α, β, y are non-zero algebraic integers, we obtain

$$\max(\overline{|\gamma|}, \overline{|\gamma_1|}, \overline{|\gamma_2|}) \leqslant c_{41}. \tag{32}$$

Consequently

$$\overline{|\alpha|}^m \leqslant c_{42}.$$

Since α is a non-unit, we see, by (A.13), that $m \leqslant c_{43}$.

Case 2: Both α and β are units. After dividing both sides of (7) by $\mu_3\alpha^m$, we apply theorem 1.3 with

$$\delta_1 = \mu_3^{-1}\alpha^{-m}\varepsilon\gamma y^q, \quad \delta_2 = -\mu_3^{-1}\left(\frac{\beta}{\alpha}\right)^m \gamma_2, \quad \delta_3 = -\mu_3^{-1}\gamma_1$$

to conclude that $\max(\overline{|\delta_1|}, \overline{|\delta_2|}, \overline{|\delta_3|}) \leqslant c_{44}$. Therefore, since $\gamma, \gamma_1, \gamma_2 \in \mathscr{S}_0$ and μ_3^{-1}, ε, α^{-1}, β/α, y are non-zero algebraic integers, we obtain (32) and $\overline{|\beta/\alpha|}^m \leqslant c_{45}$. Then, since β/α is not a root of unity, we see $m \leqslant c_{46}$. □

We apply theorem 9.3 to equation (7) to prove:

Lemma 9.6. *Suppose that the assumptions of theorem 9.5 are satisfied. Further, assume that y is not a unit. Then*

$$P(q) \leqslant c_{47} \tag{33}$$

for some computable number c_{47} depending only on τ_1, α, β, K and $\mathscr{S}_0$.

Proof. Dividing both sides of (7) by ε, we have

$$\gamma y^q = \varepsilon^{-1}\alpha^m \gamma_1 + \varepsilon^{-1}\beta^m \gamma_2. \tag{34}$$

If both α and β are units, then we apply theorem 9.3 with $\varepsilon_1 = \varepsilon^{-1}\alpha^m$ and $\varepsilon_2 = \varepsilon^{-1}\beta^m$ to equation (34) to obtain (33). If both α and β are non-units, then we apply theorem 9.3 with $\mathscr{S}_1 := \mathscr{S}_{K,v}(\pi_1, \ldots, \pi_s, \alpha\ \beta)$, γ_1 and γ_2 replaced by $\gamma_1\alpha^m$ and $\gamma_2\beta^m$, respectively, and $\varepsilon_1 = \varepsilon_2 = \varepsilon^{-1}$ to equation (34) to conclude that $P(q)$ is bounded by a computable number depending only on τ_1, α, β, K, $\mathscr{S}_1$ whence (33) follows.

Thus we may assume that exactly one of α and β is non-unit. It involves no loss of generality in assuming that α is the non-unit and β the unit. Now we apply theorem 9.3 with $\mathscr{S}_2 := \mathscr{S}_{K,v}(\pi_1, \ldots, \pi_s, \alpha)$, γ_1 replaced by $\gamma_1\alpha^m$, $\varepsilon_1 = \varepsilon^{-1}$ and $\varepsilon_2 = \varepsilon^{-1}\beta^m$ to equation (34) to conclude that $P(q)$ is bounded by a computable number depending only on τ_1, α, β, K, $\mathscr{S}_2$ whence (33) follows. □

Proof of theorem 9.5. In view of lemma 9.5, we may assume that y is a non-unit. Further, there is no loss of generality in assuming that q is prime. Then, by lemma 9.6, we see that $q \leqslant c_{47}$. Therefore it suffices to prove theorem 9.5 with C_4 depending only on q, τ_1, α, β, K and $\mathscr{S}_0$. Denote by $c_{48}, \ldots, c_{51}$ computable positive numbers depending only on q, τ_1, α, β, K and $\mathscr{S}_0$.

We apply theorem 9.4 to equation (34). It follows from (6) that there exists a unit $\varepsilon' \in \mathcal{O}_K$ (in fact, $\varepsilon' = \varepsilon$ or $\varepsilon' = \varepsilon\alpha^{-m}$) such that

$$\max(\overline{|\varepsilon'\gamma y^q|}, \overline{|\varepsilon'\varepsilon^{-1}\alpha^m\gamma_1|}, \overline{|\varepsilon'\varepsilon^{-1}\beta^m\gamma_2|}) \leqslant c_{48}$$

which, since $\gamma, \gamma_1, \gamma_2 \in \mathscr{S}_0$ and $\varepsilon', \varepsilon^{-1}, \alpha, \beta, y$ are non-zero algebraic integers, implies that

$$\max(\overline{|\gamma|}, \overline{|\gamma_1|}, \overline{|\gamma_2|}) \leqslant c_{49}.$$

Consequently, by a Liouville-type argument,

$$\overline{|\alpha/\beta|}^m \leqslant c_{50}.$$

Since α/β is not a root of unity, we find that $m \leqslant c_{51}$. □

Proof of corollary 9.1. Denote by c_{52} and c_{53} computable positive numbers depending only on τ_1, α, β, K and $\mathcal{S}_0$. In view of theorem 9.5, inequality (8) is valid. By equation (7) with $\varepsilon = 1$ and (8), we see that

$$\overline{|y|} \leqslant \overline{|y|}^q \leqslant c_{52}$$

which, since y is not a root of unity, implies that $q \leqslant c_{53}$. □

Proof of theorem 9.6. Suppose that equation (9) is valid. By § 2 of chapter C, the sequence $\{u_m\}_{m=0}^\infty$ is induced by a recurrence relation with algebraic coefficients v_0, v_1. Let α_1 and β_1 be the roots of the companion polynomial to $\{u_m\}_{m=0}^\infty$. Since $\{u_m\}_{m=0}^\infty$ is non-degenerate, α_1/β_1 is not a root of unity. Put $L = K(u_0, u_1, v_0, v_1, \alpha_1, \beta_1)$. Let $\mathcal{O}_L$ be the ring of integers of L, and h the class number of L. For $m = 0, 1, 2, \ldots$ we have

$$u_m = a_1 \alpha_1^m + b_1 \beta_1^m$$

where

$$a_1 = \frac{u_0 \beta_1 - u_1}{\beta_1 - \alpha_1} \quad \text{and} \quad b_1 = \frac{u_1 - u_0 \alpha_1}{\beta_1 - \alpha_1}.$$

Put $\mathcal{S}_3 = \mathcal{S}_{L,v}(\pi_1, \ldots, \pi_s)$. It suffices to prove that m is bounded by a computable number depending only on L, $\mathcal{S}_3$ and $\{u_m\}_{m=0}^\infty$. Denote by $c_{54}, \ldots, c_{63}$ computable positive numbers depending only on L, $\mathcal{S}_3$ and $\{u_m\}_{m=0}^\infty$.

Denote by d_1 the least positive integer such that $d_1 a_1$, $d_1 b_1$, $d_1 \alpha_1$, $d_1 \beta_1 \in \mathcal{O}_L$. Put

$$a_2 = d_1 a_1, \quad b_2 = d_1 b_1, \quad \alpha_2 = d_1 \alpha_1, \quad \beta_2 = d_1 \beta_1. \tag{35}$$

Multiplying both sides of equation (9) by d_1^{m+1}, we have

$$a_2 \alpha_2^m + b_2 \beta_2^m = \gamma_3 y^q \tag{36}$$

where

$$\gamma_3 = d_1^{m+1} \gamma. \tag{37}$$

Observe that $\gamma_3 \in \mathcal{O}_L$.

Let d_2 be the least positive integer such that $y_1 := d_2 y \in \mathcal{O}_L$. Then, since the left-hand side of (36) and γ_3 are algebraic integers, we see from (36) and (37) that

$$P(d_2) \leqslant c_{54}. \tag{38}$$

Multiplying both sides of (36) by d_2^q, we have

$$d_2^q a_2 \alpha_2^m + d_2^q b_2 \beta_2^m = \gamma_3 y_1^q. \tag{39}$$

By corollary A.3, notice that

$$([\alpha_2^h], [\beta_2^h]) = [\pi]$$

where either $\pi = 1$ or π is a non-unit in $\mathcal{O}_L$. Put

$$\alpha_3 = \pi^{-1}\alpha_2^h \quad \text{and} \quad \beta_3 = \pi^{-1}\beta_2^h. \tag{40}$$

Then $\alpha_3, \beta_3 \in \mathcal{O}_L$ satisfy $([\alpha_3], [\beta_3]) = [1]$. Further, since α_1/β_1 is not a root of unity, we see from (40) and (35) that α_3/β_3 is not a root of unity. Putting

$$m = m_1 h + m_2 \quad (0 \leqslant m_2 < h) \tag{41}$$

and

$$a_3 = \alpha_2^{m_2} a_2, \quad b_3 = \beta_2^{m_2} b_2, \tag{42}$$

we re-write (39) as

$$a_3 \alpha_3^{m_1} + b_3 \beta_3^{m_1} = d_2^{-q} \pi^{-m_1} \gamma_3 y_1^q. \tag{43}$$

Observe that $a_3, b_3 \in \mathcal{O}_L$ and, by (42) and (35),

$$\max(\overline{|a_3|}, \overline{|b_3|}) \leqslant c_{55}. \tag{44}$$

We may write

$$y_1 = y_2 y_3$$

where $y_2, y_3 \in \mathcal{O}_L$ satisfy

$$([y_3], [\pi d_2]) = [1] \tag{45}$$

and, by (38) and lemma A.12,

$$y_2 = \psi_1^{v_1} \cdots \psi_l^{v_l} \tag{46}$$

where $v_1, \ldots, v_l$ are non-negative rational integers and, for $1 \leqslant j \leqslant l$, $\psi_j \in \mathcal{O}_L$ are non-units with $|N(\psi_j)| \leqslant c_{56}$. Therefore, by corollary A.6, we may choose non-units $\psi_1, \ldots, \psi_l \in \mathcal{O}_L$ with

$$\max_{1 \leqslant j \leqslant l} \overline{|\psi_j|} \leqslant c_{57}. \tag{47}$$

Put

$$\gamma_4 = d_2^{-q} \pi^{-m_1} \gamma_3 y_2^q.$$

Then $\gamma_4 \in \mathcal{O}_L$, since the left-hand side of (43) is an algebraic integer and, by (45), the ideals $[y_3]$ and $[\pi d_2]$ are relatively prime. Further, it follows from (37), (46) and (47) that

$$P(N(\gamma_4)) \leqslant c_{58}. \tag{48}$$

Therefore, by lemma A.12 and corollary A.6, there exist non-units $\psi_1^*, \ldots,$

$\psi_\lambda^* \in \mathcal{O}_L$ such that

$$\max_{1 \leqslant j \leqslant \lambda} \overline{|\psi_j^*|} \leqslant c_{59}$$

and

$$\gamma_4 = \varepsilon_7 \gamma_5$$

where $\varepsilon_7 \in \mathcal{O}_L$ is a unit and $\gamma_5 \in \mathcal{S}_{L,1}(\psi_1^*, \ldots, \psi_\lambda^*)$. We re-write (43) as

$$a_3 \alpha_3^{m_1} + b_3 \beta_3^{m_1} = \varepsilon_7 \gamma_5 y_3^q. \tag{49}$$

Put $v_1 = \max(v, c_{55})$ and $\mathcal{S}_4 = \mathcal{S}_{L,v_1}(\pi_1, \ldots, \pi_s, \psi_1^*, \ldots, \psi_\lambda^*)$. Observe that a_3, $b_3, \gamma_5 \in \mathcal{S}_4$. We apply theorem 9.5 with $K = L, \mathcal{S}_0 = \mathcal{S}_4, m = m_1, \alpha = \alpha_3, \beta = \beta_3$, $\gamma_1 = a_3, \gamma_2 = b_3, \gamma = \gamma_5, y = y_3, \varepsilon = \varepsilon_7$ and, by (44) and $(\alpha_3, \beta_3) = [1], \tau_1 = c_{60}$ to equation (49) to conclude that m_1 is bounded by a computable number depending only on L, $\mathcal{S}_4$ and $\{u_m\}_{m=0}^\infty$. Therefore $m_1 \leqslant c_{61}$, whence, by (41), $m \leqslant (c_{61} + 1)h$. □

Proof of corollary 9.2. Let d_3 be the least positive integer such that $d_3 \delta \in \mathcal{O}_K$. Multiplying both sides of (10) by d_3, there is no loss of generality in assuming that $\delta \in \mathcal{O}_K$. Denote by $c_{62}, \ldots, c_{65}$ computable positive numbers depending only on δ, K and $\{u_m\}_{m=0}^\infty$.

We apply theorem 9.6 with $\mathcal{S}_0 = \mathcal{S}_{K,|\bar{\delta}|}(\varnothing)$ and $\gamma = \delta$ to conclude that $m \leqslant c_{62}$. Consequently, by (10), $\overline{|\delta y^q|} \leqslant c_{63}$ which implies that

$$\overline{|y|}^q \leqslant c_{64} \quad (c_{64} > 1). \tag{50}$$

Therefore $\overline{|y|} \leqslant c_{64}$. Further, since y is not a root of unity, we see from (50) that $q \leqslant c_{65}$. □

Proof of theorem 9.8. There is no loss of generality in assuming that t is prime. Denote by $c_{66}, \ldots$ computable positive numbers depending only on a_1, A_1, A_2, A_3 and S. Write

$$Q(X, Y) = A_1 X^2 + A_2 XY + A_3 Y^2.$$

Denote by α_1 and α_2 the roots of $Q(X, 1)$. Then

$$A_1 Q(x^t, y) = (A_1 x^t - A_1 \alpha_1 y)(A_1 x^t - A_1 \alpha_2 y).$$

Put $L = \mathbb{Q}(\alpha_1)$ and denote by $\mathcal{O}_L$ the ring of integers of L. In view of (13), lemma A.12 and corollary A.6, we may choose a set of non-zero non-units $\psi_1', \ldots, \psi_k' \in \mathcal{O}_L$ such that

$$k \leqslant c_{66}, \quad \max_{1 \leqslant j \leqslant k} \overline{|\psi_j'|} \leqslant c_{66} \tag{51}$$

and

$$A_1 x^t - A_1 \alpha_1 y = \varepsilon_8 \gamma_1', \quad A_1 x^t - A_1 \alpha_2 y = \varepsilon_9 \gamma_2'$$

where $\varepsilon_8, \varepsilon_9 \in \mathcal{O}_L$ are units and γ'_1, γ'_2 are products of powers of $\psi'_1, \ldots, \psi'_k$ with non-negative exponents. Further, since $(x^t, y) \leqslant a_1$, we have

$$\min(\operatorname{ord}_{\mathfrak{p}}(\gamma'_1), \operatorname{ord}_{\mathfrak{p}}(\gamma'_2)) \leqslant c_{67}$$

for every prime ideal $\mathfrak{p}$ in $\mathcal{O}_L$. It follows from the above equations that

$$A_1(\alpha_2-\alpha_1)y = \varepsilon_8\gamma'_1 - \varepsilon_9\gamma'_2 \tag{52}$$

and

$$A_1^2(\alpha_2-\alpha_1)x^t = \varepsilon_8 A_1\alpha_2\gamma'_1 - \varepsilon_9 A_1\alpha_1\gamma'_2. \tag{53}$$

Observe that $\alpha_1 \neq \alpha_2$, since $A_2^2 - 4A_1A_3 \neq 0$. Put

$$v_2 = A_1^2 \max(\overline{|\alpha_2-\alpha_1|}, \overline{|\alpha_1|}, \overline{|\alpha_2|})$$

and

$$\mathscr{S}_5 = \mathscr{S}_{L,v_2}(\psi'_1, \ldots, \psi'_k).$$

In view of (51) and corollary A.2, we see that $\mathscr{S}_5$ belongs to a computable finite collection of sets. We apply theorem 9.3 with $\gamma_1 = A_1\alpha_2\gamma'_1$, $\gamma_2 = -A_1\alpha_1\gamma'_2$, $\gamma = A_1^2(\alpha_2-\alpha_1)$ and $\mathscr{S}_0 = \mathscr{S}_5$ to equation (53) to conclude that $t \leqslant c_{68}$. Let t $(\geqslant 2)$ be fixed. By applying theorem 9.4 to equation (53), it follows that

$$\max(\overline{|\gamma'_1|}, \overline{|\gamma'_2|}) \leqslant c_{69} \tag{54}$$

and

$$\overline{|\varepsilon_i^{-1}x^t|} \leqslant c_{70} \quad (i=8,9). \tag{55}$$

Observe that $\overline{|\varepsilon_i^{-1}x^t|} = \overline{|\varepsilon_i^{-1}|}\,|x^t|$ for $i=8,9$. Therefore, by (55) and the fact that $\varepsilon_8^{-1}, \varepsilon_9^{-1}$ are algebraic integers, we see that

$$|x|^t \leqslant c_{70} \tag{56}$$

and

$$\max(\overline{|\varepsilon_8^{-1}|}, \overline{|\varepsilon_9^{-1}|}) \leqslant c_{70}.$$

Consequently, $\max(\overline{|\varepsilon_8|}, \overline{|\varepsilon_9|}) \leqslant c_{71}$ which, together with (52) and (54), implies that $|y| \leqslant c_{72}$. Further, by (56) and $|x| > 1$, we find that $\max(t, |x|) \leqslant c_{73}$. □

Proof of theorem 9.7. We may assume that y is non-zero; otherwise the assertion follows from (11) and $|x| > 1$. Further, observe that equation (11) implies that $(x^t, y) \leqslant B$. Now apply theorem 9.8 to complete the proof of theorem 9.7. □

Notes

Shorey and Tijdeman (1976*a*) applied theorems B.2 and 5.1 to show that there exist computable positive numbers C_9, C_{10} and C_{11}

depending only on P such that if

$$b \in S, \quad a = by^q + l \in S$$

with $l, q, y \in \mathbb{Z}, l \neq 0, q \geqslant 3, y > 1$, then

$$|l| \geqslant C_9(by^q)^\theta \tag{57}$$

where

$$\theta = \max\left(1 - \frac{C_{10} \log q}{q}, C_{11}\right).$$

If $q=2$, inequality (57) can be derived from theorem B.4 (see Turk, 1986, §4.1.2). Weaker versions of inequality (57) were given by Schinzel (1967) for $q=2$ and $q=3$ and by Langevin for general q.

For the Fibonacci sequence $\{u_m\}_{m=0}^{\infty}$, Cohn (1964) and Wyler (1964), independently, proved that u_m is a square only when $m=0$, 1, 2 and 13. Cohn (1965) and Steiner (1980) solved the equations $u_m = 2y^2$ and $u_m = 3y^2$. Cohn (1965) applied these results and the corresponding ones for Lucas sequences to determine all integer solutions of certain diophantine equations. London and Finkelstein (1969) determined all the cubes in the Fibonacci sequence. Lagarias and Weisser (1981) gave another proof. Steiner (1978) derived some partial results for higher powers. See also Robbins (1978, 1983). The proofs of these results do not depend on estimates for linear forms in logarithms. Pethö (1983, 1984) utilised the theory of linear forms in logarithms and computer calculations to determine all the cubes and the fifth powers in the Fibonacci sequence. As mentioned in the text of this chapter, Pethö (1982*a*) and Shorey and Stewart (1983) proved that there are only finitely many perfect powers in a simple non-degenerate binary recurrence sequence $\{v_m\}_{m=0}^{\infty}$ of rational integers. Pethö extended this result to the equation $v_m = by^q$ with $b \in S$, provided that the companion polynomial has relatively prime integral coefficients. Shorey and Stewart (1983, 1986) and Kiss (1986) proved the assertion for certain recurrence sequences of order >2. Shorey and Stewart (1983) applied their result to show that, under suitable conditions, there are only finitely many integers x, y, z, q with $q > 1$ and $|z| > 1$ satisfying

$$a_1x^2 + b_1xy + c_1y^2 = d_1$$

and

$$a_2x^2 + b_2xy + c_2y^2 = d_2z^q.$$

For a generalisation of this result, see Shorey and Stewart (1986) which also contains an inhomogeneous analogue of theorem 9.7. For $q=2$, see Mordell (1969, Ch. 8).

Finkelstein (1973), Williams (1975) and Steiner (1980) gave proofs for the fact that 1, 2 and 5 are the only Fibonacci numbers which are of the form y^2+1. Finkelstein (1975) proved a similar result for Lucas numbers. Robbins (1981) claims to have determined all Fibonacci numbers of the form y^2-1 and $y^3 \pm 1$. Stewart (1981) and Shorey and Stewart (1986) investigated the more general equation $v_m = y^2 + k$ where $\{v_m\}_{m=0}^{\infty}$ is a simple non-degenerate binary recurrence sequence of rational integers and k a given rational integer. Nemes and Pethö (1984, 1986) studied the equation

$$v_m = by^q + f(y) \tag{58}$$

in rational integers m, q, y where $b \in \mathbb{Z}$ and $f \in \mathbb{Z}[X]$ are fixed. Kiss (1986) and Shorey and Stewart (1986) dealt with equation (58) for recurrence sequences $\{v_m\}_{m=0}^{\infty}$ of any order, under certain conditions on $\{v_m\}_{m=0}^{\infty}$ and $\deg(f)$.

CHAPTER 10

Perfect powers at integral values of a polynomial

We consider the superelliptic equations of chapters 6 and 8, but now with m as a variable. Tijdeman (1976*a*) proved the following result.

Theorem 10.1. *Let $f(X)$ be a polynomial with rational integer coefficients and with at least two simple rational zeros. Suppose $b\neq 0$, $m\geqslant 0$, x and y with $|y|>1$ are rational integers. Then the equation*

$$f(x)=by^m \tag{1}$$

implies that m is bounded by a computable number depending only on b and f.

Schinzel and Tijdeman (1976) extended theorem 10.1 as follows.

Theorem 10.2. *Let $f(X)$ be a polynomial with rational integer coefficients and with at least two distinct roots. Suppose $b\neq 0$, $m\geqslant 0$, x and y with $|y|>1$ are rational integers satisfying* (1). *Then m is bounded by a computable number depending only on b and f.*

Let K be a finite extension of $\mathbb{Q}$ and denote by $\mathcal{O}_K$ the ring of integers of K. For given non-zero non-units $\pi_1,\dots,\pi_s$ of $\mathcal{O}_K$, denote by $\mathcal{S}$ the set of all the products of non-negative powers of $\pi_1,\dots,\pi_s$. Let $\alpha_1,\dots,\alpha_n$ be distinct elements of $\mathcal{O}_K$. Write

$$f(X,Z)=(X-\alpha_1 Z)^{r_1}\cdots(X-\alpha_n Z)^{r_n}$$

where $r_1,\dots,r_n$ are positive integers. Then theorems 10.2 and 9.3 are contained in the following result that generalises a theorem of Shorey, van der Poorten, Tijdeman and Schinzel (1977).

Theorem 10.3. *Suppose $f(X,1)$ has at least two distinct roots. Let $\varepsilon\in\mathcal{O}_K$ be a* *unit and $\gamma \in \mathscr{S}$. Suppose $\tau \geqslant 0$, $x \in \mathcal{O}_K$ and $z \in \mathscr{S}$ satisfy*

$$\min(\operatorname{ord}_{\wp}(x), \operatorname{ord}_{\wp}(z)) \leqslant \tau \tag{2}$$

for every prime ideal $\wp$ in $\mathcal{O}_K$. Assume $0 \neq y \in \mathcal{O}_K$ is not a unit and $m \geqslant 0$ is a rational integer. Then the equation

$$f(x, z) = \varepsilon \gamma y^m \tag{3}$$

implies that m is bounded by a computable number depending only on K, $\mathscr{S}$, f and τ.

If y is a unit in $\mathcal{O}_K$, we can apply theorem 1.3 to obtain the following result on equation (3).

Theorem 10.4. *Let $f, \varepsilon, \gamma, \tau, x, z$ and m be as in theorem 10.3. Suppose $y \in \mathcal{O}_K$ is a unit. Then equation (3) implies that*

$$\max(\overline{|\varepsilon y^m|}, \overline{|\gamma|}, \overline{|x|}, \overline{|z|}) \leqslant C_1$$

for some computable number C_1 depending only on K, $\mathscr{S}$, f and τ.

Combining theorems 10.3 and 10.4, we obtain the following result.

Theorem 10.5. *Let f, γ, τ, x, z and m be as in theorem 10.3. Suppose $0 \neq y \in \mathcal{O}_K$ is not a root of unity. Then equation*

$$f(x, z) = \gamma y^m \tag{4}$$

implies that m is bounded by a computable number depending only on K, $\mathscr{S}$, f and τ.

Theorem 8.1 can now be applied to find bounds for x, y, z.

Theorem 10.6. *Suppose $f(X, 1)$ has at least two simple roots. Let $\tau \geqslant 0$, $x \in \mathcal{O}_K$ and $z \in \mathscr{S}$ satisfy (2) for every prime ideal $\wp$ in $\mathcal{O}_K$. Suppose $\gamma \in \mathscr{S}$, $0 \neq y \in \mathcal{O}_K$ is not a root of unity and $m \geqslant 3$ is a rational integer. Then equation (4) implies that*

$$\max(\overline{|\gamma|}, \overline{|y|}, \overline{|x|}, \overline{|z|}, m) \leqslant C_2$$

where C_2 is a computable number depending only on K, $\mathscr{S}$, f and τ.

If $m = 2$, then we apply theorem 8.2 in place of theorem 8.1.

Theorem 10.7. *Suppose $f(X, 1)$ has at least three simple roots and $m \geqslant 2$ is a rational integer. Let τ, x, z, γ and y be as in theorem 10.6. Then equation (4)* *implies that*

$$\max(\overline{|\gamma|}, \overline{|y|}, \overline{|x|}, \overline{|z|}, m) \leqslant C_3$$

for some computable number C_3 depending only on K, $\mathcal{S}$, f and τ.

An immediate consequence of theorem 10.7 is the following extension of corollary 8.1.

Corollary 10.1. *Let $A \neq 0$, $B \neq 0$ and $n \geqslant 2$ be given rational integers. Then*

$$P(Ax^m + By^n) \to \infty \quad \textit{effectively},$$

as $\max(|x|, |y|, m)$ tends to infinity through rational integers $x \neq 0$, $y \neq 0$ and $m \geqslant 0$ satisfying $|x| > 1$, $(x, y) = 1$ and $mn \geqslant 6$.

Corollary 10.1 is due to Shorey, van der Poorten, Tijdeman and Schinzel (1977).

Proofs

The constants $c_1, \ldots, c_{23}$ in the proofs of theorems 10.1 and 10.2 are computable positive numbers depending only on b and f.

Proof of theorem 10.1. Suppose that equation (1) is satisfied. We may then assume $m \geqslant 2$. Let α_1 and α_2 be simple rational zeros of f. Denote by a_0 the leading coefficient of f and by N the degree of f. Multiplying both sides of (1) by a_0^{N-1}, we have

$$g(a_0 x) = b a_0^{N-1} y^m$$

where $g(X)$ is given by

$$g(a_0 X) = a_0^{N-1} f(X).$$

Observe that $g(X)$ is a monic polynomial with rational integer coefficients and has at least two simple rational zeros. Thus there is no loss of generality in assuming that f is monic. Consequently α_1 and α_2 are rational integers. Write

$$f_1(X) = f(X)/(X - \alpha_1)(X - \alpha_2).$$

Then $f_1(X) \in \mathbb{Z}[X]$. Put

$$D = b(\alpha_1 - \alpha_2) f_1(\alpha_1) f_1(\alpha_2).$$

Observe that $0 \neq D \in \mathbb{Z}$, since $\alpha_1, \alpha_2 \in \mathbb{Z}$ are simple zeros of f and $b \neq 0$.

Equation (1) implies that

$$x-\alpha_1=\mu_1^m\prod_{p|D}p^{a_p},\quad x-\alpha_2=\mu_2^m\prod_{p|D}p^{b_p}$$

where $a_p\geqslant 0$, $b_p\geqslant 0$, $\mu_1\neq 0$, $\mu_2\neq 0$ are rational integers and the product is taken over the rational primes dividing D. For $p\,|\,D$, we write

$$a_p\equiv A_p(\operatorname{mod} m),\quad b_p\equiv B_p(\operatorname{mod} m)$$

where $0\leqslant A_p<m$ and $0\leqslant B_p<m$. Then

$$x-\alpha_1=\mu_3^m\prod_{p|D}p^{A_p},\quad x-\alpha_2=\mu_4^m\prod_{p|D}p^{B_p}\tag{5}$$

for some non-zero rational integers μ_3,μ_4. By interchanging the suffixes of α_1 and α_2, if necessary, we may assume that $|\mu_3|\geqslant|\mu_4|$. We split the proof of theorem 10.1 in two cases.

Case 1.

$$|\mu_3|=1.\tag{6}$$

Then $|\mu_4|=1$. Hence, by (5),

$$\alpha_1-\alpha_2=\pm\prod_{p|D}p^{A_p}\pm\prod_{p|D}p^{B_p}.$$

We apply corollary 1.1 to the above equation to conclude that

$$\max_{p|D}(A_p,B_p)\leqslant c_1.$$

Consequently, by (5) and (6), $|x|\leqslant c_2$ which, together with $|y|>1$ and (1), implies that

$$2^m\leqslant|y|^m\leqslant c_3.$$

Hence $m\leqslant c_4$.

Case 2.

$$|\mu_3|>1.\tag{7}$$

Observe that

$$|x-\alpha_1|\geqslant|\mu_3|^m\tag{8}$$

and, by $\alpha_1\neq\alpha_2$ and (5),

$$0\neq\left|\frac{\alpha_1-\alpha_2}{x-\alpha_1}\right|=\left|\frac{x-\alpha_2}{x-\alpha_1}-1\right|=\left|\prod_{p|D}p^{B_p-A_p}\left(\frac{\mu_4}{\mu_3}\right)^m-1\right|.$$

We apply corollary B.1 with $n=\omega(D)+1$, $d=1$, $A_1=A_2=\cdots=A_{n-1}=c_5$,

$A_n = |\mu_3| + 1$ and $B = m$ to conclude that

$$\left|\frac{\alpha_1 - \alpha_2}{x - \alpha_1}\right| \geqslant |\mu_3|^{-c_6 \log m}. \tag{9}$$

Now combine (9), (8) and (7) to obtain $m \leqslant c_7$. □

Proof of theorem 10.2. As in the proof of theorem 10.1, there is no loss of generality in assuming that f is monic. Write

$$f(X) = (X - \alpha_1)^{r_1} \cdots (X - \alpha_n)^{r_n}$$

where $n \geqslant 2$, $\alpha_1, \ldots, \alpha_n$ are distinct algebraic integers and $r_1, \ldots, r_n$ are positive rational integers. Put $L = \mathbb{Q}(\alpha_1, \ldots, \alpha_n)$ and denote by $\mathcal{O}_L$ the ring of integers of L. Let $\eta_1, \ldots, \eta_r$ be an independent system of units for L satisfying (A.45), and h the class number of L. Put

$$\Delta = \left[b \prod_{1 \leqslant i < j \leqslant n} (\alpha_i - \alpha_j) \right].$$

Observe that Δ is a non-zero ideal in $\mathcal{O}_L$. Denote by $\mathfrak{p}_1, \ldots, \mathfrak{p}_t$ all the prime ideals in $\mathcal{O}_L$ dividing Δ. Observe that $\mathfrak{p}_i^h$ is principal and $N(\mathfrak{p}_i^h) \leqslant c_8$ for $1 \leqslant i \leqslant t$. Consequently, by lemma A.9 and corollary A.6, $\mathfrak{p}_i^h$ is generated by some $\psi_i \in \mathcal{O}_L$ satisfying

$$\overline{|\psi_i|} \leqslant c_9 \quad (1 \leqslant i \leqslant t). \tag{10}$$

Equation (1) gives

$$[f(x)] = [by^m].$$

Hence there exist non-zero ideals $\mathfrak{a}_1$ and $\mathfrak{a}_2$ in $\mathcal{O}_L$ with $(\mathfrak{a}_1 \mathfrak{a}_2, \Delta) = [1]$ such that

$$[(x - \alpha_j)^{r_j}] = \mathfrak{p}_1^{a_{1,j}} \cdots \mathfrak{p}_t^{a_{t,j}} \mathfrak{a}_j^m \quad (j = 1, 2) \tag{11}$$

where $a_{i,j}$ with $i = 1, 2, \ldots, t$ and $j = 1, 2$ are non-negative rational integers. If $\mathfrak{p}$ is a prime ideal in $\mathcal{O}_L$ such that $\mathfrak{p}^{l_1} \parallel \mathfrak{a}_1^m$ and $\mathfrak{p}^{l_2} \parallel \mathfrak{a}_2^m$, then $m \mid l_1, m \mid l_2$ and, by (11), $r_1 \mid l_1$ and $r_2 \mid l_2$. This implies

$$\frac{m}{(m, r_j)} \Bigg| \frac{l_j}{r_j} \quad (j = 1, 2).$$

Putting $\langle r_1, r_2 \rangle$ for the least common multiple of r_1 and r_2 and

$$M = \frac{m}{(m, \langle r_1, r_2 \rangle)}, \tag{12}$$

we see that $M \mid (l_j / r_j)$ for $j = 1, 2$. Now it follows from this observation and

(11) that

$$[x-\alpha_j]=\wp_1^{b_{1,j}}\cdots\wp_t^{b_{t,j}}\mathscr{b}_j^M \quad (j=1,2)$$

for some non-zero ideals $\mathscr{b}_1$ and $\mathscr{b}_2$ in $\mathcal{O}_L$ and non-negative integers $b_{i,j}$ with $i=1,\ldots,t$ and $j=1,2$. We may write

$$\mathscr{b}_1^h=[\xi_1], \quad \mathscr{b}_2^h=[\xi_2]$$

for some non-zero $\xi_1,\xi_2\in\mathcal{O}_L$. Then, for $j=1,2$, we have

$$(x-\alpha_j)^h=\rho_j\eta_1^{u_{1,j}}\cdots\eta_r^{u_{r,j}}\psi_1^{b_{1,j}}\cdots\psi_t^{b_{t,j}}\xi_j^M \tag{13}$$

where $\rho_1,\rho_2\in\mathcal{O}_L$ satisfy, by corollary A.5, $\max(\overline{|\rho_1|},\overline{|\rho_2|})\leqslant c_{10}$ and $u_{q,j}\in\mathbb{Z}$ for $q=1,\ldots,r$ and $j=1,2$. By incorporating every Mth power in ξ_j, we may assume that

$$0\leqslant u_{q,j}<M, \quad 0\leqslant b_{i,j}<M \tag{14}$$

for $1\leqslant q\leqslant r$, $1\leqslant i\leqslant t$ and $j=1,2$.

By interchanging the suffixes of α_1 and α_2, if necessary, we may assume that

$$\overline{|\xi_1|}\geqslant\overline{|\xi_2|}.$$

Put

$$\Lambda=\max(\overline{|\xi_1|},3).$$

Let σ_0 be an embedding of L such that

$$|\sigma_0(\xi_1)|=\overline{|\xi_1|}.$$

Further, set

$$\beta_{1,0}=\sigma_0(x-\alpha_1)=x-\sigma_0(\alpha_1), \quad \beta_{2,0}=\sigma_0(x-\alpha_2)=x-\sigma_0(\alpha_2).$$

By (12),

$$m/\langle r_1,r_2\rangle\leqslant M\leqslant m. \tag{15}$$

We may assume that $M\geqslant c_{11}$ with c_{11} sufficiently large; otherwise the assertion follows from (15). Consequently, by $|y|>1$ and (1),

$$\log|x|\geqslant c_{12}m\geqslant c_{12}c_{11}. \tag{16}$$

By taking c_{11} sufficiently large, it follows from (16) that

$$|\beta_{1,0}^h|\geqslant 2^{-h}|x|^h. \tag{17}$$

Also, by (13), (14) and (10),

$$|\beta_{1,0}^h|\geqslant(c_{13}^{-1}\Lambda)^M \tag{18}$$

with $c_{13}>1$.

Suppose that $(x-\alpha_1)^h=(x-\alpha_2)^h$. Then, since $\alpha_1\neq\alpha_2$, we see that $h>1$ and there exists an integer l with $0<l<h$ such that

$$x-\alpha_1=e^{2\pi il/h}(x-\alpha_2).$$

This implies that

$$|x|\leqslant \operatorname{cosec}(\pi/h)(|\alpha_1|+|\alpha_2|)$$

which, together with (16), implies $m\leqslant c_{14}$.

Thus we may assume $(x-\alpha_1)^h\neq(x-\alpha_2)^h$. Then $\beta_{1,0}^h\neq\beta_{2,0}^h$. Further observe that

$$0<|\beta_{1,0}^h-\beta_{2,0}^h|\leqslant c_{15}|x|^{h-1} \tag{19}$$

and

$$|\beta_{1,0}^h-\beta_{2,0}^h|=|\beta_{1,0}^h|\left|\left(\frac{\beta_{2,0}}{\beta_{1,0}}\right)^h-1\right|. \tag{20}$$

We apply (13) and corollary B.1 with $n=r+t+2$, $d=[L:\mathbb{Q}]$, $A_1=A_2=\cdots=A_{n-1}=c_{16}$, $A_n=\Lambda^{c_{17}}$ and $B=M$ to conclude that

$$\left|\left(\frac{\beta_{2,0}}{\beta_{1,0}}\right)^h-1\right|\geqslant\Lambda^{-c_{18}\log M}. \tag{21}$$

Combining (20), (17), (21) and (19), we obtain

$$|x|\leqslant\Lambda^{c_{19}\log M}. \tag{22}$$

Further, it follows from (20), (18), (21), (19) and (22) that

$$\Lambda^{M-c_{18}\log M}\leqslant|\beta_{1,0}^h-\beta_{2,0}^h|c_{13}^M\leqslant\Lambda^{c_{20}\log M}c_{13}^M.$$

By taking c_{11} sufficiently large, these inequalities imply that $\Lambda\leqslant c_{21}$. Consequently, by (22) and (15),

$$\log|x|\leqslant c_{22}\log m$$

which, together with (16), gives $m\leqslant c_{23}$. □

The constants c_{24}, c_{25}, ... in the proofs of theorems 10.3–10.7 are computable positive numbers depending only on K, $\mathscr{S}$, f and τ.

Proof of theorem 10.3. Suppose that equation (3) is valid. Let $\eta_1,\ldots,\eta_r$ be an independent system of units for K satisfying (A.45). Put

$$\Delta=\left[\pi_1\cdots\pi_s\prod_{1\leqslant i<j\leqslant n}(\alpha_i-\alpha_j)\right].$$

Denote by $\mathfrak{p}_1,\ldots,\mathfrak{p}_t$ all the prime ideals in $\mathcal{O}_K$ dividing Δ. As in the proof of theorem 10.2, we can find $\psi_1, \ldots, \psi_t \in \mathcal{O}_K$ such that $\not{p}_i^h = [\psi_i]$ and $\overline{|\psi_i|} \leqslant c_{24}$ for $1 \leqslant i \leqslant t$. Put

$$M = \frac{m}{(m, \langle r_1, r_2 \rangle)}.$$

Observe that

$$\frac{m}{\langle r_1, r_2 \rangle} \leqslant M \leqslant m. \tag{23}$$

Therefore we may assume that $M \geqslant c_{25}$ with c_{25} sufficiently large; otherwise the assertion follows from (23).

Equation (3) gives

$$[f(x, z)] = [\gamma y^m]$$

which, as in the proof of theorem 10.2, implies that

$$(x - \alpha_j z)^h = \rho_j \eta_1^{u_{1,j}} \cdots \eta_r^{u_{r,j}} \psi_1^{b_{1,j}} \cdots \psi_t^{b_{t,j}} \xi_j^M \quad (j = 1, 2) \tag{24}$$

where $\rho_1, \rho_2 \in \mathcal{O}_K$ satisfy $\max(\overline{|\rho_1|}, \overline{|\rho_2|}) \leqslant c_{26}$ and $0 \leqslant b_{i,j} \in \mathbb{Z}$, $u_{q,j} \in \mathbb{Z}$ for $1 \leqslant i \leqslant t$, $1 \leqslant q \leqslant r$ and $j = 1, 2$. By incorporating every Mth power in ξ_j, we may assume that

$$0 \leqslant u_{q,j} < M, \quad 0 \leqslant b_{i,j} < M \tag{25}$$

for $1 \leqslant q \leqslant r$, $1 \leqslant i \leqslant t$ and $j = 1, 2$. By interchanging the suffixes of α_1 and α_2, if necessary, we may assume $\overline{|\xi_1|} \geqslant \overline{|\xi_2|}$. Set

$$\Lambda = \max(\overline{|\xi_1|}, 3).$$

Further, put

$$\beta_1 = x - \alpha_1 z, \quad \beta_2 = x - \alpha_2 z.$$

Let σ_0 and σ_1 be embeddings of K satisfying

$$|\sigma_0(\xi_1)| = \overline{|\xi_1|}, \quad |\sigma_1(x)| = \overline{|x|}.$$

For $j = 1, 2$ and $\delta = 0, 1$, denote

$$\beta_{j,\delta} = \sigma_\delta(\beta_j).$$

By (24) and (25),

$$|\beta_{1,0}^h| \geqslant c_{27}^{-M} \Lambda^M \tag{26}$$

and

$$|\beta_{1,1}^h| \geqslant 2^{-h} \overline{|x|}^h \quad \text{if } \overline{|x|} \geqslant 2 \overline{|\alpha_1|}\, \overline{|z|}. \tag{27}$$

Suppose that

$$\beta_1^h = \beta_2^h. \tag{28}$$

Then

$$[z] \mid [h][\alpha_1 - \alpha_2][x]^{h-1}.$$

Therefore, by (2) and $z \in \mathscr{S}$, we see that

$$\overline{|z|} \leqslant c_{28}. \tag{29}$$

Putting $\sigma_1(x)=x'$, $\sigma_1(\alpha_1)=\alpha'_1$, $\sigma_1(\alpha_2)=\alpha'_2$ and $\sigma_1(z)=z'$, equation (28) implies that

$$(x'-\alpha'_1 z')^h=(x'-\alpha'_2 z')^h.$$

Then, since $\alpha'_1 \neq \alpha'_2$, it follows that $h>1$ and there exists an integer l with $0<l<h$ such that

$$x'-\alpha'_1 z' = e^{2\pi i l/h}(x'-\alpha'_2 z').$$

This implies that

$$\overline{|x|}=|x'| \leqslant \operatorname{cosec}(\pi/h)(|\alpha_1|+|\alpha_2|)\overline{|z|}$$

which, together with (29), gives $\overline{|x|} \leqslant c_{29}$. Therefore, by taking norms on both sides of (3), we obtain

$$2^m \leqslant |N(y)|^m \leqslant c_{30},$$

since y is not a unit. Hence $m \leqslant c_{31}$.

Thus we may assume that $\beta_1^h \neq \beta_2^h$ which implies $\beta_{1,\delta}^h \neq \beta_{2,\delta}^h$ for $\delta=0, 1$. We may write

$$z=\pi_1^{w_1}\cdots\pi_s^{w_s}$$

where $w_1,\ldots,w_s$ are non-negative integers. By interchanging the suffixes of $\pi_1,\ldots,\pi_s$, if necessary, we may assume that $w_1 \geqslant w_2 \geqslant \cdots \geqslant w_s$. Recall that π_1 is not a unit. Let $\wp$ be a prime ideal in $\mathcal{O}_K$ dividing π_1. Observe that

$$w_1 \leqslant \operatorname{ord}_\wp(z) \leqslant \operatorname{ord}_\wp(\beta_1^h-\beta_2^h).$$

If $\wp^{v_1} \parallel [\beta_1^h]$ and $\wp^{v_2} \parallel [\beta_2^h]$, then $h|v_1, h|v_2$ and $\wp^{v_1/h} | [\beta_1]$, $\wp^{v_2/h} | [\beta_2]$. Put $v_3=\min(v_1/h, v_2/h)$. Hence $\wp^{v_3}$ divides both $[\beta_1]$ and $[\beta_2]$. Consequently, $\wp^{v_3}$ divides both $[\alpha_2-\alpha_1][x]$ and $[\alpha_2-\alpha_1][z]$. Now we apply (2) to conclude that $v_3 \leqslant c_{32}$. For simplicity, assume that $v_3=v_2/h$. Then

$$w_1 \leqslant hc_{32}+\operatorname{ord}_\wp\left(\frac{\beta_1^h}{\beta_2^h}-1\right).$$

Now we apply (24) and theorem B.3 with $n=r+t+2$, $d=[K:\mathbb{Q}]$, $p \leqslant c_{33}$, $A_1=A_2=\cdots=A_{n-1}=c_{34}$, $A_n=\Lambda^{c_{35}}$ and $B=M$ to conclude that

$$\operatorname{ord}_\wp\left(\frac{\beta_1^h}{\beta_2^h}-1\right) \leqslant c_{36}(\log M)^2 \log \Lambda.$$

Consequently

$$w_1 \leqslant c_{37}(\log M)^2 \log \Lambda$$

whence

$$\overline{|z|} \leqslant \Lambda^{c_{38}(\log M)^2}. \tag{30}$$

Therefore, for $\delta = 0, 1$, we obtain

$$0 < |\beta_{1,\delta}^h - \beta_{2,\delta}^h| \leqslant \overline{|x|}^{h-1} \Lambda^{c_{39}(\log M)^2}.$$

We may write

$$|\beta_{1,\delta}^h - \beta_{2,\delta}^h| = |\beta_{1,\delta}^h| \left| \left(\frac{\beta_{2,\delta}}{\beta_{1,\delta}} \right)^h - 1 \right| \quad (\delta = 0, 1).$$

We apply (24) and corollary B.1 with $n = r + t + 2$, $d = [K : \mathbb{Q}]$, $A_1 = A_2 = \cdots = A_{n-1} = c_{40}$, $A_n = \Lambda^{c_{41}}$ and $B = M$ to conclude that

$$\left| \left(\frac{\beta_{2,\delta}}{\beta_{1,\delta}} \right)^h - 1 \right| \geqslant \Lambda^{-c_{42} \log M} \quad (\delta = 0, 1).$$

Thus

$$|\beta_{1,\delta}^h| \leqslant \overline{|x|}^{h-1} \Lambda^{(c_{42}+c_{39})(\log M)^2} \quad (\delta = 0, 1). \tag{31}$$

By (27), (31) with $\delta = 1$ and (30),

$$\overline{|x|} \leqslant \Lambda^{c_{43}(\log M)^2}. \tag{32}$$

Further, it follows from (26), (31) with $\delta = 0$ and (32) that

$$\Lambda^M \leqslant c_{27}^M \Lambda^{c_{44}(\log M)^2}$$

which, by taking c_{25} sufficiently large, implies that $\Lambda \leqslant c_{45}$. Consequently, by (32), (30) and (23), we obtain

$$\max(\overline{|x|}, \overline{|z|}) \leqslant e^{c_{46}(\log m)^2}.$$

Now we take the norm on both sides of (3) to conclude

$$2^m \leqslant |N(y)|^m \leqslant e^{c_{47}(\log m)^2},$$

since y is not a unit. Hence $m \leqslant c_{48}$. □

Proof of theorem 10.4. Suppose that equation (3) is valid. Then we apply theorem 1.3 to

$$(\alpha_2 - \alpha_1)z = (x - \alpha_1 z) - (x - \alpha_2 z)$$

to conclude that

$$\max(\overline{|x - \alpha_1 z|}, \overline{|x - \alpha_2 z|}, \overline{|z|}) \leqslant c_{49}$$

whence

$$\max(\overline{|x|}, \overline{|z|}) \leqslant c_{50}.$$

Therefore, by (3),

$$\overline{|\varepsilon y^m \gamma|} \leqslant c_{51}. \tag{33}$$

Consequently

$$|N(\gamma)| = |N(\varepsilon y^m \gamma)| \leqslant c_{52}$$

which, since $\gamma \in \mathscr{S}$, implies that $\overline{|\gamma|} \leqslant c_{53}$ and therefore, by (33), $\overline{|\varepsilon y^m|} \leqslant c_{54}$. □

Proof of theorem 10.5. Suppose that equation (4) is valid. In view of theorem 10.3, we may assume that y is a unit. Then, by theorem 10.4, it follows that

$$\overline{|y^m|} \leqslant c_{55}. \tag{34}$$

Further, since y is not a root of unity, we see from (A.13) that

$$\overline{|y|} \geqslant 1 + c_{56}. \tag{35}$$

Now combine (35) and (34) to conclude that $m \leqslant c_{57}$. □

Proof of theorem 10.6. By theorem 10.5, equation (4) implies that $m \leqslant c_{58}$. Therefore it suffices to prove theorem 10.6 with C_2 depending only on $K, \mathscr{S}$, f, τ and m. We allow the constants $c_{59}, \ldots, c_{64}$ in the proof of the theorem to depend on m too. We may write

$$\gamma = \gamma_1 \gamma_2^m, \quad \gamma_1, \gamma_2 \in \mathscr{S}, \quad \overline{|\gamma_1|} \leqslant c_{59}. \tag{36}$$

Re-write (4) as

$$f(x, z) = \gamma_1 (\gamma_2 y)^m. \tag{37}$$

Observe that there are only finitely many possibilities for γ_1 and they can be determined explicitly. Further, recall that $m \geqslant 3$. We apply theorem 8.1 to equation (37) to conclude that there exists a unit $\varepsilon_1 \in \mathcal{O}_K$ such that

$$\max(\overline{|\varepsilon_1 x|}, \overline{|\varepsilon_1 z|}) \leqslant c_{60}. \tag{38}$$

Since $z \in \mathscr{S}$, it follows from (38) that $\overline{|z|} \leqslant c_{61}$. Therefore, by (38), $\overline{|\varepsilon_1|} \leqslant c_{62}$ and consequently, again by (38), $\overline{|x|} \leqslant c_{63}$. Now we infer from (37), (36) and $\gamma \in \mathscr{S}$ that $\overline{|\gamma|} \leqslant c_{64}$ and $\overline{|y|} \leqslant c_{64}$. □

Proof of theorem 10.7. By theorem 10.5, equation (4) implies that $m \leqslant c_{65}$. Now apply theorem 8.2 and argue as in the proof of theorem 10.6 to obtain the assertion of theorem 10.7. □

Proof of corollary 10.1. Suppose that $x \neq 0$, $y \neq 0$ and $m \geqslant 0$ are rational integers satisfying $|x| > 1$, $(x, y) = 1$ and $mn \geqslant 6$. We assume that $\max(|x|, |y|, m)$ exceeds a sufficiently large computable number depending only on A, B and n. Then, since $|x| > 1$ and $(x, y) = 1$, we see that $Ax^m + By^n$ is non-zero. Write

$$Ax^m + By^m = \pm p_1^{a_1} \cdots p_s^{a_s}$$

where $a_1, \ldots, a_s$ are positive integers and $p_1, \ldots, p_s$ are primes not exceeding P_1. We may write

$$p_1^{a_1} \cdots p_s^{a_s} = wz^n, \quad 0 < w \leqslant P_1^{(n-1)s}.$$

Then

$$-By^n \pm wz^n = Ax^m. \tag{39}$$

Suppose $n = 2$. Then we see from $mn \geqslant 6$ that $m \geqslant 3$. We apply theorem 10.6 to equation (39) to conclude that $\max(|x|, |y|, m)$ is bounded by a computable number depending only on A, B and P_1.

Thus we may assume $n > 2$. Then we apply theorem 10.7 to equation (39) to derive that $\max(|x|, |y|, m)$ is bounded by a computable number depending only on A, B, n and P_1. □

Notes

The equations considered in this chapter are more general than the ones considered in chapter 9, but the approach in chapter 9 is different from the one followed in this chapter. In chapter 10, the proofs depend heavily on a factorisation that the equation under consideration provides, whereas this information is not utilised in chapter 9. For explicit estimates on the magnitude of the solutions, the approach followed in chapter 9 gives better bounds than are obtainable by the method of the present chapter.

The original proofs of Schinzel and Tijdeman (1976) of theorem 10.2 and its generalisation due to Shorey *et al.* (1977) depend on theorem 7.1. Shorey (1980) gave proofs of these theorems which do not depend on theorem 7.1; in fact he gave a quantitative version which implies the following.

Let $n > 1$ and A, B be non-zero integers. For integers $m > 3$, x and y with $|x| > 1$, $(x, y) = 1$ and $Ax^m + By^n \neq 0$, we have

$$P(Ax^m + By^n) \geqslant C_4((\log m)(\log\log m))^{1/2}$$

and

$$|Ax^m + By^n| \geqslant \exp(C_4((\log m)(\log\log m))^{1/2})$$

where $C_4 > 0$ is a computable number depending only on A, B and n (cf. corollary 10.1).

Explicit upper bounds for m in theorem 10.2 have been derived by Sprindžuk (1982, Ch. 7, § 2) and Turk (1982, 198x). For example, in the latter paper he proved that, under the conditions of theorem 10.2,

$$m < \exp\left\{\frac{C_5 N^5 (\log 3H)^2}{\log(N \log 3H)}\right\}(\log 3|b|)(\log\log 3|b|)^2$$

where C_5 is some computable absolute constant and N and H are the degree and height of f, respectively. Turk used such an estimate to deduce lower bounds for the greatest prime factor of the power-free part of $f(x)$. Here the power-free part $Z(n)$ of an integer n is the smallest integer b such that $|n|$ can be written as by^m for some $m>1$, $y>1$. Let $f \in \mathbb{Z}[X]$ with at least three simple zeros. One of the results of Turk (1982) is that

$$Z(f(x)) > \tfrac{1}{2}(\log\log(|f(x)|+3))^{C_6}$$

where $C_6>0$ is a constant depending only on f.

Let $p_1, \ldots, p_s$ be a set of distinct primes. Put $\mathscr{P} = \prod_{j=1}^{s} \log p_j$. Let S be the set of rational integers composed of $p_1, \ldots, p_s$. Brindza, Győry and Tijdeman (1985) extended theorem 10.2 to the equation

$$af(x) = bwy^m$$

in rational integers a, b, m, w, x and y with $ab \neq 0$, $w \in S$ and $|y|>1$. They proved that this equation implies that

$$m \leqslant (C_7 (s+1)^{s+1} \mathscr{P})^{C_8} (\log A)(\log\log A)^2$$

where $A = \max(|a|, |b|, 3)$ and C_7 and C_8 are computable constants depending only on N and H. The special case $a=b=1$ was already obtained by Turk (1982).

An immediate consequence of theorem 10.6 is that generalised Ramanujan–Nagell equations like

$$x^2+7=y^m \quad \text{in integers } m>2,\ x, y>1$$

and

$$7x^2+1=y^m \quad \text{in integers } m>2,\ x, y>1$$

have only finitely many solutions. For a discussion of such equations and their relations with algebraic number fields generated by roots of cyclotomic integers, see Ennola (1978).

Another choice of the polynomial f in equation (1) which has received special attention is

$$f(X) = 1^k + 2^k + \cdots + X^k$$

where k is a fixed positive integer. Schäffer (1956) proved that for fixed $k>0$

and $m>1$, the equation

$$1^k+2^k+\cdots+x^k=y^m \quad \text{in integers } x, y>1$$

has an infinite number of solutions only if $(k,m)=(1,2),(3,2),(3,4)$ or $(5,2)$. He conjectured that all other solutions have $x=y=1$ apart from $k=m=2$, $x=24$, $y=70$. Györy, Tijdeman and Voorhoeve (1980) extended Schäffer's result by proving that for fixed $b, k, r \in \mathbb{Z}$ with $b\neq 0$, $k\geqslant 2$ and $k\notin\{3,5\}$, the equation

$$1^k+2^k+\cdots+x^k+r=by^m \quad \text{in integers } m>1,\ x>0,\ y>1 \tag{40}$$

has only finitely many solutions and that all the solutions can be effectively determined. In Voorhoeve, Györy and Tijdeman (1979) they gave an ineffective proof that, in (40), r can be replaced by $R(x)$ where $R\in\mathbb{Z}[X]$ is a fixed polynomial. Brindza (1984*b*) gave an effective proof of this result and extended it to a certain class of equations

$$F(x, 1^k+2^k+\cdots+x^k)=by^m \quad \text{in integers } m>1,\ x>0,\ y>1$$

where $F\in\mathbb{Z}[X,Y]$ is a fixed polynomial. Dilcher (1986) showed that under general conditions the equation

$$\chi(1)1^k+\chi(2)2^k\cdots+\chi(\phi x)(\phi x)^k=by^m$$
$$\text{in integers } m>1,\ x>0,\ y>1,$$

where χ is a primitive quadratic residue class character with conductor ϕ and b is a non-zero integer, has only finitely many solutions. In particular, he showed that for any integers $b\neq 0$ and $k\geqslant 3$, $k\notin\{4,5\}$ there exist computable upper bounds for the solutions of the equation

$$1^k-3^k+5^k-\cdots+(4x-3)^k-(4x-1)^k=by^m$$

in integers $m>1$, $x>0$ and $y>1$.

Erdös (1951) showed that the equation

$$\binom{x+n}{n}=y^m \quad \text{in integers } m>1,\ n>1,\ x\geqslant 1,\ y>1 \tag{41}$$

has no solutions provided that $n\geqslant 4$. On the other hand, it is clear that there are infinitely many solutions when $m=n=2$. The only other known solution of (41) is $m=2$, $n=3$, $x=47$, $y=140$ and it is conjectured that there are no more. It is a direct consequence of theorems 10.6 and 10.7 that there are computable bounds for the solutions of (41) with $m\geqslant 3$, $n=2$ and $m\geqslant 2$, $n=3$ (cf. Tijdeman, 1976*a*).

Erdös and Selfridge (1975) proved that the related equation

$$(x+1)\cdots(x+n)=y^m \tag{42}$$

in integers $m>1,\ n>1,\ x\geqslant 1,\ y>1$

has no solution at all. That is, the product of two or more consecutive positive integers is never a perfect power. For earlier results in this direction, see Erdös and Selfridge (1975). Further, Shorey (1986*a*) applied linear forms in logarithms to give a different proof of the assertion that equation (42) implies that n is bounded by a computable absolute constant. The proof does not depend on the fact that $a_i a_j$ (see Erdös and Selfridge (1975) for the definition of the a_is) are distinct for distinct pairs i, j and this fact is crucial in the proof of Erdös and Selfridge.

Estimates for linear forms in logarithms have also been used to prove results like this: the product of two or more neighbouring positive integers cannot be a perfect power. Here a set of integers is called neighbouring if they all belong to some small interval. Turk (1983*b*) and Erdös and Turk (1984) proved this assertion for integers in an interval of the form

$$(N, N+C_9 \log\log\log N)$$

where $N\geqslant 16$ and $C_9>0$ is some computable absolute constant. Furthermore, Shorey (1986*a*) proved that for $\varepsilon>0$ and integers m, n, x with $m>3$, $x>n^m$ and n exceeding a computable number depending only on ε, any product of $(\frac{1}{2}(m-1)/(m-2)+\varepsilon)n$ distinct integers from $x+1,\ldots,x+n$ is never an mth power. If $m=3$ and $x>n^{4+\varepsilon}$, the assertion is valid for any product of $C_{10}n$ distinct integers from $x+1,\ldots,x+n$ where $0<C_{10}<1$ is a computable number depending only on ε.

In a similar way as the deduction of theorem 7.6 from lemma 7.1, it is possible to extend theorem 10.3 to the situation that the coefficients of $f(X, Y)$ belong to K, but the roots of $f(X, 1)$ are not necessarily in K. In case $K=\mathbb{Q}$, Shorey *et al.* (1977) proved such an assertion.

By using theorem 8.3 in place of theorems 8.1 and 8.2, we can easily derive a common generalisation of theorems 10.6 and 10.7. We recall that in theorems 10.3–10.7 the dependence on K can be refined to dependence on the degree and the discriminant of K only.

CHAPTER 11——

The Fermat equation

Fermat's Last Theorem states that the equation

$$x^n + y^n = z^n \tag{1}$$

in positive integers n, x, y, z with $n > 2$ has no solution, but no proof of this assertion is available. We call (1) the *Fermat equation.* We refer to Ribenboim (1979) for the history and a general treatment of the Fermat equation. Without loss of generality we may assume $x < y < z$ and $(x, y, z) = 1$ in (1) and we shall do so throughout the chapter without further mention. Hence $(x, y) = (y, z) = (z, x) = 1$.

The celebrated result of Faltings (1983) quoted in chapter 6 implies that for every $n > 2$ there are only finitely many triples of positive integers x, y, z such that (1) holds. Heath-Brown (1985) and Granville (1985), independently, used this result to prove that Fermat's Last Theorem is true for almost all exponents n. Faltings' proof is ineffective. In theorems 11.4, 11.6 and 11.7 we present some conditions under which effective proofs can be given.

The method of estimating linear forms in logarithms enables us to prove that, under suitable conditions, equation (1) implies that n is bounded. Such a result is given by theorem 11.3. On combining such results with those for fixed n, we obtain assertions that, under suitable conditions, equation (1) has only finitely many solutions n, x, y, z (see theorems 11.2 and 11.5).

Abel's conjecture says that (1) has no solution in positive integers $n > 2$, x, y, z such that at least one among x, y, z is a prime power. Even this special case of Fermat's Last Theorem is still open. In order to confirm Abel's conjecture it suffices to show that there are no solutions with $z - y = 1$. In theorem 11.1 lower bounds for $z - y$, $y - x$ and $|(z - y) - (y - x)|$ are given, but the case $z - y = 1$ is excluded. The bound for $z - y$ is proved by using ideas of Barlow and Abel. Bounds for $y - x$ were given by Stewart (1977*c*) and Inkeri and van der Poorten (1980). The former derived a slightly weaker lower bound; the latter authors restricted themselves (unnecessarily) to prime exponents.

Theorem 11.1. *Let $n>2$, x, y, z be positive integers satisfying* (1).

(*a*) *Suppose $z-y>1$ if n is odd. Then*

$$z-y \geqslant 2^n/n.$$

(*b*) *There is a computable absolute constant C_1 such that*

$$y-x \geqslant z^{1-C_1(\log n)^3/n}. \tag{2}$$

(*c*) *There is a computable absolute constant C_2 such that*

$$|(z-y)-(y-x)| \geqslant z^{1-C_2(\log n)^3/n}. \tag{3}$$

The next theorem and its corollary deal with the situation that x, y and z assume polynomial values. They are due to Brindza, Győry and Tijdeman (1985).

Theorem 11.2. *Let $E, F, G \in \mathbb{Z}[X, Y]$ be pairwise non-proportional binary forms of the same degree m. Then all solutions of the equation*

$$(E(t,u))^n+(F(t,u))^n=(G(t,u))^n \tag{4}$$

in rational integers n, t, u

with $n>2$, $(t,u)=1$ and $E(t,u)F(t,u)G(t,u)\neq 0$ satisfy $\max(n, |t|, |u|)<C_3$ *where C_3 is a computable number depending only on E, F and G.*

By taking $u=1$ we obtain the following result for polynomials in one variable.

Corollary 11.1. *Let $E, F, G \in \mathbb{Z}[X]$ be pairwise non-proportional polynomials. Then all solutions of the equation*

$$(E(t))^n+(F(t))^n=(G(t))^n \quad \text{in rational integers } n, t \tag{5}$$

with $n>2$ and $E(t)F(t)G(t)\neq 0$ satisfy $\max(n, |t|)<C_4$ *where C_4 is a computable number depending only on E, F and G.*

The remaining results deal with solutions of (1) such that values of certain polynomial expressions in x, y and z are composed of fixed primes. Let $P \geqslant 3$ be a fixed number. Denote by S the set of all rational integers composed of primes not exceeding P.

We shall apply corollary B.1 and theorem B.3 to prove

Theorem 11.3. *Let n, x, y, z be positive integers satisfying* (1). *Assume that at least one of the following conditions holds:*

$$(a)\ x \in S, \qquad (b)\ y \in S, \qquad (c)\ z \in S,$$

$$(d)\ y-x \in S, \quad (e)\ y+z \in S \text{ and } n \text{ odd},$$

$$(f)\ x+z \in S \text{ and } n \text{ odd}.$$

Then n is bounded by a computable number depending only on P.

Inkeri and van der Poorten (1980) gave quantitative results in case $y-x$ has a large factor belonging to S.

For fixed n we shall apply theorem 8.1 to derive the following generalisation of a theorem of Inkeri (1976).

Theorem 11.4. *Let x, y, z and $n>2$ be positive integers satisfying* (1). *For $A, B \in \mathbb{Z}$ put $\mathscr{L}(X, Y) =: AX+BY$. Assume that at least one of the following conditions holds:*

$$(a)\ \mathscr{L}(x, y) \in S, \quad (b)\ \mathscr{L}(y, z) \in S, \quad (c)\ \mathscr{L}(z, x) \in S.$$

Then there exists a computable number C_5 depending only on n, $\mathscr{L}$ and P such that

$$\max(x, y, z) \leqslant C_5.$$

The combination of theorems 11.3 and 11.4 yields the following result.

Theorem 11.5. *Let x, y, z and $n>2$ be positive integers satisfying* (1). *Assume that at least one of the following conditions holds:*

$$(a)\ x \in S, \quad (b)\ y \in S, \quad (c)\ z \in S, \quad (d)\ y-x \in S,$$

$$(e)\ y+z \in S \text{ and } n \text{ odd},$$

$$(f)\ x+z \in S \text{ and } n \text{ odd}.$$

Then there exists a computable number C_6 depending only on P such that

$$\max(n, x, y, z) \leqslant C_6.$$

We shall apply theorems 11.5 and 9.6 to prove that the assertion of theorem 11.5 remains valid if the linear form is replaced by a quadratic form.

Theorem 11.6. *Let x, y, z and $n>2$ be positive integers such that* (1) *holds. For $A, B, C \in \mathbb{Z}$ put $Q(X, Y) = AX^2+BXY+CY^2$. Assume that at least one* *of the following conditions holds:*

$$(a)\ Q(x,y)\in S,\quad (b)\ Q(y,z)\in S,\quad (c)\ Q(z,x)\in S.$$

Then there exists a computable number C_7 depending only on n, P and Q such that

$$\max(x,y,z)\leqslant C_7.$$

Theorems 11.4, 11.6 and 7.1 imply the following result.

Theorem 11.7. *Let x, y, z and $n>2$ be positive integers satisfying (1). Let $F\in\mathbb{Z}[X,Y]$ be a non-constant binary form. Assume that at least one of the following conditions holds:*

$$(a)\ F(x,y)\in S,\quad (b)\ F(y,z)\in S,\quad (c)\ F(z,x)\in S.$$

Then there exists a computable number C_8 depending only on n, F and P such that

$$\max(x,y,z)\leqslant C_8.$$

Proofs

The proofs of the theorems depend on the following result, essentially due to Abel (1823), on factorisations in (1).

Lemma 11.1. *Suppose n, x, y, z are positive integers satisfying (1). Then*

(a) there exist $\delta_1, \delta_2\in\{0,1\}$ and positive integers a_1, a_2, d_1, d_2 with $d_1|n$ and $d_2|n$ such that

$$z-x=2^{\delta_1}d_1^{-1}a_1^n \tag{6}$$

and

$$z-y=2^{\delta_2}d_2^{-1}a_2^n. \tag{7}$$

(b) Suppose n is odd. Then (6) and (7) hold with $\delta_1=\delta_2=0$. Further,

$$x+y=d_3^{-1}a_3^n. \tag{8}$$

for some positive integers a_3, d_3 with $d_3|n$.

Proof. We may assume $n>1$. It follows from (1) that

$$(z-x)V=y^n \tag{9}$$

where

$$V=\frac{z^n-x^n}{z-x}=z^{n-1}+z^{n-2}x+\cdots+x^{n-1}.$$

Let $0<r\in\mathbb{Z}$ and p prime. Suppose $p^r\mid(z-x)$ and $p^r\mid V$. Then $p^r\mid nz^{n-1}$. Since $(p,z)=1$, this implies that $p^r\mid n$. Thus $(z-x, V)\mid n$. For any odd prime p with $p\mid n$ and $p\mid(z-x)$ we have

$$\mathrm{ord}_p(V)=\mathrm{ord}_p(n).$$

Further, if $2\mid n$ and $4\mid(z-x)$, then

$$\mathrm{ord}_2(V)=\mathrm{ord}_2(n).$$

Formula (6) follows from (9). Clearly $\delta_1=0$ if n is odd. Formulas (7) and (8) are proved similarly.

Lemma 11.2. *Suppose x, y, z and $n>2$ are positive integers satisfying* (1). *Then $n>100$.*

This result was already known in 1926. See Ribenboim (1979, p. 200). The best lower bound for n known today, 125 000, is due to Wagstaff (1978). We shall use lemma 11.2 in two ways. Firstly it implies that (1) has no solutions such that n is a multiple of 4, a result already proved by Fermat. Secondly it is used for convenience in certain estimates.

Lemma 11.3. *Suppose x, y, z and $n>2$ are positive integers satisfying* (1). *Then*

$$(a)\quad y^2\geqslant 2^{n/2}/n,\quad (b)\quad y/n^2\geqslant 20.$$

Proof. If n is odd, then, by lemma 11.1(*b*), there exist positive integers a_3, d_3 with $d_3\leqslant n$ such that $x+y=d_3^{-1}a_3^n$. Hence $a_3>1$ and

$$y>\frac{x+y}{2}\geqslant\frac{2^n}{2n}. \tag{10}$$

If n is even, then, by lemma 11.2, $n=2v$ with v odd. By lemma 11.1(*b*) there exist positive integers a, d with $d\leqslant v$ such that $x^2+y^2=d^{-1}a^v$. Hence $a>1$ and $y^2>a^v/2v\geqslant 2^v/(2v)=2^{n/2}/n$. This estimate, together with (10), proves (*a*). Since $n>100$ by lemma 11.2, assertion (*b*) is an immediate consequence of (*a*). □

Theorem 11.1(*a*) is a direct consequence of lemma 11.1 if we know that $a_2>1$. This is obvious unless $z-y=1$ or $z-y=2$ and n even. For the cases n even, $z-y\in\{1,2\}$ we need a separate argument.

Lemma 11.4. *Let x, y, z and $v>1$ be positive integers satisfying*

$$x^{2v}+y^{2v}=z^{2v}. \tag{11}$$

Then $z-y>2$.

Proof. By lemma 11.2, we may assume that v is odd and $v>50$.

Case 1. Assume $z-y=2$. Then x is even and y is odd. By writing $(x^v)^2+(y^v)^2=(z^v)^2$, we see that there exist positive integers r, s with $r>s>0$, $(r,s)=1$ and rs even such that

$$x^v=2rs, \quad y^v=r^2-s^2, \quad z^v=r^2+s^2. \tag{12}$$

Since $(r,s)=1$ and r^2-s^2 is odd, there exist positive integers y_1, y_2 such that

$$r-s=y_1^v, \quad r+s=y_2^v. \tag{13}$$

We have, by lemma 11.3,

$$2s^2=z^v-y^v=(y+2)^v-y^v=2vy^{v-1}\left(1+\frac{v-1}{2}\cdot\frac{2}{y}+\frac{(v-1)(v-2)}{2.3}\cdot\frac{4}{y^2}+\cdots\right)$$

$$\leqslant 2vy^{v-1}\left(1+\frac{v}{y}+\frac{v^2}{y^2}+\cdots\right)<4vy^{v-1}.$$

This implies $s\leqslant 2v^{1/2}y^{(v-1)/2}$. Further, by (13) and (12), $y_2>r^{1/v}>y^{1/2}$. Hence, by (13),

$$y_2^v-y_1^v=2s\leqslant 4v^{1/2}y^{(v-1)/2}\leqslant 4v^{1/2}y_2^{v-1}.$$

On the other hand, $y_2>y_1$ and therefore, by lemma 11.3,

$$y_2^v-y_1^v\geqslant y_2^v-(y_2-1)^v=vy_2^{v-1}-\binom{v}{2}y_2^{v-2}+\binom{v}{3}y_2^{v-3}\cdots$$

$$\geqslant vy_2^{v-1}\left(1-\frac{v}{2y_2}-\left(\frac{v}{2y_2}\right)^3\cdots\right)\geqslant \tfrac{1}{2}vy_2^{v-1}.$$

On combining these inequalities we obtain $\frac{1}{2}v\leqslant 4v^{1/2}$, which is impossible.

Case 2. Assume $z-y=1$. We now have x odd, y even. Hence there exist positive integers r, s with $r>s>0$, $(r,s)=1$, rs even such that

$$x^v=r^2-s^2, \quad y^v=2rs, \quad z^v=r^2+s^2.$$

Since $r^2+s^2=z^v=(y+1)^v$, we have, by lemma 11.3,

$$y^v+vy^{v-1}<r^2+s^2=y^v+\binom{v}{1}y^{v-1}+\binom{v}{2}y^{v-2}+\cdots+1$$

$$<y^v+vy^{v-1}\left(1+\frac{v}{2y}+\left(\frac{v}{2y}\right)^2+\cdots\right)<y^v+2vy^{v-1}. \tag{14}$$

Furthermore, by lemma 11.3,

$$(r^2-s^2)^2=x^{2v}=(y+1)^{2v}-y^{2v}=\binom{2v}{1}y^{2v-1}+\binom{2v}{2}y^{2v-2}+\cdots$$

$$<2vy^{2v-1}\left(1+\frac{v}{y}+\frac{v^2}{y^2}+\cdots\right)<3vy^{2v-1}.$$

Hence

$$\sqrt{(2v)}y^{v-1/2}<r^2-s^2<\sqrt{(3v)}y^{v-1/2}. \tag{15}$$

On combining (14) and (15) we obtain, by lemma 11.3,

$$y^v<2r^2<y^v+2\sqrt{v}\,y^{v-1/2}$$

and

$$y^v-2\sqrt{v}\,y^{v-1/2}<2s^2<y^v.$$

This implies

$$\frac{1}{\sqrt{2}}y^{v/2}<r<\frac{1}{\sqrt{2}}y^{v/2}+\sqrt{v}\,y^{v/2-1/2}. \tag{16}$$

and, by lemma 11.3,

$$\frac{1}{\sqrt{2}}y^{v/2}-\sqrt{v}\,y^{v/2-1/2}<s<\frac{1}{\sqrt{2}}y^{v/2}. \tag{17}$$

Observe that $(r+s, r-s)=1$, $(r+s)(r-s)=x^v$ imply that $r+s=x_1^v$ for some positive integer x_1. Furthermore, either r is even, s is odd or r is odd, s is even. In the former instance $(2r, s)=1$, $2rs=y^v$, hence $2r=y_3^v$ for some $y_3\in\mathbb{Z}$; in the latter $(r, 2s)=1$, $2rs=y^v$, hence $2s=y_3^v$ for some $y_3\in\mathbb{Z}$. We have, by (16) and (17),

$$|x_1^v-y_3^v|=\max(|r+s-2r|, |r+s-2s|)=r-s<2\sqrt{v}\,y^{v/2-1/2}. \tag{18}$$

On the other hand, by (17),

$$\min(x_1^v, y_3^v)\geq 2s>\sqrt{2}\,y^{v/2}-2\sqrt{v}\,y^{v/2-1/2}.$$

Hence, by lemma 11.3,

$$\min(x_1^v, y_3^v)\geq y^{v/2}+((\sqrt{2}-1)\sqrt{y}-2\sqrt{v})y^{v/2-1/2}>y^{v/2}.$$

Thus $\min(x_1, y_3)>\sqrt{y}$. Since $x_1\neq y_3$, this implies

$$|x_1^v-y_3^v|\geq(\sqrt{y}+1)^v-(\sqrt{y})^v|>vy^{v/2-1/2}.$$

Since $v>4$, this yields a contradiction to (18). □

We shall use lemmas 11.1 and 11.4 to prove that (1) has no solution with $z-y=y-x=1$.

Lemma 11.5. *If $n>2$ and y are positive integers, then*

$$(y-1)^n+y^n\neq(y+1)^n.$$

Proof. Suppose $n>2$ and y satisfy $(y-1)^n+y^n=(y+1)^n$. By lemma 11.4, n is odd. Hence, by (6) with $\delta_1=0$, there are positive integers a_1 and d_1 with $d_1 \mid n$ such that $2d_1=a_1^n$. This is impossible, since d_1 is odd and $n>1$. □

Proof of theorem 11.1. (*a*) By lemma 11.1 we have $z-y\leqslant a_2^n$ if n is odd and $z-y\leqslant 2a_2^n$ if n is even. Since $z-y>1$ if n is odd and, by lemma 11.4, $z-y>2$ if n is even, we have $a_2>1$. Thus, by (7),

$$z-y\geqslant d_2^{-1}2^n\geqslant 2^n/n.$$

(*b*) Let c_1, c_2, c_3 denote computable absolute positive constants. By lemma 11.1, we have

$$z-x=2^{\delta_1}d_1^{-1}a_1^n, \quad z-y=2^{\delta_2}d_2^{-1}a_2^n \tag{19}$$

where $\delta_1, \delta_2\in\{0,1\}$ and d_1, d_2, a_1, a_2 are positive integers with $d_1|n$ and $d_2|n$. Hence

$$0\neq\frac{y-x}{z-x}=\left|\frac{z-y}{z-x}-1\right|=\left|2^{\delta_2-\delta_1}\frac{d_1}{d_2}\left(\frac{a_2}{a_1}\right)^n-1\right|. \tag{20}$$

From (19) we deduce that $\max(a_1, a_2)\leqslant(nz)^{1/n}$. By lemma 11.1, we have $z> z-x\geqslant 2^n/n>2^{n/3}$. On applying corollary B.1 we obtain

$$\left|2^{\delta_2-\delta_1}\frac{d_1}{d_2}\left(\frac{a_2}{a_1}\right)^n-1\right|>\exp(-c_1(\log n)^3(\log z^{1/n})). \tag{21}$$

Since $x<y<z$, we see from (1) that $2x^n<z^n$, hence $z-x>(1-2^{-1/n})z>z/2n$. We therefore obtain, by (20) and (21),

$$y-x>\frac{1}{2n}z^{1-c_1(\log n)^3/n}.$$

Hence

$$\frac{1}{2n}>2^{-4\log n}>z^{-c_2(\log n)^3/n}.$$

Thus $y-x>z^{1-c_3(\log n)^3/n}$.

(*c*) In lemma 11.6 we shall prove slightly more than we need, namely that if $(z-y)/(y-x)$ is almost equal to some rational number r^*, then it equals r^*.

Lemma 11.6. *Let x, y, z and $n>2$ be positive integers satisfying (1). Let r be some rational integer. Then there exists a computable number C_9 depending* *only on r such that if*

$$|r(z-y)-(y-x)| \leqslant z^{1-C_9(\log n)^3/n}, \tag{22}$$

then $y-x=r(z-y)$.

Proof. Put $s=r+1$. Assume that $r(z-y)\neq y-x$. Then $s(z-y)\neq z-x$. As in the proof of theorem 11.1(*b*) we have (19), hence

$$\left|s\frac{z-y}{z-x}-1\right| = \left|s\,.\,2^{\delta_2-\delta_1}\frac{d_1}{d_2}\left(\frac{a_2}{a_1}\right)^n-1\right| \tag{23}$$

where $\delta_1, \delta_2 \in \{0, 1\}$ and d_1, d_2, a_1, a_2 are positive integers with $d_1 \mid n$ and $d_2 \mid n$ and $\max(a_1, a_2) \leqslant (nz)^{1/n}$ with $z^{1/n} > 2^{1/3}$. Suppose $s(z-y)\neq z-x$. On applying corollary B.1 we obtain that the right-hand side of (23) exceeds

$$\exp(-c_4(\log n)^3(\log z^{1/n}))$$

where c_4 is a computable constant depending only on r. Thus, using $z-x>z/2n$ and arguing as before, we find

$$|r(z-y)-(y-x)| = |s(z-y)-(z-x)| > z^{1-c_5(\log n)^3/n}$$

where c_5 is a computable number depending only on r. □

Proof of theorem 11.1(*c*) (continued). In view of lemma 11.6 with $r=1$, we may assume that $z-y=y-x$. It was proved by Goldziher (1913) and rediscovered by Mihaljinec (1952) and Rameswar Rao (1969) that (1) has no solutions with x, y, z in arithmetical progression. We give a proof which is an immediate consequence of lemma 11.5. Put $\Delta=z-y$. Hence, by (1),

$$(y-\Delta)^n+y^n=(y+\Delta)^n. \tag{24}$$

It is clear from equation (24) that every prime divisor of Δ is a prime divisor of y. By $(y, z)=1$, we must have $(y, \Delta)=1$. Thus $\Delta=1$ in contradiction to lemma 11.5.

Proof of theorem 11.3. We have $n>6$ by lemma 11.2. If (*a*), (*b*) or (*c*) holds, then the assertion follows from theorem 2.2. Thus we may assume that (*d*), (*e*) or (*f*) holds. Denote by $c_6, c_7, \ldots, c_{13}$ computable positive numbers depending only on P.

(*d*) Suppose $y-x\in S$. By lemma 11.1 there exist $\delta_1, \delta_2 \in \{0, 1\}$ and positive integers a_1, a_2, d_1, d_2 with $d_1 \mid n, d_2 \mid n$ such that (6) and (7) hold. By lemma 11.5, we have $z-x>2$. By a deduction similar to that of (20) and (21), we obtain, by corollary B.1,

$$\frac{y-x}{z-x} \geqslant a^{-c_6(\log n)^3} \tag{25}$$

where $a = \max(a_1, a_2) > 1$. We have $y-x=(z-x)-(z-y)$. By (1) and $(x, y, z) = 1$, it follows that $z-x$ and $z-y$ are relatively prime. Let p be a prime dividing $y-x$. Then $p \leqslant P$, since $y-x \in S$. Further, either $\mathrm{ord}_p(z-x)=0$ or $\mathrm{ord}_p(z-y)=0$. Assume, for simplicity, that $\mathrm{ord}_p(z-x)=0$. Now we apply (19) and theorem B.3 with $n=3$, $d=1$, $p \leqslant P$, $A_1=3$, $A_2=n$, $A_3=a+1$ and $B=n$ to conclude that

$$\mathrm{ord}_p(y-x) = \mathrm{ord}_p\left(1 - \frac{z-y}{z-x}\right) \leqslant c_7 \log a (\log n)^3.$$

The above inequality is also valid when $\mathrm{ord}_p(z-y)=0$. Consequently,

$$\log(y-x) = \sum_{p \leqslant P} \mathrm{ord}_p(y-x) \log p \leqslant c_8 \log a (\log n)^3. \tag{26}$$

Combining (25) and (26), we obtain

$$\log(z-x) \leqslant c_9 \log a (\log n)^3. \tag{27}$$

On the other hand, it follows from (6), (7) and $x<y<z$ that

$$\log(z-x) \geqslant n \log a - \log n. \tag{28}$$

Now (28), (27) and $a>1$ imply that $n \leqslant c_{10}$.

(*e*) Suppose $y+z \in S$ and n is odd. By lemma 11.1, there exist positive integers a_1, a_3, d_1, d_3 with $d_1 \mid n$, $d_3 \mid n$ such that (6) with $\delta_1 = 0$ and (8) are valid. Observe that $a_3 > 1$. Put $b = \max(a_1, a_3)$. By (6) and (8),

$$\frac{y+z}{x+y} = 1 - \frac{x-z}{x+y} = 1 - \left(-\frac{d_3}{d_1}\right)\left(\frac{a_1}{a_3}\right)^n.$$

Apply corollary B.1 with $n=2$, $d=1$, $A_1=n$, $A_2=b+1$ and $B=n$ to obtain

$$\frac{y+z}{x+y} \geqslant b^{-c_{11}(\log n)^3}.$$

As in the proof of (*d*), it follows from $y+z \in S$ and theorem B.3 that

$$\log(y+z) \leqslant c_{12} \log b (\log n)^3. \tag{29}$$

By $x<z$ this implies $\log(x+y) \leqslant c_{12} \log b (\log n)^3$. On the other hand, it follows from (8) that $\log(x+y) \geqslant n \log a_3 - \log n$. Therefore we may assume

that $b=a_1$. Then, by (6) and (29),

$$n \log b - \log n \leqslant \log(z-x) \leqslant \log(y+z) \leqslant c_{12} \log b(\log n)^3$$

which implies that $n \leqslant c_{13}$.

(*f*) The proof is similar to that of (*e*). □

Proof of theorem 11.4. Since $0 \notin S$, we may assume that at least one of A and B is non-zero. By theorem 7.1 we may suppose that none of x, y and z is a member of S. Consequently we may assume that $AB \neq 0$. Denote by c_{14}, c_{15}, c_{16} computable positive numbers depending only on n, $\mathscr{L}$ and P.

(*a*) Suppose $\mathscr{L}(x, y) = Ax + By \in S$. Put

$$Ax + By = k. \tag{30}$$

Observe that $(x, k) \mid B$, since $(x, y) = 1$. Further, by (1) and (30),

$$B^n z^n = B^n(x^n + y^n) = (Bx)^n + (k - Ax)^n. \tag{31}$$

Set

$$U = B^n + (-A)^n$$

and

$$f(X, Y) = \begin{cases} (BX)^n + (Y - AX)^n & \text{if } U \neq 0, \\ \{(BX)^n + (Y - AX)^n\}/Y & \text{if } U = 0. \end{cases} \tag{32}$$

Observe that $f \in \mathbb{Z}[X, Y]$ is a binary form of degree $N \geqslant n-1 \geqslant 2$ with $f(1, 0) \neq 0$. By (32) and (31),

$$B^n z^n = \begin{cases} f(x, k) & \text{if } U \neq 0, \\ kf(x, k) & \text{if } U = 0. \end{cases}$$

If $U = 0$, then $B^n z^n / k \in \mathbb{Z}$ and we may write $B^n z^n / k = B_1 z_1^n$ for some integers B_1, z_1 with $|B_1| \leqslant c_{14}$. Thus

$$f(x, k) = B_2 z_2^n \tag{33}$$

where $z_2 = z_1$ or Bz and $B_2 = B_1$ or 1. Notice that $|B_2| \leqslant c_{14}$. Put

$$F(X) = f(X, 1).$$

Notice that $F \in \mathbb{Z}[X]$ has degree greater than or equal to $n - 1 \geqslant 2$. Further, observe that

$$nAF(X) + (1 - AX)F'(X) = nB^n X^{n-1}$$

which, together with $F(0) \neq 0$, implies that all the roots of F are simple.

Apply theorem 8.1 to (33) to conclude that

$$\max(x, |k|) \leqslant c_{15}$$

which, together with (30) and (1), implies that $\max(y, z) \leqslant c_{16}$.

The proofs for (*b*) and (*c*) are similar. □

Proof of theorem 11.5. Apply first theorem 11.3 then theorem 11.4. □

Proof of theorem 11.6. We may assume that at least one of A, B, C is non-zero. In view of theorem 11.4 we may further assume that $Q(X, Y)$ is irreducible over the rationals. Put $\Delta = B^2 - 4AC$ and denote by α, β the roots of $Q(X, 1)$. Thus

$$Q(X, Y) = A(X - \alpha Y)(X - \beta Y).$$

Set $L = \mathbb{Q}(\alpha)$. Observe that $[L:\mathbb{Q}] = 2$. Let $\mathcal{O}_L$ be the ring of integers of L. Define

$$\varepsilon = \begin{cases} \text{fundamental unit in } L & \text{if } \Delta > 0, \\ 1 & \text{if } \Delta < 0. \end{cases}$$

Denote by $c_{17}, \ldots, c_{28}$ computable positive numbers depending only on A, B, C, n and P. It follows from lemma A.12 and corollary A.6 that there exists a c_{17} such that for every pair x_0, y_0 of rational integers with $Q(x_0, y_0) \in S$, we have

$$A(x_0 - \alpha y_0) = \rho \pi_1^{l_1} \cdots \pi_s^{l_s} \varepsilon^M \tag{34}$$

where $\rho \in \mathcal{O}_L$ is a root of unity, $M, l_1 \geqslant 0, \ldots, l_s \geqslant 0$ are rational integers and $\pi_1, \ldots, \pi_s$ are all non-units of $\mathcal{O}_L$ satisfying $\left|\overline{\pi_v}\right| \leqslant c_{17}$ for $1 \leqslant v \leqslant s$. Denote by $\mathscr{S}$ the set of all the products of non-negative powers of $\pi_1, \ldots, \pi_s$.

(*b*) Assume that $Q(y, z) \in S$. By lemma 11.1, there exist a $\delta_2 \in \{0, 1\}$ and positive integers a_2, d_2 with $d_2 \mid n$ such that

$$z - y = 2^{\delta_2} d_2^{-1} a_2^n. \tag{35}$$

By (34) with $x_0 = y$ and $y_0 = z$,

$$A(y - \alpha z) = \rho_1 \phi_1 \varepsilon^m \tag{36}$$

where $m \in \mathbb{Z}$, $\phi_1 \in \mathscr{S}$ and ρ_1 is a root of unity in $\mathcal{O}_L$. Let σ be an embedding of L such that $\sigma(\beta) = \alpha$. Observe that $\sigma(\varepsilon) = \pm \varepsilon^{-1}$. We prove the theorem when $\sigma(\varepsilon) = \varepsilon^{-1}$. If $\sigma(\varepsilon) = -\varepsilon^{-1}$, the proof is similar. Put $\sigma(\rho_1) = \rho_2$ and $\sigma(\phi_1) = \phi_2$. Notice that $\phi_2 \in \mathscr{S}$. By taking images under σ on both sides in (36), we obtain

$$A(y - \beta z) = \rho_2 \phi_2 \varepsilon^{-m}. \tag{37}$$

Solving for y and z in equations (36) and (37), we obtain

$$A(\beta-\alpha)y=\rho_1\beta\phi_1\varepsilon^m-\rho_2\alpha\phi_2\varepsilon^{-m} \tag{38}$$

and

$$A(\beta-\alpha)z=\rho_1\phi_1\varepsilon^m-\rho_2\phi_2\varepsilon^{-m}. \tag{39}$$

Consequently

$$A(\beta-\alpha)(z-y)=\rho_1(1-\beta)\phi_1\varepsilon^m-\rho_2(1-\alpha)\phi_2\varepsilon^{-m}$$

which, together with (35), implies that

$$2^{\delta_2}A(\beta-\alpha)a_2^n=\phi_3\varepsilon^m-\phi_4\varepsilon^{-m} \tag{40}$$

where

$$\phi_3=\rho_1 d_2(1-\beta)\phi_1, \quad \phi_4=\rho_2 d_2(1-\alpha)\phi_2. \tag{41}$$

Notice that $\sigma(\phi_3)=\phi_4$. Since $\beta\neq 1$, we have $\phi_3\neq 0$ and consequently $\phi_4\neq 0$. Further, it follows from (38), (39) and $(y, z)=1$ that

$$\min(\operatorname{ord}_{\mathfrak{p}}(\phi_1), \operatorname{ord}_{\mathfrak{p}}(\phi_2))\leqslant c_{18},$$

and consequently, by (41),

$$\min(\operatorname{ord}_{\mathfrak{p}}(\phi_3), \operatorname{ord}_{\mathfrak{p}}(\phi_4))\leqslant c_{19}$$

for every prime ideal $\mathfrak{p}$ in $\mathcal{O}_L$. Now we apply theorem 9.4 to equation (40). It follows from (6) of chapter 9 that there exists a unit $\varepsilon_1\in\mathcal{O}_L$ such that

$$\max(\overline{|\varepsilon_1\phi_3\varepsilon^m|}, \overline{|\varepsilon_1\phi_4\varepsilon^{-m}|})\leqslant c_{20}.$$

Since $\sigma(\phi_3\varepsilon^m)=\phi_4\varepsilon^{-m}$ and ε_1 is an algebraic integer, we obtain $\overline{|\varepsilon_1|}\leqslant c_{21}$ and

$$\max(\overline{|\phi_3\varepsilon^m|}, \overline{|\phi_4\varepsilon^{-m}|})\leqslant c_{22}.$$

By (41) this implies $\max(\overline{|\phi_1\varepsilon^m|}, \overline{|\phi_2\varepsilon^{-m}|})\leqslant c_{23}$. Now it follows from (38), (39) and $\alpha\neq\beta$ that $\max(y, z)\leqslant c_{24}$. Hence, by (1), $x\leqslant c_{25}$.

(*c*) The proof is similar to that of (*b*).

(*a*) Assume $Q(x, y)\in S$. If n is odd, proceed in a similar way to (*b*) to conclude that $\max(x, y, z)\leqslant c_{26}$. Thus we may assume that n is even. Further, we may suppose that $n=2v$ with v odd and $v>2$. Re-write (1) as

$$(x^2)^v+(y^2)^v=(z^2)^v.$$

In view of theorem 11.4, we may assume that $Q(X, Y)\neq A(X^2+Y^2)$. Further, by lemma 11.1, there exist positive integers a_5, d_5 with $d_5\mid v$ such that

$$x^2+y^2=d_5^{-1}a_5^v.$$

Thus

$$(x+iy)(x-iy)=d_5^{-1}a_5^v.$$

Therefore, since $(x, y)=1$, we obtain

$$x+iy=d_6^{-1}a_6^v$$

where $a_6, d_6 \in \mathbb{Z}[i]$ with $\overline{|d_6|} \leqslant c_{27}$. Now proceed as in (*b*) to conclude that $\max(x, y, z) \leqslant c_{28}$. □

Proof of theorem 11.7

(*a*) If F contains a linear factor $\mathscr{L}$ in its factorisation over $\mathbb{Q}$, then $\mathscr{L}(x, y) \in S$ and we apply theorem 11.4. If F contains a quadratic factor in its factorisation over $\mathbb{Q}$, then we apply theorem 11.6. Otherwise F has at least three simple roots and, by theorem 7.1, it follows already from the fact that $F(x, y) \in S$, that $\max(x, y, z)$ is bounded. The proofs of (*b*) and (*c*) are similar. □

Now we turn to the proof of theorem 11.2. The proof of theorem 11.2 depends on the following lemma.

Lemma 11.7. *Let $M \in \mathbb{R}$. If x, y, z and $n>2$ satisfy* (1) *and $y<M^n$, then $n<C_{10}$ where C_{10} is a computable number depending only on M.*

Proof. Since there are no solutions of (1) with $4 \mid n$, it suffices to prove the theorem when n is odd. By c_{29}, c_{30}, c_{31} we shall denote positive computable numbers depending only on M. By (1), we have $z^n<2y^n$, hence $z<2M^n$. By lemma 11.1, we can write

$$z-x=d_1^{-1}a_1^n, \quad z-y=d_2^{-1}a_2^n, \quad x+y=d_3^{-1}a_3^n \tag{42}$$

where a_i and d_i are positive integers satisfying $d_i \mid n$ for $i=1,2,3$. We deduce $a_i^n \leqslant 4nM^n$, hence

$$a_i \leqslant 3M, \; d_i \leqslant n \quad (i=1,2,3).$$

We may therefore assume that a_i is fixed for $i=1,2,3$. By (42)

$$\left.\begin{aligned} 2x &= -d_1^{-1}a_1^n+d_2^{-1}a_2^n+d_3^{-1}a_3^n \\ 2y &= d_1^{-1}a_1^n-d_2^{-1}a_2^n+d_3^{-1}a_3^n \\ 2z &= d_1^{-1}a_1^n+d_2^{-1}a_2^n+d_3^{-1}a_3^n \end{aligned}\right\}. \tag{43}$$

Let $n>c_{29}$ with c_{29} sufficiently large. Then, by $x<y$ and $d_2 \leqslant$ n, we have $d_2^{-1}a_2^n<d_1^{-1}a_1^n$ and $a_2 \leqslant a_1$. Further, since $x>0$ and $d_1 \leqslant n$, it follows that $a_1 \leqslant a_3$.

Case 1. $a_3 > a_1$. Observe that $(X-A)^n + (X+A)^n \geqslant 2X^n$ for all real A, X with $X > A > 0$. Using this inequality, (1) and (43) we obtain

$$2(d_3^{-1}a_3^n)^n \leqslant (2x)^n + (2y)^n = (2z)^n \leqslant (d_3^{-1}a_3^n + (d_1^{-1} + d_2^{-1})a_1^n)^n.$$

Hence

$$(d_2^{-1} + d_1^{-1})a_1^n \geqslant d_3^{-1}a_3^n(\sqrt[n]{2} - 1)$$

which implies $n \leqslant c_{30}$.

Case 2. $a_3 = a_1$. By (42) we have $d_1(z-x) = d_3(x+y)$. By (1) and $(z, x) = 1$, we see that $(x+y,\ z-x) = 1$. Thus $y < x+y \leqslant d_1 \leqslant n$, which, together with lemma 11.3(*a*), implies that $n \leqslant c_{31}$. □

In the proof of theorem 11.2 we shall also apply the following lemma. This lemma is well known and can be found, for example, in the introduction of Shafarevich (1977).

Lemma 11.8. *Let $e(X), f(X), g(X)$ be relatively prime non-trivial polynomials in $\mathbb{C}[X]$, not all constant. Let $n > 2$ be a rational integer. Then*

$$(e(X))^n \pm (f(X))^n \neq (g(X))^n.$$

Proof. Suppose $(e(X))^n \pm (f(X))^n = (g(X))^n$. Without loss of generality we may assume that $\deg(g) \geqslant \max(\deg(e), \deg(f))$ and that e, f and g are pairwise relatively prime. We have

$$\left(\frac{e(X)}{f(X)}\right)^n \pm 1 = \left(\frac{g(X)}{f(X)}\right)^n.$$

Hence, by differentiation,

$$(e(X))^{n-1}(e'(X)f(X) - e(X)f'(X)) = (g(X))^{n-1}(g'(X)f(X) - g(X)f'(X)).$$

Since e and g are relatively prime, we obtain

$$(g(X))^{n-1} \mid (e'(X)f(X) - e(X)f'(X)).$$

Hence

$$(n-1)\deg(g) \leqslant \deg(e) + \deg(f) - 1 \leqslant 2\deg(g) - 1.$$

Consequently $n \leqslant 2$, a contradiction. □

Proof of theorem 11.2. By $c_{32}, \ldots, c_{37}$ we shall denote computable positive numbers depending only on E, F and G. Suppose n, t, u is a solution of (4) as specified in the theorem. There exists a rational integer b with $|b| \leqslant 3m$ such that the coefficients of X^m in $E(X, bX+Y)$, $F(X, bX+Y)$ and $G(X, bX+Y)$ are non-zero (cf. the proof of theorem 5.5). Hence we may assume without loss of generality that the coefficients of X^m in E, F and G are non-zero. By applying a similar argument to $E(X+aY, Y), F(X+aY, Y)$ and $G(X+aY, Y)$ with an appropriate rational integer a with $|a| \leqslant 3m$, we may further assume that the coefficients of Y^m in E, F and G are also non-zero. Let

$$\left.\begin{aligned} E(X,Y) &= e_0X^m + e_1X^{m-1}Y + \cdots + e_mY^m \\ F(X,Y) &= f_0X^m + f_1X^{m-1}Y + \cdots + f_mY^m \\ G(X,Y) &= g_0X^m + g_1X^{m-1}Y + \cdots + g_mY^m \end{aligned}\right\}. \tag{44}$$

Then $e_0e_mf_0f_mg_0g_m \neq 0$.

First we shall prove that n is bounded. If $e_0^n + f_0^n = g_0^n$ or $e_m^n + f_m^n = g_m^n$ then $n \leqslant c_{33}$ by corollary 1.2. If

$$(e_0^n + f_0^n - g_0^n)(e_m^n + f_m^n - g_m^n) \neq 0,$$

then, by (4) and (44),

$$u \mid (e_0^n + f_0^n - g_0^n)t^{mn} \quad \text{and} \quad t \mid (e_m^n + f_m^n - g_m^n)u^{mn},$$

hence, by $(t, u) = 1$,

$$\max(|t|, |u|) \leqslant \max(|e_0^n + f_0^n - g_0^n|, |e_m^n + f_m^n - g_m^n|) \leqslant c_{34}^n.$$

Thus

$$\max(|E(t,u)|, |F(t,u)|, |G(t,u)|) \leqslant c_{35}^n.$$

By (4) and lemma 11.7 with $M = c_{35}$, we obtain $n \leqslant c_{36}$.

In the sequel we may assume that n is fixed. Observe that $E^n + F^n - G^n$ is a binary form of degree mn or vanishes. In the former case, $(E(X,Y))^n + (F(X,Y))^n - (G(X,Y))^n$ can be decomposed into linear factors $\alpha_iX + \beta_iY$ over $\mathbb{C}$ and we find $\alpha_it + \beta_iu = 0$ for some i. Since the coefficients α_i, β_i are constants and $(t, u) = 1$, we obtain $\max(|t|, |u|) \leqslant c_{37}$ in this case. Now suppose $E^n + F^n = G^n$. We may divide by any common factor. By the conditions of the theorem it follows that it is no restriction to assume that E, F and G are relatively prime non-trivial binary forms in $\mathbb{Z}[X, Y]$, not all constant. Applying lemma 11.8 with $e(X) = E(X, 1)$, $f(X) = F(X, 1)$ and $g(X) = G(X, 1)$ we derive a contradiction. □

Notes

In these notes we restrict ourselves to publications which are of special interest in connection with the results proved in this chapter. For other results on the Fermat equation, see Ribenboim (1979). Other books on Fermat's Last Theorem are Bachmann (1919) and Edwards (1977).

Inkeri (1953) derived lower bounds for the solutions of

$$x^p + y^p = z^p \quad \text{with } (x, y, z) = 1,\ 0 < x < y \text{ and } p > 2, \tag{45}$$

in terms of prime p. He proved that if $p \nmid xyz$ then $x > ((2p^3+p)/\log(3p))^p$ and if $p \mid xyz$ then $x > p^{3p-4}$ and $y > \frac{1}{2}p^{3p-1}$ (see Rotkiewicz, 1960). Brindza, Győry and Tijdeman (1985) proved a similar bound for general exponents n (cf. lemma 11.7). They showed that (1) implies $x > n^{n/3}$. Turk (1983*a*) showed that every solution of (1) implies $n < \mathscr{P}$ and $z^n < \exp(\exp(C_{11}\mathscr{P}^3))$ where $\mathscr{P} = \min(P(x), P(y), P(z))$ and C_{11} is a computable absolute constant.

Inkeri (1946) proved that, for given p, there are only finitely many solutions (x, y, z) of (45) for which $y-x$ or $z-y$ is less than a given number M (cf. Everett, 1973; Inkeri, 1976). Stewart (1977*c*) proved that there is some computable upper bound C_{12} for all solutions n, x, y, z of (1) with $y-x<M$. Inkeri and van der Poorten (1980) refined Inkeri's result as follows: Let $p_1, \ldots, p_m$ be distinct primes less than p such that $p_i^{w_i} \mid y-x$ for $i=1,\ldots,m$. Then (45) implies that

$$(y-x)\Big/\prod_{i=1}^{m} p_i^{w_i} > (z-x)^{1-C_{13}(\log p)^3/(p-1)}$$

where C_{13} equals $1+p_1+\cdots+p_m$ multiplied by some computable absolute constant. Furthermore, they proved that (45) implies $z-x>p^{2p}$.

Lemma 11.4 is the main result of Tijdeman (1986). It is remarkable that Fermat's Last Theorem has not even been established for even exponents. Terjanian (1977) proved that for an odd prime p the equation $x^{2p}+y^{2p}=z^{2p}$ in positive integers x, y, z implies $2p \mid x$ or $2p \mid y$.

Consider the equation (cf. (5))

$$(E(t))^n + (F(t))^n = z^n \quad \text{in rational integers } n, t, z \tag{46}$$

where $E, F \in \mathbb{Z}[X]$ are non-constant relatively prime polynomials. Brindza (1984*c*) improved on a result of Inkeri (1976) by showing that, for given $n>2$, all solutions t, z of (46) with $E(t)F(t)z \neq 0$ satisfy $\max(|t|, |z|) \leqslant C_{14}$ where C_{14} is a computable number depending only on n, E and F. Brindza, Győry and Tijdeman (1985) generalised another result of Brindza (1984*c*) by proving that all solutions n, t, z of (46) with $n>2$ and $E(t)F(t)z \neq 0$ satisfy $\max(n, |t|, |z|) < C_{15}$ where C_{15} is a computable number depending only on E and F, provided that at least one of the following conditions holds:

(*a*) $E+F$ has at least two distinct zeros,
(*b*) the degrees of E and F are different,
(*c*) the leading coefficients of E and F are equal.

For Fermat's equation over function fields, see lemma 11.8, Shanks (1962, pp. 144–7), Gross (1966*a*, *b*), Greenleaf (1969) and Albis Gonzales (1975).

CHAPTER 12

The Catalan equation and related equations

Catalan (1844) conjectured that 8 and 9 are the only two consecutive positive integers which are both perfect powers. Here and elsewhere in this chapter we use integers for rational integers unless stated otherwise. Catalan's conjecture says that the equation

$$x^m - y^n = 1 \quad \text{in } m, n, x, y \in \mathbb{Z} \tag{1}$$
$$\text{with } m > 1, n > 1, x > 1, y > 1$$

has only one solution, $m = y = 2$, $n = x = 3$. Pillai (1945) conjectured that for given non-zero integers a, b and k, the more general equation

$$ax^m - by^n = k \quad \text{in } m, n, x, y \in \mathbb{Z} \tag{2}$$
$$\text{with } m > 1, n > 1, x > 1, y > 1$$

and $mn > 4$ has only finitely many solutions. Both conjectures are still open. By the results in this tract we can show that equation (2) has only finitely many solutions if m, n, x or y is fixed and $mn > 4$. This is a straightforward consequence of theorems 12.1 and 12.2.

Let $P \geqslant 2$ and denote by S the set of all integers which are composed of primes less than or equal to P. The first theorem is an extension of Shorey and Tijdeman (1976*a*, theorem 4(iii)).

Theorem 12.1. *Let $\tau > 0$. There exists a computable number C_1 depending only on P and τ such that the equation*

$$ax^m - by^n = k \tag{3}$$
$$\text{in } a \in S, b \in S, k \in S, x \in S, y \in \mathbb{Z}, m \in \mathbb{Z}, n \in \mathbb{Z}$$

with $m>1$, $n>1$, $x>1$, $y>1$ *and* $(ax^m,k)\leqslant\tau$ *implies that*

$$\max(|a|,|b|,|k|,m,n,x,y)\leqslant C_1.$$

The second theorem is an extension of Shorey *et al.* (1977, theorem 3).

Theorem 12.2. *Let* $\tau>0, m\in\mathbb{Z}, m>1$. *There exists a computable number* C_2 *depending only on m, P and* τ *such that the equation*

$$ax^m-by^n=k \tag{4}$$

in $a\in S$, $b\in S$, $k\in S$, $n\in\mathbb{Z}$, $x\in\mathbb{Z}$, $y\in\mathbb{Z}$

with $n>1$, $|x|>1$, $|y|>1$, $mn>4$ *and* $(ax^m,k)\leqslant\tau$ *implies that*

$$\max(|a|,|b|,|k|,n,x,y)\leqslant C_2.$$

The only values of a, b and k for which Pillai's conjecture has been proved are $a=b=k=1$, the case of Catalan's equation (1).

Theorem 12.3 (Tijdeman, 1976*b*). *There exists a computable absolute constant* C_3 *such that* (1) *implies that* $\max(m,n,x,y)\leqslant C_3$.

Van der Poorten (1977*b*) generalised this result to the equation $x^m-y^n=z^{\langle m,n\rangle}$ in positive integers $m>1, n>1, x>1, y>1, z\in S$ where $\langle m,n\rangle$ denotes the least common multiple of m and n. We shall give a further extension.

Theorem 12.4. *There exists a computable constant* C_4 *depending only on P such that the equation*

$$\left(\frac{x}{v}\right)^m-\left(\frac{y}{w}\right)^n=1 \tag{5}$$

in positive integers $m>1$, $n>1$, v,w,x,y

with $(x,v)=(y,w)=1$, $mn>4$ *and at least one of* v,w,x,y *in S implies that* $\max(m,n,v,w,x,y)\leqslant C_4$.

We conjecture that (5) with $(x,v)=(y,w)=1$ and $mn>4$ has only finitely many solutions. For a historical survey on consecutive powers, see Ribenboim (1984).

An equation which is closely related to (1) and has been studied by several authors is

$$\frac{x^m-1}{x-1}=y^n \quad \text{in integers } m>2, n>1, x>1, y>1. \tag{6}$$

Known solutions are $(m,n,x,y)=(4,2,7,20), (5,2,3,11)$ and $(3,3,18,7)$. We conjecture that (6) has only finitely many solutions, but we can only prove the following weaker result (cf. Shorey and Tijdeman, 1976*a*).

Theorem 12.5. *The equation* (6) *has only finitely many solutions if at least one of the following conditions is satisfied:*

(i) *x is fixed,*
(ii) *m has a fixed prime divisor,*
(iii) *y has a fixed prime divisor.*

Bounds for the solutions can be computed in each of the three instances.

The number-theoretical interpretation of (6) is which perfect powers in the x-adic number system are written as a repetition of digits 1. A more general question is which perfect powers in the x-adic number system are written by repeating the same digit a again and again. Obláth (1956) proved that the latter question has no solution for $x=10$, $1<a<x$. Inkeri (1972) determined all solutions for $1<a<x\leqslant 10$. The following result is a generalisation of one from Shorey and Tijdeman (1976*a*). It shows that for given x there are only finitely many perfect powers of the form $baa\cdots a$ in the x-adic system.

Theorem 12.6. *Let c and x be integers with $x>1$. There exists a computable constant C_5 depending only on c and x such that*

$$a\frac{x^m-1}{x-1}=y^n+c \quad \text{in integers } a\geqslant 1, m>2, n>1, y>1 \tag{7}$$

subject to $a<x$ and $a\neq -c(x-1)$ implies that $\max(m,n,y)\leqslant C_5$.

Note that if $1\leqslant a<x$ and $a=-c(x-1)$, then $a=x-1$, $c=-1$. Hence (7) becomes $x^m=y^n$. This equation has infinitely many solutions m,n,y.

Goormaghtigh posed the question which numbers have identical digits in two different number systems. The only known solutions of $x^{m-1}+x^{m-2}+\cdots+1=y^{n-1}+y^{n-2}+\cdots+1$ in integers $m>n>2$, $y>x>1$ are $(m,n,x,y)=(5,3,2,5)$ and $(13,3,2,90)$. The following result is a straightforward application of theorem 1.2.

Theorem 12.7. *Let x and y be integers with $y>x>1$. There exists a computable constant C_6 depending only on x and y such that the equation*

$$a\frac{x^m-1}{x-1}=b\frac{y^n-1}{y-1} \tag{8}$$

in integers $1\leqslant a<x,\ 1\leqslant b<y,\ m>1,\ n>1$

with $a(y-1)\neq b(x-1)$ *implies that* $\max(m,n)\leqslant C_6$.

Balasubramanian and Shorey (1980) generalised theorem 12.7 as follows.

Theorem 12.8. *There exists a computable constant* C_7 *depending only on* P *such that the equation*

$$a\frac{x^m-1}{x-1}=b\frac{y^n-1}{y-1} \tag{9}$$

in integers $a\geqslant 1, b\geqslant 1, m>1, n>1, x>1, y>1$

with a, b, x, y *in* S, $(a,b)=1$ *and* $a(y-1)\neq b(x-1)$ *implies that* $\max(a,b,m,n,x,y)\leqslant C_7$.

Proofs

Proof of theorem 12.1. The constants c_1, c_2, c_3 occurring in the proof are computable and depend only on P and τ. Assume that (3) holds with $m>1$, $n>1, x>1, y>1$ and $(ax^m,k)\leqslant\tau$. Without loss of generality we may assume that $d:=(ax^m,k)$ is fixed. Since $d\in S$, there exist integers a_1, b_1, k_1, x_1 in S and $y_1>0$ such that $a_1x_1^m=ax^m/d$, $b_1y_1^n=by^n/d$, $k_1=k/d$, $(a_1x_1^m,k_1)=1$ and

$$a_1x_1^m-b_1y_1^n=k_1. \tag{10}$$

If $y_1=1$, then we can apply corollary 1.2. If $y_1>1$, then we apply theorem 9.2 with $u_1=a_1x_1^m, u_2=-k_1, u=b_1, q=n, y=y_1$. In either case $|a_1|, |b_1|, |k_1|$, $|x_1^m|$ and $|y_1^n|$ are bounded by some constant c_1. Since $d\leqslant c_2$, this implies that a, b, k, m, n, x and y are bounded by some constant c_3. □

Proof of theorem 12.2. Without loss of generality we may assume that (4) holds with $n>1$, $x>1$, $y>1$, $mn>4$ and $(ax^m,k)\leqslant\tau$. If $y\in S$, then the assertion follows from theorem 12.1. We consider the case $y\notin S$. Hence, following the argument of the previous proof, we see that we may assume without loss of generality that $(ax^m,by^n)=(ax^m,k)=1$ and $y>1$. We may further assume that a is m-free. Write $k=k_1k_2^m$ where k_1 is m-free and $k_2\in S$. Since there are only finitely many possibilities for a and k_1, we may assume that they are fixed. We apply theorem 10.6 if $n>2$ and theorem 10.7 if $n=2$, with $\tau=1$ and $f(X,Z)=aX^m-k_1Z^m$, to the equation

$$ax^m-k_1k_2^m=-by^n.$$

We obtain that $|b|$, y, x, $|k_2|$ and n are bounded by a computable constant depending only on m, P and τ. This implies the statement of the theorem. □

Proof of theorem 12.3. In the proof $c_4, c_5, \ldots, c_{11}$ denote computable absolute constants. We shall deal with the equivalent equation

$$x^p - y^q = \varepsilon \quad \text{in integers } p \geqslant q > 1,\ x > 1,\ y > 1,\ \varepsilon \in \{-1, 1\}. \tag{11}$$

Without loss of generality we may assume that p and q are distinct primes. Moreover, by theorem 12.1 applied with x fixed and by theorem 12.2 applied with p fixed, it is no restriction to assume that $x > c_4$, $y > c_4$, $p > c_4$, $q > c_4$ where c_4 is some suitable large constant, and that p and q are odd. By (11) and $p > q$ we have

$$(x, y) = 1, \quad x < y \tag{12}$$

and

$$x^p = y^q + \varepsilon = (y + \varepsilon)(y^{q-1} - \varepsilon y^{q-2} + \varepsilon^2 y^{q-3} \cdots + \varepsilon^{q-1}).$$

Let $d = (y + \varepsilon, y^{q-1} - \varepsilon y^{q-2} + \cdots + \varepsilon^{q-1})$. Then $y \equiv -\varepsilon \pmod d$, hence $d \mid q$. If $q \mid (y + \varepsilon)$, then $(y^q + \varepsilon)/(y + \varepsilon)$ contains exactly one factor q. Thus there are integers $\delta \in \{-1, 0\}$ and $s > 0$ such that

$$y + \varepsilon = q^\delta s^p. \tag{13}$$

In a similar way we derive from

$$y^q = x^p - \varepsilon = (x - \varepsilon)(x^{p-1} + \varepsilon x^{p-2} + \cdots + \varepsilon^{p-1}),$$

that there exist integers $\gamma \in \{-1, 0\}$ and $r > 0$ such that

$$x - \varepsilon = p^\gamma r^q. \tag{14}$$

Note that $r > 1$, $s > 1$, if $\gamma = -1$ then $p \mid r$, if $\delta = -1$ then $q \mid s$. Hence $p^\gamma r^q \geqslant 2^{q-1}$ and $q^\delta s^p \geqslant 2^{p-1}$. On substituting (13) and (14) into (11) we obtain

$$(p^\gamma r^q + \varepsilon)^p - (q^\delta s^p - \varepsilon)^q = \varepsilon. \tag{15}$$

The crucial point for the proof is that r and s are nearly equal. We shall use the following estimates. From (14), (11) and (13) we infer that

$$2^p r^{pq} \geqslant (r^q + 1)^p + 1 \geqslant x^p + 1 \geqslant y^q \geqslant (q^\delta s^p - 1)^q \geqslant \frac{s^{pq}}{(2q)^q}.$$

These inequalities are also valid when p and q, r and s, x and y are interchanged. Hence, by $p > q \geqslant c_4$,

$$s \leqslant (4q)^{1/q} r \leqslant 2r, \quad r \leqslant (4p)^{1/q} s. \tag{16}$$

We shall first prove that a constant c_5 exists such that

$$q \leqslant c_5(\log p)^4. \tag{17}$$

It follows from (13), (14) and

$$\max((x-1)^p, (y-1)^q) < x^p = y^q + \varepsilon < \min((x+1)^p, (y+1)^q) \tag{18}$$

that

$$p^{\gamma p} r^{pq} - q^{\delta q} s^{pq} = (x-\varepsilon)^p - (y+\varepsilon)^q \neq 0. \tag{19}$$

In order to prove (17) we may assume that $p^\gamma r^q \geqslant 2^{q-1} > 12p^3$. We have, by (14), (11) and (13) respectively,

$$\left|\frac{x}{p^\gamma r^q} - 1\right| = \frac{1}{p^\gamma r^q}, \quad \left|\frac{y^q}{x^p} - 1\right| = \frac{1}{x^p}, \quad \left|\frac{y}{q^\delta s^p} - 1\right| = \frac{1}{q^\delta s^p}.$$

Recalling that if $|\alpha| \leqslant \frac{1}{2}$ then $|\log(1+\alpha)| \leqslant 2|\alpha|$, we obtain, by $-1 \leqslant \gamma \leqslant 0$, $-1 \leqslant \delta \leqslant 0$, $p > q$, (14) and (16),

$$|p \log(p^\gamma r^q) - p \log x| \leqslant 2p^{1-\gamma} r^{-q} \leqslant 2p^2 r^{-q},$$

$$|p \log x - q \log y| \leqslant 2x^{-p} \leqslant 2pr^{-q}, \tag{20}$$

$$|q \log y - q \log(q^\delta s^p)| \leqslant 2q^{1-\delta} s^{-p} \leqslant 2q^2 s^{-q} \leqslant 8p^3 r^{-q}. \tag{21}$$

Hence,

$$\Lambda_1 := |p \log(p^\gamma r^q) - q \log(q^\delta s^p)| \leqslant 12p^3 r^{-q}. \tag{22}$$

By (19), $\Lambda_1 = |p\gamma \log p - q\delta \log q + pq \log(r/s)| \neq 0$. On applying theorem B.1 to Λ_1 with $d=1$, $n=3$, $A_1 = A_2 = p$, $A_3 = 2r$, $B = p^2$ we obtain, by (16),

$$\Lambda_1 \geqslant \exp(-c_6(\log p)^4 \log r). \tag{23}$$

Comparing (23) with (22) we see that

$$r^q \leqslant 12p^3 r^{c_6(\log p)^4} < r^{c_7(\log p)^4}$$

which implies the assertion (17).

By another application of theorem B.1 we shall show that p is bounded. It follows from (14), (13) and (18) that

$$(p^\gamma r^q + \varepsilon)^p - q^{\delta q} s^{pq} = x^p - (y+\varepsilon)^q \neq 0. \tag{24}$$

We have, by (20) and (21),

$$|p \log x - q \log(q^\delta s^p)| \leqslant 2x^{-p} + 2q^2 s^{-p}.$$

Further, by (11) and (13), $x^p \geqslant y^q - 1 > 2^{q/2} y > 2qy > s^p$. Hence

$$\Lambda_2 := \left| -q\delta \log q + p \log \frac{p^\gamma r^q + \varepsilon}{s^q} \right| \leqslant 4q^2 s^{-p}. \tag{25}$$

We apply theorem B.1 to Λ_2 with $n=2$, $A_1=p$, $A_2=5ps^q$, $B=p$. Since $\Lambda_2 \neq 0$ in view of (24), we obtain, by (16),

$$\Lambda_2 \geqslant \exp(-c_8(\log p)^3 \log(5ps^q)). \tag{26}$$

Comparing (26) and (25) we find, by (17),

$$s^p \leqslant 4q^2(5ps^q)^{c_8(\log p)^3} < s^{c_9(\log p)^7}.$$

Hence $p \leqslant c_9(\log p)^7$. This implies that p and q are bounded by a computable number c_{10}.

We may now assume that p and q are fixed. It is a direct consequence of theorem 6.1 that there is a constant c_{11} such that $x \leqslant c_{11}$, $y \leqslant c_{11}$ for every solution of (11). Thus the total number of solutions of (1) is finite and there is a computable constant C_3 such that $\max(m, n, x, y) \leqslant C_3$. □

Proof of theorem 12.4. In this proof $c_{12}, c_{13}, \ldots, c_{28}$ denote computable constants which depend only on P. Suppose that (5) holds with $(x, v)=(y, w)=1$, $mn>4$ and at least one of v, w, x, y in S. Then

$$x^m w^n - y^n v^m = v^m w^n. \tag{27}$$

By $(x, v)=(y, w)=1$, we have $v^m \mid w^n$ and $w^n \mid v^m$, hence $v^m = w^n$. Choose $z>0$ such that $z^{\langle m,n\rangle} = v^m = w^n$. Then, on dividing by $v^m = w^n$ in (27),

$$x^m - y^n = z^{\langle m,n\rangle}. \tag{28}$$

By theorem 12.3 we may assume $z>1$. If p is a prime such that $p \mid x$ and $p \mid y$, then $p \mid z$, hence $p \mid v$. But $(x, v)=1$. Thus $(x, y)=1$. Further, at least one of x, y, z belongs to S.

If $x \in S$, then (28) can be written as

$$y^n + w^n \in S.$$

It follows from theorem 2.2 with $A=B=1$ that $n \leqslant c_{12}$. Then we apply theorem 12.2 to the equation $y^n + z^{\langle m,n\rangle} = x^m$ to conclude that $\max(m, v, w, x, y) \leqslant c_{13}$. If $y \in S$, then we can apply a similar argument.

The only remaining case is $z \in S$, which is in fact van der Poorten's theorem. If m is even and $n>2$, then the claim follows from theorem 12.2 applied to $(x^{m/2})^2 - y^n = z^{\langle m,n\rangle}$. If n is even and $m>2$, then the claim follows from theorem 12.2 applied to $(y^{n/2})^2 - x^m = -z^{\langle m,n\rangle}$. Thus we may assume, without loss of generality, that m and n are odd. It follows that m has an odd prime factor p and n has an odd prime factor q. If $p=q$, then (28) implies

$$x_1^q + y_1^q = z_1^q$$

with $x_1 = y^{n/q}$, $y_1 = z^{\langle m,n\rangle/q}$, $z_1 = x^{m/q}$, $y_1 \in S$ and the result follows from theorem 11.3(*b*) and theorem 7.1. Without loss of generality we may consider in the sequel

$$x^p - y^q = z^{pq} \quad \text{in integers } p>q>2,\ x>1,\ y>1,\ z\in S \tag{29}$$

with $(x, y)=1$, p and q prime. It will suffice to show that $\max(p, q, x, y, |z|)\leqslant c_{14}$ for some constant c_{14}.

We have

$$y^q + z^{pq} = (y+z^p)\left(\frac{y^q+z^{pq}}{y+z^p}\right).$$

Let d be the greatest common divisor of the two factors on the right-hand side. Then $y\equiv -z^p \pmod d$, hence, by $(y, z)=1$, $d\,|\,q$. If $q\,|\,(y+z^p)$, then $(y^q+z^{pq})/(y+z^p)$ contains at most one factor q. Thus there are integers $\delta\in\{-1,0\}$ and $s\neq 0$ such that

$$y+z^p = q^\delta s^p. \tag{30}$$

Similarly, we derive the existence of integers $\gamma\in\{-1,0\}$ and $r\neq 0$ such that

$$x - z^q = p^\gamma r^q. \tag{31}$$

Note, by distinguishing $z>0$ and $z<0$, that $|p^\gamma q^\delta r^q s^p|\geqslant 2$, whence $|rs|\geqslant 2$. Furthermore, $q^{-\delta}\,|\,s$ and $p^{-\gamma}\,|\,r$. On substituting (30) and (31) into (29) we obtain

$$(p^\gamma r^q + z^q)^p - (q^\delta s^p - z^p)^q = z^{pq}. \tag{32}$$

Our first object is to show by p-adic methods that $|z|$ is relatively small. Suppose $p_1^l \,\|\, z$, $l>0$, p_1 prime. Plainly, by (32),

$$z^q\,|\,(p^{\gamma p} r^{pq} - q^{\delta q} s^{pq}), \tag{33}$$

hence,

$$\operatorname{ord}_{p_1}(p^{\gamma p} r^{pq} - q^{\delta q} s^{pq}) \geqslant lq. \tag{34}$$

It follows from (29) and

$$\max((x-|z|^q)^p, (y-|z|^p)^q) < x^p = y^q + z^{pq} < \min((x+|z|^q)^p, (y+|z|^p)^q) \tag{35}$$

that

$$p^{\gamma p} r^{pq} - q^{\delta q} s^{pq} = (x-z^q)^p - (y+z^p)^q \neq 0. \tag{36}$$

We apply theorem B.3 to $\Lambda_3 := p^{\gamma p} q^{-\delta q} (r/s)^{pq} - 1$ with $n=3$, $A_1=A_2=p$, $A_3=2|rs|$, $B=p^2$. Since $\Lambda_3\neq 0$ in view of (36), we obtain that

$$\operatorname{ord}_{p_1}(\Lambda_3) \leqslant c_{15} (\log p)^4 \log(|rs|).$$

Comparing this with (34) we see that

$$lq \leqslant c_{15} (\log p)^4 \log(|rs|).$$

Let $|z| = \prod_j p_j^{l_j}$. Then

$$|z|^q = \prod_j p_j^{l_j q} \leqslant (|rs|)^{c_{16}(\log p)^4}. \tag{37}$$

Next we intend to show that

$$q \leqslant c_{17}(\log p)^4 \tag{38}$$

for some computable constant c_{17}. Thus we may assume that

$$q > c_{18}(\log p)^4 \quad \text{and} \quad p > c_{19}, \tag{39}$$

where $c_{18} > 6c_{16}$ and c_{19} are suitable large constants. In order to derive inequalities like (16) we use (39) repeatedly without further reference. Assume $|r| \leqslant |s|$. Then $|s| > 1$ and, by (37) and (39),

$$|z| \leqslant |s|^{2c_{16}(\log p)^4/q} \leqslant |s|^{1/3}. \tag{40}$$

Hence, by (30) and $s > 1$,

$$y \geqslant q^\delta s^p - |z|^p > q^\delta s^p - s^{p/3} > \tfrac{1}{2} q^\delta s^p.$$

By (31) and (29) it follows that

$$(p^\gamma r^q + z^q)^p = x^p = y^p + z^{pq} > (\tfrac{1}{2} q^\delta s^p)^q - s^{pq/3} > (\tfrac{1}{2} q^\delta)^p s^{pq}.$$

Hence

$$p^\gamma r^q > \tfrac{1}{2} q^\delta s^q - z^q > \tfrac{1}{2} q^\delta s^q - s^{q/3} > \tfrac{1}{4} q^\delta s^q.$$

In particular,

$$s \leqslant (4q)^{1/q} r < 2r \tag{41}$$

and $r > 1$. By a similar reasoning we find that if $|s| \leqslant |r|$, then $r > 1$ and $r \leqslant (4p)^{1/q} s$. Combining this with (41) we see that in both cases

$$\begin{gathered} r > 1, \quad s > 1, \quad |z|^3 \leqslant \max(r, s), \\ (4p)^{-1/q} r \leqslant s \leqslant (4q)^{1/q} r < 2r. \end{gathered} \tag{42}$$

We have, by (31) and (42),

$$\left| \frac{x}{p^\gamma r^q} - 1 \right| = \left| \frac{z^q}{p^\gamma r^q} \right| \leqslant 4pq r^{q/3} r^{-q} \leqslant 4p^2 r^{-2q/3},$$

by (29), (42) and (31),

$$\left| \frac{y^q}{x^p} - 1 \right| = \left| \frac{z^{pq}}{x^p} \right| \leqslant \frac{(4q)^p r^{pq/3}}{r^{2pq/3}} = (4q)^p r^{-pq/3},$$

and by (30), (40) and (42),

$$\left|\frac{y}{q^\delta s^p}-1\right|=\left|\frac{z^p}{q^\delta s^p}\right|\leqslant q(4p)^{p/q}r^{-2p/3}.$$

Recalling that if $|\alpha|\leqslant\frac{1}{2}$ then $|\log(1+\alpha)|\leqslant 2|\alpha|$, we obtain

$$|p\log(p^\gamma r^q)-p\log x|\leqslant 8p^3r^{-2q/3},$$

$$|p\log x-q\log y|\leqslant 2\frac{|z|^{pq}}{x^p}\leqslant(8pr^{-q/3})^p,$$

$$|q\log y-q\log(q^\delta s^p)|\leqslant 2\frac{|z|^p}{q^\delta s^p}\leqslant(8p^3r^{-2q/3})^{p/q}.$$

Hence,

$$\Lambda_4:=|\gamma p\log p-\delta q\log q+pq\log(r/s)|\leqslant 24p^3r^{-q/3}. \tag{43}$$

By (36), $\Lambda_4\neq 0$. On applying theorem B.1 to Λ_4 with $n=3$, $A_1=A_2=p$, $A_3=2r$, $B=p^2$ we obtain, by (41),

$$\Lambda_4\geqslant\exp(-c_{20}(log\ p)^4\log r). \tag{44}$$

Comparing (44) with (43) we find

$$r^{q/3}\leqslant 24p^3r^{c_{20}(\log p)^4}<r^{c_{21}(\log p)^4},$$

which implies the assertion (38).

Our third object is to improve upon estimate (37). Suppose $p_1^l\parallel z, l>0$. By (32), we have

$$z^p\,|\,((p^\gamma r^q+z^q)^p-q^{\delta q}s^{pq}).$$

Put $\Lambda_5=(p^\gamma r^q+z^q)^pq^{-\delta q}s^{-pq}-1$. We have, by $(x,z)=1$,

$$\mathrm{ord}_{p_1}(\Lambda_5)\geqslant lp. \tag{45}$$

Suppose $\Lambda_5=0$. Then, by (32), $(q^\delta s^p-z^p)^q=(q^\delta s^p)^q-z^{pq}$. Since $q^\delta s^p-z^p=y>0$, this is impossible. Thus $\Lambda_5\neq 0$. We apply theorem B.3 to $\Lambda_5=q^{-\delta q}((p^\gamma r^q+z^q)/s^q)^p-1$ with $n=2$, $A_1=p$, $A_2=x|s|^q$, $B=p$. By (31) we obtain

$$\mathrm{ord}_{p_1}(\Lambda_5)\leqslant c_{22}(\log p)^3\log(x|s|^q). \tag{46}$$

Comparing (46) with (45) we see that

$$p_1^{lp}\leqslant(x|s|^q)^{c_{22}(\log p)^3\log p_1}.$$

On taking the product of these inequalities for all prime divisors of z, we obtain that

$$|z|^p\leqslant(x|s|^q)^{c_{23}(\log p)^3}. \tag{47}$$

We now want to show that p is bounded. Suppose $|s|^q \leqslant x$. Then, by (29) and (30),

$$x^p = y^q + z^{pq} = (q^\delta s^p - z^p)^q + z^{pq} \leqslant (|s|^p + |z|^p)^q.$$

Here we use that q is odd and therefore the terms $-z^{pq}$ and z^{pq} cancel. Hence, by (47) and (38).

$$x^{p/q} - |s|^p \leqslant |z|^p \leqslant x^{2c_{23}(\log p)^3} < \tfrac{1}{2} x^{p/q}.$$

Here and in the sequel we suppose that p is sufficiently large. It follows that $x^{p/q} < 2|s|^p$, which implies that $x < 2|s|^q$. Thus unconditionally

$$x \leqslant 2|s|^q. \tag{48}$$

Hence, by (47) and (38),

$$|z|^p \leqslant (2|s|)^{2c_{23}q(\log p)^3} \leqslant (2|s|)^{c_{24}(\log p)^7}. \tag{49}$$

By $|z| > 1$, we have $|s| > 1$. Since $y > 0$ and p is odd, we infer from (30) that $s > 1$. We have, by (29), (49) and (30),

$$\left|\frac{x^p}{y^q} - 1\right| = \left|\frac{z^{pq}}{y^q}\right| \leqslant \frac{(2s)^{c_{24}q(\log p)^7}}{(q^\delta s^p - z^p)^q}$$

$$\leqslant \left(\frac{(2s)^{c_{24}(\log p)^7}}{s^{3p/4}}\right)^q < \tfrac{1}{4} s^{-pq/2},$$

and, by (30),

$$\left|\frac{y}{q^\delta s^p} - 1\right| = \frac{|z^p|}{q^\delta s^p} \leqslant q(2s)^{c_{24}(\log p)^7} s^{-p} < \frac{1}{4q} s^{-p/2}.$$

Hence,

$$|p \log x - q \log y| < \tfrac{1}{2} s^{-pq/2},$$

and

$$|q \log y - q \log(q^\delta s^p)| < \tfrac{1}{2} s^{-p/2}.$$

So we obtain

$$\Lambda_6 := |-\delta q \log q + p \log(x/s^q)| < s^{-p/2}. \tag{50}$$

We apply theorem B.1 to Λ_6 with $n = 2$, $A_1 = p$, $A_2 = 2s^q$, $B = p$. By (32) we have $\Lambda_6 \neq 0$. It follows, by (48), that

$$\Lambda_6 \geqslant \exp(-c_{25}(\log p)^3 q \log s). \tag{51}$$

Comparing (51) with (50) we see that, by (38),

$$s^{p/2} \leqslant s^{c_{25}q(\log p)^3} \leqslant s^{c_{26}(\log p)^7}.$$

Thus $p \leqslant 2c_{26}(\log p)^7$, which implies that $p \leqslant c_{27}$. Hence $q \leqslant c_{27}$.

We have reduced (29) to finitely many cases with p, q fixed. On applying theorem 7.1 we find that for each pair p, q there are only finitely many integer solutions x, y of (29). Thus $\max(p, q, x, y, |z|) \leqslant c_{28}$. □

Proof of theorem 12.5. Assume that (6) holds.

(i) Suppose x is fixed. Apply theorem 9.2 with $u_1 = x^m, u_2 = -1, u = x-1$. It follows that m, n and therefore y are bounded.

(ii) Let $d > 1$ be a fixed divisor of m. Put

$$A = \frac{x^m - 1}{x^{m/d} - 1}, \quad B = \frac{x^{m/d} - 1}{x - 1}. \tag{52}$$

Then $AB = y^n$ and $(A, B) \mid d$. Hence there exist positive integers r, s, y_1, y_2 with $(r, s) = 1$, $rs \leqslant d$ such that

$$A = \frac{r}{s} y_1^n, \quad B = \frac{s}{r} y_2^n. \tag{53}$$

Without loss of generality we may assume that r and s are fixed. Put $z = x^{m/d}$. Then

$$s\frac{z^d - 1}{z - 1} = s(z^{d-1} + z^{d-2} + \cdots + 1) = ry_1^n. \tag{54}$$

We distinguish three cases.

Case 1. $d \geqslant 4$. On applying theorem 10.7 with $f(X, Z) = s(X^{d-1} + X^{d-2}Z + \cdots + Z^{d-1})$, $\gamma = r$, $m = n$, $X = z$, $Y = y_1$, $Z = 1$, we obtain that there are only finitely many solutions n, $y_1 > 1$, $z = x^{m/d}$ of (54). Obviously there are only finitely many solutions with $y_1 = 1$. Thus there are only finitely many solutions of (6).

Case 2. $d = 2$. If m is divisible by 4, then the assertion follows from case 1 with $d = 4$. So we may assume that m is an odd multiple of 2. If $(A, B) = 1$, then $r/s = 1$ and $A = x^{m/2} + 1 = y_1^n$. Theorem 12.3 implies that there are only finitely many solutions of this equation. If $(A, B) > 1$, then $(A, B) = 2$ and x is odd. By (52) we have $4 \mid B(x-1) = x^{m/2} - 1$, hence $x \equiv 1 \pmod 4$. We have

$$\frac{x^m - 1}{x^2 - 1}(x + 1) = y^n.$$

Since $x + 1 \equiv 2 \pmod 4$ and $n > 1$, the quotient $(x^m - 1)/(x^2 - 1)$ is even.

However,

$$\frac{x^m-1}{x^2-1}=1+x^2+\cdots+x^{m-2}\equiv\frac{m}{2}\pmod 2$$

is odd.

Case 3. $d=3$. On applying theorem 10.6 we obtain that (54) has only finitely many solutions $n>2$, $y_1>1$ and $z=x^{m/d}$. Obviously there are only finitely many solutions with $y_1=1$. We may therefore restrict our attention to the case $n=2$. Since case 2 can be applied if m is even, we may also assume that m is odd. Since case 1 can be applied if $9\mid m$, we may further assume that $m/3$ is not divisible by 3. If $(A,B)=1$, then $r=s=1$ and $A=1+z+z^2=y_1^2$. Since $z^2<z^2+z+1<(z+1)^2$, this is impossible. If $(A,B)>1$, then $(A,B)=3$ and $r=3$, $s=1$. Hence

$$B=\frac{x^{m/3}-1}{x-1}=3^{n-1}\left(\frac{y_2}{3}\right)^n.$$

If $x\equiv1\pmod 3$, then $3\mid(m/3)$ because of $3\mid B$, a contradiction. If $x\equiv2\pmod 3$, then $2\mid(m/3)$, again a contradiction. Since $3\mid x$ is plainly impossible, there are no solutions at all if $d=3$, $n=2$, $2\nmid m$, $9\nmid m$.

(iii) Suppose that y has a fixed prime factor p. If $p\mid(x-1)$, then

$$\frac{x^m-1}{x-1}=x^{m-1}+x^{m-2}+\cdots+1\equiv m\pmod p.$$

Since $p\mid(x^m-1)/(x-1)$, we see that m has a fixed prime factor and we can apply (ii). If $p\nmid(x-1)$, then let $p\mid(x^t-1)$ with $t>1$ minimal. Since $t\mid m$ and $t\mid(p-1)$, t is a bounded divisor of m. As $t>1$, we can apply (ii) again.

Proof of theorem 12.6. By c_{29}, c_{30}, c_{31} we shall denote computable positive constants depending only on c and x. Suppose (7) holds subject to $a<x$ and $a\neq-c(x-1)$. We may assume that a is fixed. By (7),

$$ax^m-(cx+a-c)=(x-1)y^n.$$

We apply theorem 9.2 with $u_1=ax^m$, $u_2=-(cx+a-c)$, $u=x-1$ and $q=n$. Note that $(ax^m, cx+a-c)\leqslant c_{29}$, since $cx+a-c$ is fixed and non-zero. We infer $\max(m,n)\leqslant c_{30}$. Since the left-hand side of (7) is bounded, we obtain $y\leqslant c_{31}$. □

Proof of theorem 12.7. Assume that (8) holds with $a(y-1)\neq b(x-1)$. We may assume that a and b are fixed. We have

$$a(y-1)x^m-b(x-1)y^n-(a(y-1)-b(x-1))=0.$$

The common factor of the three terms is bounded, since the third term is bounded and non-zero. On dividing by the common factor and applying corollary 1.2 we obtain that m and n are bounded. □

Proof of theorem 12.8. Assume that (9) holds with a, b, x, y in S, $(a, b) = 1$ and $a(y-1) \neq b(x-1)$. Write

$$x = p_1^{x_1} \cdots p_s^{x_s}, \quad y = p_1^{y_1} \cdots p_s^{y_s},$$
$$a = p_1^{a_1} \cdots p_s^{a_s}, \quad b = p_1^{b_1} \cdots p_s^{b_s}.$$

Here $p_1, \ldots, p_s$ are primes less than or equal to P and the exponents of $p_1, \ldots, p_s$ in the factorisations of x, y, a, b are non-negative integers not exceeding $2 \log x$, $2 \log y$, $2 \log(a+1)$, $2 \log(b+1)$, respectively. It involves no loss of generality to assume that $m \geqslant n$. Further, (9) implies

$$ax^{m-1} < nby^{n-1}, \quad by^{n-1} < max^{m-1}. \tag{55}$$

By $c_{32}, c_{33}, \ldots, c_{48}$ we shall denote computable positive constants depending only on P.

Lemma 12.1. $\max(\log a, \log b) \leqslant c_{32}(\log(m \log x))^2$.

Proof. We prove the inequality for $\log b$. Suppose $p \mid b$. Then, by (9), we have

$$\operatorname{ord}_p(b) \leqslant \operatorname{ord}_p\left(a \frac{x^m - 1}{x - 1}\right) \leqslant \operatorname{ord}_p(x^m - 1) = \operatorname{ord}_p(p_1^{mx_1} \cdots p_s^{mx_s} - 1).$$

On applying theorem B.3 with $n = s \leqslant P$, $A_1 = A_2 = \cdots = A_n = P$ and $B = 2m \log x$ to the right-hand side of the above inequality, we obtain

$$\operatorname{ord}_p(b) \leqslant c_{33}(\log(m \log x))^2,$$

hence

$$\log b = \sum_{p \mid b} \operatorname{ord}_p(b) \log p \leqslant c_{34}(\log(m \log x))^2.$$

Similarly

$$\log a \leqslant c_{35}(\log(n \log y))^2.$$

In view of (55) the lemma follows immediately. □

Lemma 12.2. $\min(\log x, \log y) \leqslant c_{36} \log m$.

Proof. We prove the lemma for $x \leqslant y$. The proof for the case $x > y$ is similar. Let δ be the smallest positive integer such that $ax^{m-\delta} \neq by^{n-\delta}$. Observe that $\delta \leqslant 2$, since $a(y-1) \neq b(x-1)$ and $(a, b) = 1$. Now it follows from (9) that

$$ax^{m-\delta} + ax^{m-\delta-1} + \cdots + a = by^{n-\delta} + by^{n-\delta-1} + \cdots + b,$$

hence, by (55),

$$0<|ax^{m-\delta}-by^{n-\delta}|\leqslant max^{m-\delta-1}+nby^{n-\delta-1}\leqslant 2m^2ax^{m-\delta-1}.$$

Thus we have

$$0<|p_1^{u_1}\cdots p_s^{u_s}-1|\leqslant 2m^2x^{-1}, \tag{56}$$

where $u_i=b_i-a_i+(n-\delta)y_i-(m-\delta)x_i$ for $i=1,\ldots,s$. By lemma 12.1 and (55), we find that the integers $|u_i|$ do not exceed $c_{37}m\log x$. Now apply theorem B.1 with $n=s\leqslant P$, $A_1=A_2=\cdots=A_n=P$ and $B=c_{37}m\log x$ to obtain

$$|p_1^{u_1}\cdots p_s^{u_s}-1|>(m\log x)^{-c_{38}}. \tag{57}$$

Comparing (57) with (56) we obtain lemma 12.2 by transferring secondary factors. □

Lemma 12.3. $\max(\log x, \log y)\leqslant c_{39}(\log m)^2$.

Proof. We prove the lemma for $x\leqslant y$. The proof for the case $x>y$ is similar. From (9), lemma 12.1 and (55) we have

$$0<\frac{ax^m}{x-1}-by^{n-1}=\frac{a}{x-1}+b(y^{n-2}+\cdots+1)$$

$$\leqslant a+nby^{n-2}\leqslant y^{n-2}\exp(c_{40}(\log(n\log y))^2).$$

Thus

$$0\neq(p_1^{v_1}\cdots p_s^{v_s}(x-1)^{-1}-1)\leqslant y^{-1}\exp(c_{40}(\log(n\log y))^2) \tag{58}$$

where $v_i=a_i-b_i+mx_i-(n-1)y_i$ for $i=1,\ldots,s$. From lemma 12.1 and (55) we observe that the absolute values of v_i with $i=1,\ldots,s$ do not exceed $c_{41}m\log x$ which, by lemma 12.2, is less than $c_{42}m\log m$. Now apply theorem B.1 with $n=s+1\leqslant P+1$, $A_1=\cdots=A_{n-1}=P$, $A_n=x\leqslant m^{c_{36}}$ and $B=c_{42}m\log m$ to conclude that

$$p_1^{v_1}\cdots p_s^{v_s}(x-1)^{-1}-1\geqslant\exp(-c_{43}(\log m)^2). \tag{59}$$

Now lemma 12.3 follows immediately from (59), (58), (55) and lemma 12.1. □

From (9) and lemma 12.1 we have

$$0\neq\left|\frac{ax^m}{x-1}-\frac{by^n}{y-1}\right|=\left|\frac{a}{x-1}-\frac{b}{y-1}\right|\leqslant\exp(c_{44}(\log(m\log x))^2).$$

Thus, by lemma 12.3,

$$0<\left|p_1^{w_1}\cdots p_s^{w_s}\frac{x-1}{y-1}-1\right|\leqslant x^{-m+1}\exp(c_{45}(\log m)^2). \tag{60}$$

where $w_i=b_i-a_i+ny_i-mx_i$. Observe that $|w_i|\leqslant c_{46}m(\log m)^2$ for $i=1,\ldots,s$. Now apply theorem B.1 with $n=s+1\leqslant P+1$, $A_1=\cdots=A_{n-1}=P$, $A_n=\max(x-1,\ y-1)\leqslant\exp(c_{39}(\log m)^2)$ and $B=c_{46}m(\log m)^2$. We obtain

$$\left|p_1^{w_1}\cdots p_s^{w_s}\frac{x-1}{y-1}-1\right|>\exp(-c_{47}(\log m)^3). \tag{61}$$

Combining (61) and (60) we find that $m\leqslant c_{48}$. This completes the proof of theorem 12.8 in view of lemma 12.3 and lemma 12.1. □

Notes

Theorem 12.1 implies that (1) has only finitely many solutions if x or y is fixed. Hyyrö (1964*a*) improved upon a result of Rotkiewicz (1961) by showing that for all solutions of (1) apart from

$$(m,n,x,y)=(2,3,3,2) \tag{62}$$

we have $\min(x,y)>10^{11}$. The theory of linear forms in logarithms can be applied to show that if $m>1$, $n>1$, $x\geqslant 1$, $y\geqslant 1$ are positive integers with $x^m\neq y^n$, then

$$|x^m-y^n|>(x^m)^{1-C_8(\log m)/m}$$

and

$$|x^m-y^n|>C_9(x^m)^{C_{10}}$$

where C_8, C_9 and C_{10} are certain computable positive numbers depending only on x. See Shorey and Tijdeman (1976*a*, theorems 1, 2) and Turk (1986).

It follows from a result of LeVeque (1952) that (1) has no solutions with $|x-y|=1$ apart from (62). Schinzel (1956) gave a simpler proof of this fact. Rotkiewicz (1956) generalised this result to the equation $x^m-y^n=a^n$ subject to $|x-y|=a$. By estimating linear forms in logarithms it can be proved that, for given non-zero integers a,b,k, equation (2) has only finitely many solutions for which $|x-y|$ is bounded. If $m=n$, then the result follows (unconditionally) from theorem 2.1. If $m\neq n$, then a simple estimation yields that x/m and y/n are bounded from above. By estimating $|ax^m/by^n-1|$ from below by corollary B.1 and from above in a trivial way and comparing both estimates it follows that m and x, hence n and y, are bounded. Theorem 12.2 implies that (1) has only finitely many solutions if m or n is fixed. Lebesgue (1850) proved that (1) has no solutions when $n=2$. Chao Ko

(1965) proved that (1) has no solutions apart from (62) when $m=2$. Nagell (1921) showed that (62) is the only solution of (1) if $m=3$ or $n=3$. Hence there is no solution with $\min(m,n)<5$ other than (62). Inkeri (1964) proved that (1) is unsolvable for a large number of pairs m,n. Evertse (1983*b*) improved upon an estimate of Hyyrö (1964*a*) by showing that for fixed m and n the number of solutions of (1) is at most $(mn)^{\min(m,n)}$. Turk (1986) applied linear forms in logarithms to prove that there exists a computable absolute constant $C_{11}>0$ such that, for all integers $m>1, n>1, x>1, y>1$ with $x^m \neq y^n$,

$$|x^m - y^n| > \frac{1}{n}\exp\left(\frac{C_{11}}{n^2}(\log m \log\log(m+1))^{1/2}\right).$$

For lower estimates of $|ax^m - by^n|$, see the notes of chapter 10. Shorey and Tijdeman (1976*a*) proved that for given non-zero integers a, b, k equation (2) has only finitely many solutions if $m \mid n$ and $mn>4$. Ribenboim (198x) proved that, for any $k \neq 0$, the density of pairs (m,n) such that $x^m - y^n = k$ has a non-trivial solution x, y is zero. For integers $a>0, b>0, k\neq 0, x>1$ and $y>1$, Shorey (1986*c*) proved that there are at most nine distinct pairs (m,n) in positive integers satisfying (2) and $\max(ax^m, by^n) > 953\,k^6$. If $x \geqslant 4$ and $y \geqslant 4$, Shorey (1986*c*) derived from the above result that there are at most nine distinct pairs (m,n) in integers $m \geqslant 3$ and $n \geqslant 3$ satisfying (2) with $k=1$.

Langevin (1976*b*) elaborated the proof of theorem 12.3 to show that if (1) holds, then

$$x^m < \exp\exp\exp\exp(730) \quad \text{and} \quad P(mn) < \exp(241).$$

Cassels (1960*a*) proved that if (1) holds for x, y and primes m, n then $m \mid y$ and $n \mid x$. This result was used by Makowski (1962) and Hyyrö (1963) to prove that no three consecutive positive integers can be all perfect powers. Turk (1980*a*) used estimates of linear forms in logarithms to derive an upper bound for the number of perfect powers in the interval $[n, n+\sqrt{n}]$. Loxton (198x) used his simultaneous version of theorem B.1 to improve on this bound by showing that there are at most $\exp(40(\log_2 n \log_3 n)^{1/2})$ perfect powers in the interval $[n, n+\sqrt{n}]$ for $n \geqslant 20$. For comparable results on numbers ax^m with $|a|$ small, see Turk (1984) and Loxton (1986).

In theorem 12.4 the condition that x and y are positive is superfluous (see Tijdeman, 1985). In this paper it is also shown that if (5) holds and $t=\max(m,n)$, then

$$|x| \leqslant C_{12} v^{1+C_{13}(\log t)^4/t^{1/2}}, \quad |y| \leqslant C_{12} w^{1+C_{13}(\log t)^4/t^{1/2}}$$

where C_{12} and C_{13} are certain (ineffective) absolute constants.

Obláth (1956) showed that $a.x^{m-1} + a.x^{m-2} + \cdots + a$ is never a perfect power y^n if $m \geqslant 2$, $x = 10$ and $1 < a < 10$. If there is some solution with $a = 1$, then $2 \nmid m$, $2 \nmid n$, $3 \nmid m$, $3 \nmid n$. Shorey and Tijdeman (1976*a*) proved that in this case $n \geqslant 23$. Inkeri (1972) determined all solutions for $1 < a < x \leqslant 10$ and many other pairs (a, x). All solutions of (6) have been determined if $4 \mid m$ by Nagell (1920) and if $2 \mid n$ or $3 \mid m$ by Ljunggren (1943). Further, Ljunggren (1943) determined all the solutions of (6) if $n = 3$ and $m \not\equiv 5 \pmod 6$. Shorey (1986*a*, 1986*b*) proved that (6) has only finitely many solutions if n is prime and $\omega(m) > n - 2$. This, together with theorem 12.5(ii), implies that (6) has only finitely many solutions if $n = 3$ and $m \not\equiv 5 \pmod 6$ whenever $\omega(m) = 1$. Richter (1982) considered equation (6) with $m = n$. For fixed integers a and b with $(a, b) = 1$ and a n-free, Shorey and Tijdeman (1976*a*) gave a number of conditions under which the equation

$$a\frac{x^m - 1}{x - 1} = by^n \quad \text{in integers } m > 2,\ n > 1,\ x > 1,\ y > 1$$

has only finitely many solutions. One would like to show that (6) has only finitely many solutions. For this, it suffices to restrict the variable m to prime powers (see Shorey, 1986*a*). For integers X, A with $1 \leqslant A < X$ and a prime $p \geqslant 5$, Shorey (1986*b*) showed that the number of pth powers whose all the digits are equal to A in the X-adic number system is at most $p + C_{14}$ where C_{14} is a computable absolute constant. If $A = 1$, the restriction $p \geqslant 5$ is not necessary.

Győry, Kiss and Schinzel (1981) and Győry (1982*a*) considered the equation

$$\frac{x^m - y^m}{x - y} = z \quad \text{in integers } m > 3,\ x > y \geqslant 1,\ z > 0$$

with $(x, y) = 1$ (cf. chapter 3, notes). Győry (1982*a*) proved that $m \leqslant P$ and

$$\max(x, y, z) < \exp\{s^{sP(P+30)/2}(20P^2)^{sP(P+6)+14(P+2)}\}$$

where $P = P(z)$ and $s = \omega(z)$. Loxton (1986) proved that for any $\varepsilon > 0$ and any positive integer z the equation $(x^m - 1)/(x - 1) = z$ has at most $C_{15}(\log z)^{(1/2)+\varepsilon}$ solutions in integers m, x with $m > 2$ and $x > 1$ where C_{15} is a computable number depending only on ε.

Davenport, Lewis and Schinzel (1961) proved that equation (9) has only finitely many solutions if $a = b = 1$ and m and n are fixed. If in (9) a, b, n and x are fixed, then it follows from theorem 10.4 that there is only a finite number of solutions. Shorey (1984*b*) proved, effectively, that equation (9) has only finitely many solutions in integers $x > 1, y > 1, m \geqslant n > 1, m > 2, 1 \leqslant a < x, 1 \leqslant b < y$ with $(a, b) = 1$, $a(y-1) \neq b(x-1)$ and $|x - y|$ bounded. Further, if $a =$

$b=1$, equation (9) has only finitely many solutions in integers $m>1, n>1$, $x>1$ and $y>1$ with $x\neq y$ and $P(x(y-x))$ bounded. It follows from a result of Shorey (1984*c*) that equation (9) with $m=n$, $a<b$ and $x>y$ implies that either $m\leqslant C_{16}$ or $m=[\alpha+1]$ where C_{16} is a computable number depending only on a and b and $\alpha=\log(b/a)/\log(x/y)$. For given integers $a\geqslant 1$, $b\geqslant 1$ and l, the arguments in the proof of theorem 12.8 allow us to show that equation

$$a\frac{x^m-1}{x-1}=b\frac{y^n-1}{y-1}+l \quad \text{in integers } m>1,\ n>1,\ x>1,\ y>1$$

with $x, y\in S$ and $a(x-1)^{-1}-b(y-1)^{-1}\neq l$ implies that $\max(m, n, x, y)$ is bounded by a computable number depending only on a, b, l and P.

Let $x>1$ and $y>1$ be fixed distinct integers. Senge and Straus (1973) proved that the number of integers, the sum of whose digits in each of the bases x and y lies below a fixed bound, is finite if and only if $(\log x)/\log y$ is irrational. Their proof depends on a p-adic version of the ineffective Thue–Siegel–Roth method. Stewart (1980) used estimates of linear forms in logarithms to exhibit an explicit lower bound for the sum of the digits of n in base x plus the sum of the digits of n in base y, which tends to infinity as n tends to infinity provided that $(\log x)/\log y$ is irrational. For integers X and A with $1\leqslant A<X$, denote by $S_X(A)$ the set of integers whose digits are all equal to A in their X-adic expansions. For integers X, Y, A and B with $1\leqslant A<X$, $1\leqslant B<Y$ and $A(Y-1)\neq B(X-1)$, Shorey (1986*c*) showed that the number of elements in $S_X(A)\cap S_Y(B)$ and $S_X(1)\cap S_Y(1)$ is at most 24 and 17, respectively.

Brindza, Györy and Tijdeman (1986) obtained the following generalisation of theorem 12.3 to algebraic number fields. *Let K be an algebraic number field with ring of integers* $\mathcal{O}_K$. *There exists a computable number* C_{17} *depending only on K such that all solutions of the equation*

$$x^m-y^n=1 \quad \text{in } x, y\in\mathcal{O}_K,\ m, n\in\mathbb{Z}_+$$

with x, y not roots of unity and $m>1$, $n>1$, $mn>4$ *satisfy*

$$\max(\overline{|x|}, \overline{|y|}, m, n)<C_{17}.$$

The assumptions made on x, y, m and n are necessary.

Catalan's equation has also been considered over function fields. Let K be any field. Let m and n be integers greater than 1 and not divisible by the characteristic of K. Nathanson (1974) and Albis González (1975) proved, independently, that the equation $x^m-y^n=1$ has no non-constant solution x, y in the polynomial ring $K[t]$. Nathanson further proved that if $m>2$ and

$n > 2$, then $x^m - y^n = 1$ has no non-constant solution in the rational function field $K(t)$ either. Silverman (1982*b*) studied the Pillai equation $ax^m + by^n = c$ over function fields of projective varieties.

References

The figures within square brackets refer to the chapters where the publication is quoted.

Abel, N. H. (1823), Extraits de quelques lettres à Holmboe, *Oeuvres Complètes*, Grondahl, Christiania, 1881, Vol. II, pp. 254–5. [11]

Agrawal, M. K., Coates, J. H., Hunt, D. C. and van der Poorten, A. J. (1980), Elliptic curves of conductor 11, *Math. Comp.* **35**, 991–1002. [7]

Albis González, V. S. (1975), The equations of Fermat and Catalan in $K[t]$ (Spanish), *Bol. Mat.* **9**, 217–20. [11, 12]

Alex, L. J. (1973), On simple groups of order $2^a3^b7^cp$, *J. Algebra* **25**, 113–24. [1]

Alex, L. J. (1976), Diophantine equations related to finite groups, *Comm. Algebra* **4**, 77–100. [1]

Alex, L. J. and Foster, L. L. (1983), On diophantine equations of the form $1+2^a=p^bq^c+2^dp^eq^f$, *Rocky Mountain J. Math.* **13**, 321–31. [1]

Alter, R. and Kubota, K. K. (1973), Multiplicities of second order linear recurrences, *Trans. Am. Math. Soc.* **178**, 271–84. [3]

Apéry, R. (1960*a*), Sur une équation diophantienne, *C.r. hebd. Séanc. Acad. Sci., Paris* **251**, 1263–4. [7]

Apéry, R. (1960*b*), Sur une équation diophantienne, *C.r. hebd. Séanc. Acad. Sci., Paris* **251**, 1451–2. [7]

Bachmann, P. (1919), *Das Fermatproblem in seiner bisherigen Entwicklung*, Walter de Gruyter, Berlin. Reprint: Springer-Verlag, Berlin, 1976. [11]

Baker, A. (1964*a*), Rational approximations to certain algebraic numbers, *Proc. Lond. Math. Soc.* (3) **14**, 385–98. [5]

Baker, A. (1964*b*), Rational approximations to $\sqrt[3]{2}$ and other algebraic numbers, *Q. Jl Math. Oxford* (2) **15**, 375–83. [5]

Baker, A. (1966), Linear forms in the logarithms of algebraic numbers, *Mathematika* **13**, 204–16. [B]

Baker, A. (1967), Simultaneous rational approximations to certain algebraic numbers, *Proc. Camb. Phil. Soc.* **63**, 693–702. [5]

Baker, A. (1968*a*), Linear forms in the logarithms of algebraic numbers IV, *Mathematika* **15**, 204–16. [B]

Baker, A. (1968*b*), Contributions to the theory of diophantine equations. I, On the representation of integers by binary forms. II, The diophantine equation $y^2=x^3+k$, *Phil. Trans. R. Soc. London.* A **263**, 173–208. [Introduction, A, 1, 5, 6]

Baker, A. (1968*c*), The diophantine equation $y^2=ax^3+bx^2+cx+d$, *J. Lond. Math. Soc.* **43**, 1–9. [6, 7]

Baker, A. (1969), Bounds for the solutions of the hyperelliptic equation, *Proc. Camb. Phil. Soc.* **65**, 439–44. [5, 6]

Baker, A. (1972), A sharpening of the bounds for linear forms in logarithms I, *Acta Arith.* **21**, 117–29. [Introduction]

Baker, A. (1973), A sharpening of the bounds for linear forms in logarithms II, *Acta Arith.* **24**, 33–6. [B, 5]

Baker, A. (1975), *Transcendental Number Theory*, Cambridge University Press (2nd edn 1979). [6]

Baker, A. (1977), The theory of linear forms in logarithms, *Transcendence Theory: Advances and Applications*, Academic Press, London, pp. 1–27. [B]

Baker, A. and Coates, J. (1970), Integer points on curves of genus 1, *Proc. Camb. Phil. Soc.* **67**, 595–602. [5, 6]

Baker, A. and Davenport, H. (1969), The equations $3x^2-2=y^2$ and $8x^2-7=z^2$, *Q. Jl Math. Oxford* (2) **20**, 129–37. [6]

Balasubramanian, R. and Shorey, T. N. (1980), On the equation $a(x^m-1)/(x-1)=b(y^n-1)/(y-1)$. *Math. Scand.* **46**, 177–82. [12]

Bannai, E. (1979), On tight spherical designs, *J. Comb. Th.* A **26**, 38–47. [6]

Berstel, J. (1974), Sur le calcul des termes d'une suite récurrente linéaire, Exposé fait à l'I.R.I.A. (Rocquencourt) en mars 1974. [4]

Berstel, J. and Mignotte, M. (1976), Deux propriétés décidables des suites récurrentes linéaires, *Bull. Soc. Math. France* **104**, 175–84. [4]

Bertrand, D. (1978), Approximations diophantiennes p-adiques sur les courbes elliptiques admettant une multiplication complexe, *Compositio Math.* **37**, 21–50. [8]

Beukers, F. (1979), The Generalised Ramanujan–Nagell Equation, Dissertation, R.U. Leiden. [7]

Beukers, F. (1980), The multiplicity of binary recurrences, *Compositio Math.* **40**, 251–67. [3]

Beukers, F. (1981), On the generalized Ramanujan–Nagell equation, *Acta Arith.* I: **38**, 389–410; II: **39**, 113–23. [7]

Beukers, F. (1982), The zero-multiplicity of ternary recurrences, unpublished. [4]

Beukers, F. and Tijdeman, R. (1984), On the multiplicities of binary complex recurrences, *Compositio Math.* **51**, 193–213. [4]

Birch, B. J. and Merriman, J. R. (1972), Finiteness theorems for binary forms with given discriminant, *Proc. Lond. Math. Soc.* (3) **24**, 385–94. [1]

Birkhoff, G. D. and Vandiver, H. S. (1904), On the integral divisors of a^n-b^n, *Ann. Math.* (2) **5**, 173–80. [3]

Blanksby, P. E. (1969), A note on algebraic integers, *J. Number Theory* **1**, 155–60. [A]

Blanksby, P. E. and Montgomery, H. L. (1971), Algebraic integers near the unit circle, *Acta Arith.* **18**, 355–69. [A]

Bombieri, E. (1982), On the Thue–Siegel–Dyson theorem, *Acta Math.* **148**, 255–96. [5]

Bombieri, E. and Schmidt, W. M. (1986), On Thue's equation, to appear. [5]

Bombieri, E. and Mueller, J. (1983), On effective measures of irrationality for $\sqrt[n]{(a/b)}$ and related numbers, *J. reine angew. Math.* **342**, 173–96. [5]

Bombieri, E. and Mueller, J. (1986), Remarks on the approximation to an algebraic number by algebraic numbers, *Mich. Math. J.* **33**, 83–93. [5]

Borevich, Z. I. and Shafarevich, I. R. (1964), *Number Theory* (Russian). English edn: *Pure Appl. Math.* Vol. 20, Academic Press, New York, 1966. [A]

Boyd, D. W. and Kisilevsky, H. H. (1972), The diophantine equation $u(u+1)(u+2)(u+3)=v(v+1)(v+2)$, *Pacific J. Math.* **40**, 23–32. [6]

Brauer, R. (1968), On simple groups of order $5.3^a.2^b$, *Bull. Am. Math. Soc.* **74**, 900–3. [1]

Bremner, A. (1979), A diophantine equation arising from tight 4-designs, *Osaka J. Math.* **16**, 353–6. [6]

Bremner, A., Calderbank, R., Hanlon, P., Morton, P. and Wolfskill, J. (1983), Two-weight ternary codes and the equation $y^2=4.3^a+13$, *J. Number Theory* **16**, 212–34. [7]

Brenner, J. L. and Foster, L. L. (1982), Exponential diophantine equations, *Pacific J. Math.* **101**, 263–301. [1]

Brindza, B. (1984*a*), On S-integral solutions of the equation $y^m=f(x)$, *Acta Math. Hung.* **44**, 133–9. [6, 8]

Brindza, B. (1984*b*), On some generalizations of the diophantine equation $1^k+2^k+\cdots+x^k=y^z$, *Acta Arith.* **44**, 99–107. [10]

Brindza, B. (1984*c*), On a diophantine equation connected with the Fermat equation, *Acta Arith.* **44**, 357–63. [11]

Brindza, B., Győry, K. and Tijdeman, R. (1985), The Fermat equation with polynomial values as base variables, *Invent. Math.* **80**, 139–51. [10, 11]

Brindza, B., Győry, K. and Tijdeman, R. (1986), On the Catalan equation over algebraic number fields, *J. reine angew. Math.* **367**, 90–102. [12]

Browkin, J. and Schinzel, A. (1956), Sur les nombres de Mersenne qui sont triangulaires, *C.r. hebd. Séanc. Acad. Sci., Paris* **242**, 1780–2. [7]

Browkin, J. and Schinzel, A. (1960), On the equation $2^n - D = y^2$, *Bull. Acad. Polon. Sci. Sér. Math. Astron. Phys.* **8**, 311–18. [7]

Cantor, D. C. and Straus, E. G. (1982), On a conjecture of D. H. Lehmer, *Acta Arith.* **42**, 97–100. Corr. **42**, 327. [A]

Carmichael, R. D. (1913), On the numerical factors of the arithmetic forms $\alpha^n \pm \beta^n$, *Ann. Math.* (2) **15**, 30–70. [3]

Cassels, J. W. S. (1953), On the equation $a^x - b^y = 1$, *Am. J. Math.* **75**, 159–62. [1]

Cassels, J. W. S. (1957), *An Introduction to Diophantine Approximation*, Cambridge University Press, No. 45. [A]

Cassels, J. W. S. (1959), *An Introduction to the Geometry of Numbers*, Springer-Verlag, Berlin. [A]

Cassels, J. W. S. (1960*a*), On the equation $a^x - b^y = 1$ II, *Proc. Camb. Phil. Soc.* **56**, 97–103. Corr. **57** (1961), 187. [12]

Cassels, J. W. S. (1960*b*), On a class of exponential equations, *Ark. Mat.* **4**, 231–3. [1]

Catalan, E. (1844), Note extraite d'une lettre adressée à l'éditeur, *J. reine angew. Math.* **27**, 192. [12]

Cerlienco, L., Mignotte, M. and Piras, F. (1984), *Suites Récurrentes Linéaires*, Propriétés algébriques et arithmétiques, Publ. I.R.M.A., Université L. Pasteur, Strasbourg. [C]

Chowla, P. (1969), Remarks on a previous paper, *J. Number Theory* **1**, 522–4. [3]

Chowla, P., Chowla, S. D., Dunton, M. and Lewis, D. J. (1959), Some diophantine equations in quadratic number fields, *Norske Vid. Selsk. Forh.* (Trondheim) **31**, 181–3. [3]

Chowla, P., Chowla, S. D., Dunton, M. and Lewis, D. J. (1963), Diophantine equations in quadratic number fields, *Golden Jubilee Commemoration Volume* (1958/1959), Calcutta Mathematical Society, Part II, pp. 317–32. [3]

Chowla, S. (1935), The greatest prime factor of $x^2 + 1$, *J. Lond. Math. Soc.* **10**, 117–20. [7]

Chowla, S. (1961), Proof of a conjecture of Julia Robinson, *Norske Vid. Selsk. Forh.* (Trondheim) **34**, 100–1. [1]

Chowla, S., Dunton, M. and Lewis, D. J. (1960), All integral solutions of $2^n - 7 = x^2$ are given by $n = 3, 4, 5, 7, 15$, *Norske Vid. Selsk. Forh.* (Trondheim) **33**, 37–8. [7]

Chowla, S., Dunton, M. and Lewis, D. J. (1961), Linear recurrences of order two, *Pacific J. Math.* **11**, 833–45. [3]

Chudnovsky, G. V. (1983*a*), On the method of Thue–Siegel, *Ann. Math.* (2) **117**, 325–82. [5]

Chudnovsky, G. V. (1983*b*), The Thue–Siegel–Roth theorem for values of algebraic functions, *Proc. Japan Acad.* **59** A, 281–4. [5]

Cijsouw, P. L. and Tijdeman, R. (1973), Distinct prime factors of consecutive integers, *Diophantine Approximation and its Applications*, Proceedings of the Conference at Washington D.C. 1972, Academic Press, New York, pp. 59–76. [7]

Coates, J. (1969), An effective *p*-adic analogue of a theorem of Thue, *Acta Arith.* **15**, 279–305. [1, 7]

Coates, J. (1970*a*), An effective *p*-adic analogue of a theorem of Thue II: The greatest prime factor of a binary form, *Acta Arith.* **16**, 399–412. [1, 7, 8]

Coates, J. (1970*b*), An effective *p*-adic analogue of a theorem of Thue III: The diophantine equation $y^2 = x^3 + k$, *Acta Arith.* **16**, 425–35. [8]

Cohen, E. L. (1978), Sur une équation diophantienne de Ljunggren, *Ann. Sci. Math. Québec* **2**, 109–12. [7]

Cohn, J. H. E. (1964), On square Fibonacci numbers, *J. Lond. Math. Soc.* **39**, 537–40. [9]

Cohn, J. H. E. (1965), Lucas and Fibonacci numbers and some diophantine equations, *Proc. Glasgow Math. Assoc.* **7**, 24–8. [9]

Cohn, J. H. E. (1971), The diophantine equation $Y(Y+1)(Y+2)(Y+3) = 2X(X+1)(X+2)(X+3)$, *Pacific J. Math.* **37**, 331–5. [6]

Danilov, L. V. (1982), The diophantine equation $x^3 - y^2 = k$ and a conjecture of M. Hall (Russian), *Mat. Zametki* **32**, 273–5, 425. Corr. **36** (1984), 457–8. Engl. transl.: *Math. Notes* **32**, 617–18; **36**, 726. [6]

Davenport, H., Lewis, D. J. and Schinzel, A. (1961), Equations of the form $f(x) = g(y)$, *Q. Jl Math. Oxford* (2) **12**, 304–12. [12]

Davenport, H. and Roth, K. F. (1955), Rational approximations to algebraic numbers, *Mathematika* **2**, 160–7. [5]
Dilcher, K. (1986), On a diophantine equation involving quadratic characters, *Compositio Math.* **57**, 383–403. [10]
Dobrowolski, E. (1979), On a question of Lehmer and the number of irreducible factors of a polynomial, *Acta Arith.* **34**, 391–401. [A]
Dubois, E. and Rhin, G. (1976), Sur la majoration de formes linéaires à coefficients algébriques réels et p-adiques, Démonstration d'une conjecture de K. Mahler, *C.r. hebd. Séanc. Acad. Sci., Paris* A **282**, 1211–14. [1]
Durst, L. K. (1959), Exceptional real Lehmer sequences, *Pacific J. Math.* **9**, 437–41. [3]
Dyson, F. J. (1947), The approximation to algebraic numbers by rationals, *Acta Math.* **79**, 225–40. [5]
Edwards, H. M. (1977), *Fermat's Last Theorem: A Genetic Introduction to Algebraic Number Theory*, Springer-Verlag, New York. [11]
Ellison, W. J. (1971*a*), Recipes for solving diophantine problems by Baker's method, Sém. Théorie des Nombres, Université de Bordeaux I, 1970–1, Lab. Th. Nombr. C.N.R.S., Exp. 11, 10 pp. [1]
Ellison, W. J. (1971*b*), On a theorem of S. Sivasankaranarayana Pillai, Sém. Théorie des Nombres, Université de Bordeaux I, 1970–1, Lab. Th. Nombr. C.N.R.S., Exp. 12, 10 pp. [1]
Ellison, W. J., Ellison, F., Pesek, J., Stahl, C. E. and Stall, D. S. (1972), The diophantine equation $y^2+k=x^3$, *J. Number Theory* **4**, 107–17. [6]
Ennola, V. (1978), *J*-fields generated by roots of cyclotomic integers, *Mathematika* **25**, 242–50. [10]
Erdös, P. (1934), A theorem of Sylvester and Schur, *J. Lond. Math. Soc.* **9**, 282–8. [7]
Erdös, P. (1951), On a diophantine equation, *J. Lond. Math. Soc.* **26**, 176–8. [10]
Erdös, P. (1955), On consecutive integers, *Nieuw Arch. Wiskunde* (3) **3**, 124–8. [7]
Erdös, P. (1965), Some recent advances and current problems in number theory, *Lectures on Modern Mathematics*, Vol. III, Wiley, New York, pp. 196–244. [1]
Erdös, P. and Mahler, K. (1939), Some arithmetical properties of the convergents of a continued fraction, *J. Lond. Math. Soc.* **14**, 12–18. [5]
Erdös, P. and Selfridge, J. L. (1971), Some problems on the prime factors of consecutive integers II, *Proceedings of the Washington State University Conference on Number Theory*, Pullman, Washington, pp. 13–21. [7]
Erdös, P. and Selfridge, J. L. (1975), The product of consecutive integers is never a power, *Illinois J. Math.* **19**, 292–301. [10]
Erdös, P. and Shorey, T. N. (1976), On the greatest prime factor of 2^p-1 for a prime p and other expressions, *Acta Arith.* **30**, 257–65. [3, 7]
Erdös, P. and Turk, J. (1984), Products of integers in short intervals, *Acta Arith.* **44**, 147–74. [10]
Everett, C. J. (1973), Fermat's conjecture, Roth's theorem, Pythagorean triangles, and Pell's equation, *Duke Math. J.* **40**, 801–4. [11]
Evertse, J.-H. (1983*a*), On the representation of integers by binary cubic forms of positive discriminant, *Invent. Math.* **73**, 117–38. Corr. **75**, 379. [5]
Evertse, J.-H. (1983*b*), *Upper Bounds for the Numbers of Solutions of Diophantine Equations*, Math. Centre Tract 168, Centr. Math. Comp. Sci., Amsterdam. [1, 5, 7, 12]
Evertse, J.-H. (1984*a*), On equations in *S*-units and the Thue–Mahler equation, *Invent. Math.* **75**, 561–84. [1, 5, 7]
Evertse, J.-H. (1984*b*), On sums of *S*-units and linear recurrences, *Compositio Math.* **53**, 225–44. [1, 4]
Evertse, J.-H. (198x), On equations in two *S*-units over function fields of characteristic zero, *Acta Arith.*, to appear. [1, 5]
Evertse, J.-H. and Györy, K. (1985), On unit equations and decomposable form equations, *J. reine angew. Math.* **358**, 6–19. [1, 5, 7]
Evertse, J.-H. and Györy, K. (1986*a*), On the number of polynomials and integral elements of given discriminant, *Acta Math. Hung.*, to appear. [1]
Evertse, J.-H. and Györy, K. (1986*b*), Finiteness theorems for decomposable form equations, to appear. [5, 7]

Evertse, J.-H., Győry, K., Shorey, T. N. and Tijdeman, R. (1986), Equal values of binary forms at integral points, *Acta Arith.*, to appear. [7]

Faddeev, D. K. (1966), On a paper of A. Baker (Russian), *Zap. Naučn. Sem. Leningrad. Otdel. Mat. Inst. Steklov.* (LOMI) **1**, 128–39. [5]

Faltings, G. (1983), Endlichkeitssätze für abelsche Varietäten über Zahlkörpern, *Invent. Math.* **73**, 349–66. [6, 11]

Fatou, P. (1906), Séries trigonométriques et séries de Taylor, *Acta Math.* **30**, 335–400. [C]

Feldman, N. I. (1968*a*), Estimate for a linear form of logarithms of algebraic numbers (Russian), *Mat. Sb.* **76** (118), 304–19. English trans.: *Math. USSR Sb.* **5**, 291–307. [5]

Feldman, N. I. (1968*b*), Improved estimate for a linear form of logarithms of algebraic numbers (Russian), *Mat. Sb.* **77** (119), 423–36. English trans.: *Math. USSR Sb.* **6**, 393–406. [5]

Feldman, N. I. (1970), Effective bounds for the size of the solutions of certain diophantine equations (Russian), *Mat. Zametki* **8**, 361–71. English trans.: *Math. Notes* **8**, 674–9. [5]

Feldman, N. I. (1971*a*), An effective sharpening of the exponent in Liouville's theorem (Russian), *Izv. Akad. Nauk SSSR* Ser. mat. **35**, 973–90. English trans.: *Math. USSR Izv.* **5**, 985–1002. [5]

Feldman, N. I. (1971*b*), Certain diophantine equations with a finite number of solutions (Russian), *Vestnik Moskov. Univ.* Ser. I **26**, 52–8. [5]

Feldman, N. I. (1979), Effective bounds of the solutions of certain diophantine equations, *J. Austral. Math. Soc.* A **28**, 129–35. [5]

Finkelstein, R. (1973), On Fibonacci numbers which are one more than a square, *J. reine angew. Math.* **262/263**, 171–8. [9]

Finkelstein, R. (1975), On Lucas numbers which are one more than a square, *Fibonacci Quart.* **13**, 340–2. [9]

Gaál, I. (1986), Norm form equations with several dominating variables and explicit lower bounds for inhomogeneous linear forms with algebraic coefficients I, II, *Studia Sci. Math. Hung.*, to appear. [5]

Gelfond, A. O. (1934), Sur le septième problème de D. Hilbert, *Dokl. Akad. Nauk SSSR* (N.S.) **2**, 1–3 (Russian), 4–6 (French). *Izv. Akad. Nauk SSSR* Ser. Mat. 7, 623–30 (Russian). [B]

Gelfond, A. O. (1940), Sur la divisibilité de la différence des puissances de deux nombres entiers par une puissance d'un idéal premier, *Mat. Sb.* **7** (49), 7–25. [1]

Gelfond, A. O. (1952), *Transcendental and Algebraic Numbers* (Russian). English trans.: Dover, New York, 1960. [5]

Glass, J. P., Loxton, J. H. and van der Poorten, A. J. (1981), Identifying a rational function, *C.r. Math. Rep. Acad. Sci. Canada* **3**, 279–84. [4]

Glass, J. P., Loxton, J. H. and van der Poorten, A. J. (1986), On the total multiplicity of recurrence sequences, to appear. [4]

Goldziher, K. (1913), Decomposition of perfect powers into the sum of perfect powers (Hungarian), *Köz. Mat. Lapok* **21**, 177–84. [11]

Granville, A. (1985), The set of exponents, for which Fermat's Last Theorem is true, has density one, *C.r. Math. Rep. Acad. Sci. Canada* **7**, 55–60. [11]

Greenleaf, N. (1969), On Fermat's equation in $\mathscr{C}(t)$, *Am. Math. Monthly* **76**, 808–9. [11]

Grimm, C. A. (1969), A conjecture on consecutive composite numbers, *Am. Math. Monthly* **76**, 1126–8. [7]

Grinstead, C. M. (1978), On a method of solving a class of diophantine equations, *Math. Comp.* **32**, 936–40. [6]

Gross, F. (1966*a*), On the functional equation $f^n+g^n=h^n$, *Am. Math. Monthly* **73**, 1093–6. [11]

Gross, F. (1966*b*), On the equation $f^n+g^n=1$, *Bull. Am. Math. Soc.* **72**, 86–8. [11]

Győry, K. (1972), Sur l'irréductibilité d'une classe des polynômes II, *Publ. Math. Debrecen* **19**, 293–326. [1]

Győry, K. (1973), Sur les polynômes à coefficients entiers et de discriminant donné, *Acta Arith.* **23**, 419–26. [1]

Győry, K. (1974), Sur les polynômes à coefficients entiers et de discriminant donné II, *Publ. Math. Debrecen* **21**, 125–44. [1]

Győry, K. (1975), Sur une classe des corps de nombres algébriques et ses applications, *Publ. Math. Debrecen* **22**, 151–75. [1]

Györy, K. (1976), Sur les polynômes à coefficients entiers et de discriminant donné III, *Publ. Math. Debrecen* **23**, 141–65. [1, 5, 7]
Györy, K. (1978*a*), On polynomials with integer coefficients and given discriminant IV, *Publ. Math. Debrecen* **25**, 155–67. [1]
Györy, K. (1978*b*), On polynomials with integer coefficients and given discriminant V, p-adic generalizations, *Acta Math. Acad. Sci. Hung.* **32**, 175–90. [1]
Györy, K. (1979*a*), On the number of solutions of linear equations in units of an algebraic number field, *Comment. Math. Helv.* **54**, 583–600. [1]
Györy, K. (1979*b*), On the greatest prime factors of decomposable forms at integer points, *Ann. Acad. Sci. Fenn.* Ser. AI **4**, 341–55. [7]
Györy, K. (1980*a*), On the solutions of linear diophantine equations in algebraic integers of bounded norm, *Ann. Univ. Sci. Budapest. Eötvös* Sect. Math. **22/23**, 225–33. [1]
Györy, K. (1980*b*), On certain graphs composed of algebraic integers of a number field and their applications I, *Publ. Math. Debrecen* **27**, 229–42. [1]
Györy, K. (1980*c*), Explicit upper bounds for the solutions of some diophantine equations, *Ann. Acad. Sci. Fenn.* Ser. AI **5**, 3–12. [7]
Györy, K. (1980*d*), Sur une généralisation de l'équation de Thue–Mahler, *C.r. hebd. Séanc. Acad. Sci., Paris* Sér. A **290**, 633–5. [7]
Györy, K. (1980*e*), Résultats Effectifs sur la Représentation des Entiers par des Formes Décomposables, Queen's papers in pure and applied mathematics, No. 56, Kingston, Canada. [1, 5, 7]
Györy, K. (1980*f*), Corps de nombres algébriques d'anneau d'entiers monogène, Sém. Delange-Pisot-Poitou 1978/79, Paris, Exp. 26, 7 pp. [1]
Györy, K. (1980*g*), Sur certaines généralisations de l'équation de Thue–Mahler, *Enseign. Math.* **26**, 247–55. [7]
Györy, K. (1980*h*), Explicit lower bounds for linear forms with algebraic coefficients, *Arch. Math.* **35**, 438–46. [5, 7]
Györy, K. (1981*a*), On the representation of integers by decomposable forms in several variables, *Publ. Math. Debrecen* **28**, 89–98. [5, 7]
Györy, K. (1981*b*), On *S*-integral solutions of norm form, discriminant form and index form equations, *Studia Sci. Math. Hung.* **16**, 149–61. [5, 7]
Györy, K. (1981*c*), On discriminants and indices of integers of an algebraic number field, *J. reine angew. Math.* **324**, 114–26. [1]
Györy, K. (1982*a*), On some arithmetical properties of Lucas and Lehmer numbers, *Acta Arith.* **40**, 369–73. [3, 12]
Györy, K. (1982*b*), On the irreducibility of a class of polynomials III, *J. Number Theory* **15**, 164–81. [1]
Györy, K. (1982*c*), On certain graphs associated with an integral domain and their applications to diophantine problems, *Publ. Math. Debrecen* **29**, 79–94. [1]
Györy, K. (1983), Bounds for the solutions of norm form, discriminant form and index form equations in finitely generated integral domains, *Acta Math. Hung.* **42**, 45–80. [1, 5, 7]
Györy, K. (1984*a*), Effective finiteness theorems for polynomials with given discriminant and integral elements with given discriminants over finitely generated domains, *J. reine angew. Math.* **346**, 54–100. [1]
Györy, K. (1984*b*), On norm form, discriminant form and index form equations, *Topics in Classical Number Theory*, Proceedings of the Conference at Budapest 1981, Colloq. Math. Soc. János Bolyai 34, North-Holland, Amsterdam, pp. 617–76. [5, 7]
Györy, K. (1984*c*), Sur les générateurs des ordres monogènes des corps de nombres algébriques, Sém. Théorie des Nombres, Université de Bordeaux I 1983–4, Lab. Th. Nombr. C.N.R.S., Exp. 32, 12 pp. [7]
Györy, K., Kiss, P. and Schinzel, A. (1981), On Lucas and Lehmer sequences and their applications to diophantine equations, *Colloq. Math.* **45**, 75–80. [3, 12]
Györy, K. and Lovász, L. (1970), Representation of integers by norm forms II, *Publ. Math. Debrecen* **17**, 173–81. [5]
Györy, K. and Papp, Z. Z. (1977), On discriminant form and index form equations, *Studia Sci. Math. Hung.* **12**, 47–60. [7]
Györy, K. and Papp, Z. Z. (1978), Effective estimates for the integer solutions of norm form and discriminant form equations, *Publ. Math. Debrecen* **25**, 311–25. [5]

Győry, K. and Papp, Z. Z. (1983), Norm form equations and explicit lower bounds for linear forms with algebraic coefficients, *Studies in Pure Mathematics to the Memory of Paul Turán*, Birkhäuser, Basel, pp. 245–57. [5]

Győry, K. and Pethő, A. (1977), Ueber die Verteilung der Lösungen von Normformen Gleichungen II, *Acta Arith.* **32**, 349–63. [5, 7]

Győry, K. and Pethő, A. (1980), Ueber die Verteilung der Lösungen von Normformen Gleichungen III, *Acta Arith.* **37**, 143–65. [5, 7]

Győry, K., Tijdeman, R. and Voorhoeve, M. (1980), On the equation $1^k+2^k+\cdots+x^k=y^z$, *Acta Arith.* **37**, 233–40. [10]

Hall, M. Jr (1971), The diophantine equation $x^3-y^2=k$, *Computers in Number Theory*, Proceedings of the Science Research Council Atlas Symposium No. 2, Oxford 1969, Academic Press, London, pp. 173–98. [6]

Hanson, D. (1973), On a theorem of Sylvester and Schur, *Can. Math. Bull.* **16**, 195–9. [7]

Hasse, H. (1966), Ueber eine diophantische Gleichung von Ramanujan–Nagell und ihre Verallgemeinerung, *Nagoya Math. J.* **27**, 77–102. [7]

Heath-Brown, D. R. (1985), Fermat's Last Theorem for 'almost all' exponents, *Bull. Lond. Math. Soc.* **17**, 15–16. [11]

Hecke, E. (1923), *Lectures on the Theory of Algebraic Numbers* (German). English trans.: Springer-Verlag, New York, 1981. [A]

Herzberg, N. P. (1975), Integer solutions of $by^2+p^n=x^3$, *J. Number Theory* **7**, 221–34. [8]

Hilliker, D. L. (1982), An algorithm for solving a certain class of diophantine equations I, *Math. Comp.* **38**, 611–26. [6]

Hilliker, D. L. and Straus, E. G. (1983), Determination of bounds for the solutions to those binary diophantine equations that satisfy the hypotheses of Runge's theorem, *Trans. Am. Math. Soc.* **280**, 637–57. [6]

Hooley, C. (1976), On the greatest prime factor of a quadratic polynomial, *Acta Math.* **117**, 281–99. [7]

Hurwitz, A. (1917), Ueber die diophantischen Gleichungen dritten Grades, *Vierteljahrschrift Naturf. Ges. Zürich* **62**. [6]

Hyyrö, S. (1963), On the Catalan problem (Finnish), *Arkhimedes* 1963, 53–4. [12]

Hyyrö, S. (1964*a*), Ueber das Catalansche Problem, Ann. Univ. Turku Ser. AI No. 79, 10 pp. [12]

Hyyrö, S. (1964*b*), Ueber die Gleichung $ax^n-by^n=z$ und das Catalansche Problem, Ann. Acad. Sci. Fenn. Ser. AI No. 355, 50 pp. [5]

Inkeri, K. (1946), Untersuchungen über die Fermatsche Vermutung, Ann. Acad. Sci. Fenn. Ser. AI No. 33, 60 pp. [11]

Inkeri, K. (1953), Abschätzungen für eventuelle Lösungen der Gleichung im Fermatschen Problem, Ann. Univ. Turku. Ser. A 16, No. 1, 9 pp. [11]

Inkeri, K. (1964), On Catalan's problem, *Acta Arith.* **9**, 285–90. [12]

Inkeri, K. (1972), On the diophantine equation $a(x^n-1)/(x-1)=y^m$, *Acta Arith.* **21**, 299–311. [12]

Inkeri, K. (1976), A note on Fermat's conjecture, *Acta Arith.* **29**, 251–6. [11]

Inkeri, K. (1979), On the diophantine equations $2y^2=7^k+1$ and $x^2+11=3^n$, *Elem. Math.* **34**, 119–21. [7]

Inkeri, K. and van der Poorten, A. J. (1980), Some remarks on Fermat's conjecture, *Acta Arith.* **36**, 107–11. [11]

Jones, B. W. (1976), A variation on a problem of Davenport and Diophantus, *Q. Jl Math. Oxford* (2) **27**, 349–53. [6]

Jones, B. W. (1978), A second variation on a problem of Diophantus and Davenport, *Fibonacci Quart.* **16**, 155–65. [6]

Jutila, M. (1974), On numbers with a large prime factor II, *J. Indian Math. Soc.* (N.S.) **38**, 125–30. [7]

Kanagasabapathy, P. and Ponnudurai, T. (1975), The simultaneous diophantine equations $y^2-3x^2=-2$ and $z^2-8x^2=-7$, *Q. Jl Math. Oxford* (2) **26**, 275–8. [6]

Keates, M. (1969), On the greatest prime factor of a polynomial, *Proc. Edin. Math. Soc.* (2) **16**, 301–3. [7]

Kiss, P. (1979), Zero terms in second order linear recurrences, *Math. Sem. Notes Kobe Univ.* (Japan) **7**, 145–52. [3]

Kiss, P. (1982), On common terms of linear recurrences, *Acta Math. Acad. Sci. Hung.* **40**, 119–23. [4]

Kiss, P. (198x), Differences of the terms of linear recurrences, *Studia Sci. Math. Hung.*, to appear. [4, 9]

Kleiman, H. (1976), On the diophantine equation $f(x, y)=0$, *J. reine angew. Math.* **286/287**, 124–31. [6]

Ko, Chao (1965), On the diophantine equation $x^2=y^n+1$, $xy\neq 0$, *Sci. Sinica* **14**, 457–60. [12]

Kotov, S. V. (1973*a*), The greatest prime factor of a polynomial (Russian), *Mat. Zametki* **13**, 515–22. English trans.: *Math. Notes* **13**, 313–17. [7]

Kotov, S. V. (1973*b*), The law of the iterated logarithm for binary forms with algebraic coefficients (Russian), *Dokl. Akad. Nauk. BSSR* **17**, 591–4. [7]

Kotov, S. V. (1975), The Thue–Mahler equation in relative fields (Russian), *Acta Arith.* **27**, 293–315. [7]

Kotov, S. V. (1976), Ueber die maximale Norm der Idealteiler des Polynoms $\alpha x^m+\beta y^n$ mit den algebraischen Koeffizienten, *Acta Arith.* **31**, 219–30. [8]

Kotov, S. V. (1977), The effectivisation of a theorem of Mahler on the rational points on a curve of genus one (Russian), *Dokl. Akad. Nauk BSSR* **21**, 101–4, 187. [8]

Kotov, S. V. (1979), Die arithmetische Struktur der rationalen Punkte und Kurven vom Geschlecht Eins, *Acta Arith.* **35**, 103–15. [8]

Kotov, S. V. (1980*a*), On diophantine equations of norm form type I (Russian), Inst. Math. Akad. Nauk BSSR, Preprint No. 9, Minsk. [5]

Kotov, S. V. (1980*b*), On diophantine equations of norm form type II (Russian), Inst. Math. Akad. Nauk BSSR, Preprint No. 10, Minsk. [7]

Kotov, S. V. (1981), Effective bounds for a linear form with algebraic coefficients in the archimedean and p-adic metrics (Russian), Inst. Math. Akad. Nauk BSSR, Preprint No. 24, Minsk. [5, 7]

Kotov, S. V. (1983), Effective bounds for the sizes of the solutions of a class of diophantine equations of norm form type (Russian), *Mat. Zametki* **33**, 801–6. English trans.: *Math. Notes* **33**, 411–14. [5]

Kotov, S. V. and Sprindžuk, V. G. (1977), The Thue–Mahler equation in a relative field and the approximation of algebraic numbers by algebraic numbers (Russian), *Izv. Akad. Nauk SSSR* Ser. Mat. **41**, 723–51, 799. English transl.: *Math. USSR Izv.* **11**, 677–707. [5, 7]

Kotov, S. V. and Trelina, L. A. (1979), *S*-ganze Punkte auf elliptischen Kurven, *J. reine angew. Math.* **306**, 28–41. [1, 8]

Kubota, K. K. (1977*a*), On a conjecture of Morgan Ward I, II, *Acta Arith.* **33**, 11–48. [3]

Kubota, K. K. (1977*b*), On a conjecture of Morgan Ward III, *Acta Arith.* **33**, 99–109. [4]

Lagarias, J. C. and Weisser, D. P. (1981), Fibonacci and Lucas cubes, *Fibonacci Quart.* **19**, 39–43. [9]

Landau, E. (1918), Verallgemeinerung eines Pólyaschen Satzes auf algebraische Zahlkörper, *Nachr. Ges. Wiss. Göttingen*, 478–88. [A]

Lang, S. (1960), Integral points on curves, *Publ. Math. I.H.E.S.* **6**, 27–43. [1, 5, 6, 8]

Lang, S. (1965), Report on diophantine approximations, *Bull. Soc. Math. France* **93**, 177–92. [5]

Lang, S. (1978), *Elliptic Curves Diophantine Analysis*, Grundl. Math. Wissensch. 231, Springer-Verlag, Berlin. [1, 5]

Lang, S. (1983), *Fundamentals of Diophantine Geometry*, Springer-Verlag, New York. (An earlier version was published by Wiley-Interscience in 1962 under the title *Diophantine Geometry*.) [1, 5]

Langevin, M. (1975*a*), Plus grand facteur premier d'entiers consécutifs, *C.r. hebd. Séanc. Acad. Sci., Paris* A **280**, 1567–70. [7]

Langevin, M. (1975*b*), Plus grand facteur premier d'entiers voisins, *C.r. hebd. Séanc. Acad. Sci., Paris* A **281**, 491–3. [7]

Langevin, M. (1975*c*), Sur la fonction plus grand facteur premier, Sém. Delange-Pisot-Poitou 1974/75, Paris, Exp. G3, G7, G10, G12, G22, 29 pp. [7]

Langevin, M. (1976*a*), Méthodes élémentaires en vue du théorème de Sylvester, Sém. Delange-Pisot-Poitou 1975/76, Paris, Exp. G2, 9 pp. [7]

Langevin, M. (1976*b*), Quelques applications de nouveaux résultats de van der Poorten, Sém. Delange-Pisot-Poitou 1975/76, Paris, Exp. G12, 11 pp. [7, 12]
Langevin, M. (1978), Facteurs premiers d'entiers en progression arithmétique, Sém. Delange-Pisot-Poitou 1977/78, Paris, Exp. 4, 7 pp. [7]
Langevin, M. (1979), Facteurs premiers des coefficients binomiaux, Sém. Delange-Pisot-Poitou 1978/79, Paris, Exp. 27, 15 pp. [7]
Langevin, M. (1981), Facteurs premiers d'entiers en progression arithmétique, *Acta Arith.* **39**, 241–9. [7]
Laurent, M. (1984), Equations diophantiennes exponentielles, *Invent. Math.* **78**, 299–327. [1, 4, 7]
Laxton, R. R. (1967), Linear recurrences of order two, *J. Austral. Math. Soc.* **7**, 108–14. [3]
Laxton, R. R. (1974), On a problem of M. Ward, *Fibonacci Quart.* **12**, 41–4. [4]
Lebesgue, V. A. (1850), Sur l'impossibilité, en nombres entiers, de l'équation $x^m = y^2 + 1$, *Nouv. Ann. Math.* **9**, 178–81. [12]
Lech, C. (1953), A note on recurring series, *Ark. Mat.* **2**, 417–21. [C, 4]
van Leeuwen, J. (1980), Reconsidering a problem of M. Ward, *A Collection of Manuscripts Related to the Fibonacci Sequence*, Fibonacci Association, Santa Clara, California, pp. 45–6. [4]
Lehmer, D. H. (1964), On a problem of Störmer, *Illinois J. Math.* **8**, 57–79. [1]
Lenstra, H. W., Jr (1977), Euclidean number fields of large degree, *Invent. Math.* **38**, 237–54. [1]
LeVeque, W. J. (1952), On the equation $a^x - b^y = 1$, *Am. J. Math.* **74**, 325–31. [1, 12]
LeVeque, W. J. (1961), Rational points on curves of genus greater than 1, *J. reine angew. Math.* **206**, 45–52. [8]
LeVeque, W. J. (1964), On the equation $y^m = f(x)$, *Acta Arith.* **9**, 209–19. [6, 8]
LeVeque, W. J. (1974), *Reviews in Number Theory*, American Mathematical Society, Providence, R.I. [C, 1]
Lewis, D. J. (1961), Two classes of diophantine equations, *Pacific J. Math.* **11**, 1063–76. [7]
Lewis, D. J. (1969), Diophantine equations: *p*-adic methods, *Studies in Number Theory*, Mathematical Association of America, pp. 25–75. [C]
Lewis, D. J. and Mahler, K. (1960), On the representation of integers by binary forms, *Acta Arith.* **6**, 333–63. [5, 7]
Lewis, D. J. and Turk, J. (1985), Repetitiveness in binary recurrences, *J. reine angew. Math.* **356**, 19–48. [3, 4]
Liouville, J. (1844), Sur des classes très étendues de quantités dont la valeur n'est ni algébrique, ni même réductible à des irrationelles algébriques. *C.r. hebd. Séanc. Acad. Sci., Paris* **18**, 883–5, 910–11. *J. Math. Pures Appl.* **16** (1851), 133–42. [A]
Ljunggren, W. (1943), Some theorems on indeterminate equations of the form $(x^n - 1)/(x - 1) = y^q$ (Norwegian), *Norsk Mat. Tidsskr.* **25**, 17–20. [12]
London, H. and Finkelstein, R. (1969), On Fibonacci and Lucas numbers which are perfect powers, *Fibonacci Quart.* **7**, 476–81, 487. Corr. **8** (1970), 248. [9]
London, H. and Finkelstein, R. (1973), *On Mordell's Equation* $y^2 - k = x^3$, Bowling Green State University, Ohio. [6]
Louboutin, R. (1983), Sur la mesure de Mahler d'un nombre algébrique, *C.r. hebd. Séanc. Acad. Sci., Paris* I **296**, 707–8. [A]
Loxton, J. H. (1986), Some problems involving powers of integers, *Acta Arith.* **46**, 113–23. [B, 12]
Loxton, J. H. and van der Poorten, A. J. (1977), On the growth of recurrence sequences, *Math. Proc. Camb. Phil. Soc.* **81**, 369–76. [C, 4]
MacLeod, R. A. and Barrodale, I. (1970), On equal products of consecutive integers, *Can. Math. Bull.* **13**, 255–9. [6]
Mahler, K. (1933*a*), Zur Approximation algebraischer Zahlen, I: Ueber den grössten Primteiler binärer Formen, *Math. Ann.* **107**, 691–730. [1, 7]
Mahler, K. (1933*b*), Zur Approximation algebraischer Zahlen, II: Ueber die Anzahl der Darstellungen grosser Zahlen durch binäre Formen, *Math. Ann.* **108**, 37–55. [7]
Mahler, K. (1934*a*), Ueber die rationalen Punkte auf Kurven vom Geschlecht Eins, *J. reine angew. Math.* **170**, 168–78. [8]

Mahler, K. (1934*b*), Eine arithmetische Eigenschaft der rekurrierenden Reihen, *Mathematica* B (Leiden) **3**, 153–6. [3]

Mahler, K. (1935*a*), Eine arithmetische Eigenschaft der Taylor–Koeffizienten rationaler Funktionen, *Proc. Akad. Wetensch. Amsterdam* **38**, 50–60. [4]

Mahler, K. (1935*b*), Ueber den grössten Primteiler spezieller Polynome zweiten Grades, *Arch. Math. Naturvid.* B **41**, No. 6, 26 pp. [7]

Mahler, K. (1936), Ein Analogon zu einem Schneiderschen Satz, *Proc. Akad. Wetensch. Amsterdam* **39**, 633–40, 729–37. [5]

Mahler, K. (1950), On algebraic relations between two units of an algebraic field, *Algèbre et Théorie des Nombres*, Colloq. Intern. C.N.R.S. No. 24, Paris, pp. 47–55. [1]

Mahler, K. (1953), On the greatest prime factor of $ax^m + by^n$, *Nieuw Arch. Wiskunde* (3) **1**, 113–22. [8]

Mahler, K. (1956), On the Taylor coefficients of rational functions, *Proc. Camb. Phil. Soc.* **52**, 39–48. Addendum: **53**, 544. [4]

Mahler, K. (1966), A remark on recursive sequences, *J. Math. Sci.* **1**, 12–17. [3]

Mahler, K. (1984), On Thue's theorem, *Math. Scand.* **55**, 188–200. [5]

Makowski, A. (1962), Three consecutive integers cannot be powers, *Colloq. Math.* **9**, 297. [12]

Mason, R. C. (1981), On Thue's equation over function fields, *J. Lond. Math. Soc.* (2) **24**, 414–26. [1, 5]

Mason, R. C. (1983), The hyperelliptic equation over function fields, *Math. Proc. Camb. Phil. Soc.* **93**, 219–30. [1, 6]

Mason, R. C. (1984*a*), *Diophantine Equations over Function Fields*, LMS Lecture Notes No. 96, Cambridge University Press [1, 5, 6]

Mason, R. C. (1984*b*), Equations over function fields, *Number Theory, Noordwijkerhout* 1983 Proceedings, Lecture Notes in Mathematics 1068, Springer-Verlag, Berlin, pp. 149–57. [1, 5]

Mason, R. C. (1986), Norm form equations I, *J. Number Theory* **22**, 190–207. [1,5]

Mason, R. C. and Brindza, B. (1986), LeVeque's superelliptic equation over function fields, *Acta Arith.*, to appear. [6]

Matveev, E. M. (1979), Some diophantine equations in several variables (Russian), *Vestnik Moskov. Univ.* Ser. I, No. 2, 22–30, 101. [7]

Matveev, E. M. (1980), Effective bounds for the solutions of some diophantine equations (Russian), *Vestnik Moskov. Univ.* Ser. I, No. 6, 23–7, 116. [7]

Matveev, E. M. (1981), An effective estimate for some polynomials in several variables (Russian), *Vestnik Moskov. Univ.* Ser. I, No. 1, 29–33, 107. [7]

Mignotte, M. (1974), Suites récurrentes linéaires, Sém. Delange-Pisot-Poitou 1973/74, Paris, Exp. G 14, 9 pp. [4]

Mignotte, M. (1975), A note on linear recursive sequences, *J. Austral. Math. Soc.* **A20**, 242–4. [4]

Mignotte, M. (1978), Some effective results about linear recursive sequences, *Automata, Languages and Programming*, Fifth International Colloquium, Udine 1978, Lecture Notes in Computer Science 62, Springer-Verlag, Berlin, pp. 322–9. [4]

Mignotte, M. (1979), Une extension du théorème de Skolem–Mahler, *C.r. hebd. Séanc. Acad. Sci., Paris* A **288**, 233–5. [4]

Mignotte, M. (1984), On the automatic resolution of certain diophantine equations, *Proceedings of EUROSAM 84*, Lecture Notes in Computer Science 174, Springer-Verlag, Berlin, pp. 378–85. [7]

Mignotte, M., Shorey, T. N. and Tijdeman, R. (1984), The distance between terms of an algebraic recurrence sequence, *J. reine angew. Math.* **349**, 63–76. [4]

Mignotte, M. and Waldschmidt, M. (1978), Linear forms in two logarithms and Schneider's method, *Math. Ann.* **231**, 241–67. [B]

Mihaljinec, M. (1952), A contribution to Fermat's problem (Serbo-Croatian), *Glasnik Mat. Fiz. Astr.* (2) **7**, 12–18. [11]

Mohanty, S. P. and Ramasamy, A. M. S. (1984), The simultaneous diophantine equations $5Y^2-20=X^2$ and $2Y^2+1=Z^2$, *J. Number Theory* **18**, 356–9. [6]

Montel, P. (1957), *Leçons sur les Récurrences et leurs Applications*, Gauthier-Villars, Paris. [C]

Mordell, L. J. (1914), Indeterminate equations of the third and fourth degrees, *Q. Jl Pure Appl. Math.* **45**, 170–81. [6]
Mordell, L. J. (1922), Note on the integer solutions of the equation $Ey^2 = Ax^3 + Bx^2 + Cx + D$, *Messenger Math.* **51**, 169–71. [6]
Mordell, L. J. (1923), On the integer solutions of the equation $ey^2 = ax^3 + bx^2 + cx + d$, *Proc. Lond. Math. Soc.* (2) **21**, 415–19. [6]
Mordell, L. J. (1962), The diophantine equation $2^n = x^2 + 7$, *Ark. Mat.* **4**, 455–60. [7]
Mordell, L. J. (1963), On the integer solutions of $y(y+1) = x(x+1)(x+2)$, *Pacific J. Math.* **13**, 1347–51. [6]
Mordell, L. J. (1969), *Diophantine Equations*, Academic Press, London. [9]
Mueller, J. (1984), On Thue's principle and its applications, *Number Theory, Noordwijkerhout* 1983 Proc., Lect. Notes Math. 1068, Springer-Verlag, Berlin, pp. 158–66. [5]
Ram Murty, M., Kumar Murty, V. and Shorey, T. N. (1986), Odd values of the Ramanujan τ-function, to appear. [3]
Nagell, T. (1920), Note sur l'équation indéterminée $(x^n - 1)/(x - 1) = y^q$, *Norsk Mat. Tidsskr.* **2**, 75–8. [12]
Nagell, T. (1921), Des équations indéterminées $x^2 + x + 1 = y^n$ et $x^2 + x + 1 = 3y^n$, *Norsk. Mat. Forenings Skr.* (1) **2**, 14 pp. [12]
Nagell, T. (1937), Ueber den grössten Primteiler gewisser Polynome dritten Grades, *Math. Ann.* **114**, 284–92. [7]
Nagell, T. (1948), (Norwegian) *Norsk Mat. Tidsskr.* **30**, 62–4. [7]
Nagell, T. (1955), Contributions to the theory of a category of diophantine equations of the second degree with two unknowns, *Nova Acta Regiae Soc. Sc. Upsaliensis* (4) **16**, No. 2. [7]
Nagell, T. (1958), Sur une classe d'équations exponentielles, *Ark. Mat.* **3**, 569–82. [1]
Nagell, T. (1961), The diophantine equation $x^2 + 7 = 2^n$, *Ark. Mat.* **4**, 185–7. [7]
Nagell, T. (1964), Sur une propriété des unités d'un corps algébrique, *Ark. Mat.* **5**, 343–56. [1]
Nagell, T. (1969), Sur un type particulier d'unités algébriques, *Ark. Mat.* **8**, 163–84. [1]
Nair, M. (1978), A note on the equation $x^3 - y^2 = k$, *Quart. Jl Math. Oxford* (2) **29**, 483–7. [6]
Narkiewicz, W. (1974), *Elementary and Analytic Theory of Algebraic Numbers*, Monogr. Matem. Tom 57, PWN, Warsaw. [A]
Nathanson, M. B. (1974), Catalan's equation in $K(t)$, *Am. Math. Monthly* **81**, 371–3. [12]
Nemes, I. and Pethö, A. (1984), Polynomial values in linear recurrences, *Publ. Math. Debrecen* **31**, 229–33. [9]
Nemes, I. and Pethö, A. (1986), Polynomial values in linear recurrences II, *J. Number Theory*, to appear. [9]
Obláth, R. (1956), Une propriété des puissances parfaites, *Mathesis* **65**, 356–64. [12]
Osgood, C. F. (1970*a*), The simultaneous diophantine approximation of certain kth roots, *Proc. Camb. Phil. Soc.* **67**, 75–86. [5]
Osgood, C. F. (1970*b*), The diophantine approximation of roots of positive integers, *J. Res. Nat. Bur. Standards* **74 B**, 241–4. [5]
Osgood, C. F. (1971), On the simultaneous diophantine approximation of values of certain algebraic functions, *Acta Arith.* **19**, 343–86. [5]
Parnami, J. C. and Shorey, T. N. (1982), Subsequences of binary recursive sequences, *Acta Arith.* **40**, 193–6. [3]
Parry, C. J. (1950), The $\mathfrak{p}$-adic generalisation of the Thue–Siegel theorem, *Acta Math.* **83**, 1–100. [1, 7]
Perelli, A. and Zannier, U. (1982), On periodic mod p sequences, *J. Number Theory* **15**, 77–82. [1]
Pethö, A. (1982*a*), Perfect powers in second order linear recurrences, *J. Number Theory* **15**, 5–13. [9]
Pethö, A. (1982*b*), Ueber die Verteilung der Lösungen von S-Normformen Gleichungen, *Publ. Math. Debrecen* **29**, 1–17. [7]
Pethö, A. (1983), Full cubes in the Fibonacci sequence, *Publ. Math. Debrecen* **30**, 117–27. [9]

Pethö, A. (1984), Perfect powers in second order recurrences, *Topics in Classical Number Theory*, Proceedings of the Conference in Budapest 1981, Colloq. Math. Soc. János Bolyai 34, North-Holland, Amsterdam, pp. 1217–27. [9]

Pethö, A. (1985), On the solution of the diophantine equation $G_n = p^z$, *Proceedings of EUROCAL '85*, Linz, Lecture Notes in Computer Science 204, Springer-Verlag, Berlin, pp. 503–12. [3]

Picon, P. A. (1978), Sur certaines suites récurrentes cubiques ayant deux ou trois termes nuls, *Discrete Math.* **21**, 285–96. [4]

Pillai, S. S. (1931), On the inequality $0 < a^x - b^y \leqslant n$, *J. Indian Math. Soc.* (1) **19**, 1–11. [1]

Pillai, S. S. (1936), On $a^x - b^y = c$, *J. Indian Math. Soc.* (2) **2**, 119–22. Corr. **2**, 215. [1]

Pillai, S. S. (1945), On the equation $2^x - 3^y = 2^X + 3^Y$, *Bull. Calcutta Math. Soc.* **37**, 15–20. [1, 12]

Pollard, H. (1950), *The Theory of Algebraic Numbers*, Mathematical Association of America, Wiley, New York. [A]

Pólya, G. (1918), Zur arithmetischen Untersuchung der Polynome, *Math. Z.* **1**, 143–8. [1, 7]

Pólya, G. (1921), Arithmetische Eigenschaften der Reihenentwicklungen rationaler Funktionen, *J. reine angew. Math.* **151**, 1–31. [C, 3, 4]

Pólya, G. and Szegö, G. (1925), *Problems and Theorems in Analysis II* (German). English edn: Springer-Verlag, New York, 1976. [C]

Ponnudurai, T. (1975), The diophantine equation $Y(Y+1)(Y+2)(Y+3) = 3X(X+1)(X+2)(X+3)$, *J. Lond. Math. Soc.* (2) **10**, 232–40. [6]

van der Poorten, A. J. (1977*a*), Linear forms in logarithms in the p-adic case, *Transcendence Theory: Advances and Applications*, Academic Press, London, pp. 29–57. [B]

van der Poorten, A. J. (1977*b*), Effectively computable bounds for the solutions of certain diophantine equations, *Acta Arith.* **33**, 195–207. [2, 7, 12]

van der Poorten, A. J. and Loxton, J. H. (1976), Computing the effectively computable bound in Baker's inequality for linear forms in logarithms, *Bull. Austral. Math. Soc.* **15**, 33–57. Corr. **17** (1977), 151–5. [B]

van der Poorten, A. J. and Schlickewei, H. P. (1982), The growth conditions for recurrence sequences, Macquarie University Mathematical Report 82-0041, North-Ryde, Australia. [1, 4]

Postnikova, L. P. and Schinzel, A. (1968), Primitive divisors of the expression $a^n - b^n$ in algebraic number fields (Russian), *Mat. Sb.* **75**, 171–7. English trans.: *Math. USSR Sb.* **4**, 153–9. [3]

Ramachandra, K. (1969), A note on Baker's method, *J. Austral. Math. Soc.* **10**, 197–203. [B]

Ramachandra, K. (1970), A note on numbers with a large prime factor II, *J. Indian Math. Soc.* (N.S.) **34**, 39–48. [7]

Ramachandra, K. (1971), A note on numbers with a large prime factor III, *Acta Arith.* **19**, 49–62. [7]

Ramachandra, K. (1973), Application of Baker's theory to two problems considered by Erdös and Selfridge, *J. Indian Math. Soc.* (N.S.) **37**, 25–34. [7]

Ramachandra, K. and Shorey, T. N. (1973), On gaps between numbers with a large prime factor, *Acta Arith.* **24**, 99–111. [B, 7]

Ramachandra, K., Shorey, T. N. and Tijdeman, R. (1975), On Grimm's problem relating to factorisation of a block of consecutive integers, *J. reine angew. Math.* **273**, 109–24. [B, 7]

Ramachandra, K., Shorey, T. N. and Tijdeman, R. (1976), On Grimm's problem relating to factorisation of a block of consecutive integers II, *J. reine angew. Math.* **288**, 192–201. [B, 7]

Ramanujan, S. (1913), Question 464, *J. Indian Math. Soc.* **5**, 120. *Collected Papers*, Cambridge University Press, 1927, p. 327. [7]

Rameswar Rao, D. (1969), Some theorems on Fermat's Last Theorem, *Math. Student* **37**, 208–10. [11]

Ribenboim, P. (1979), *13 Lectures on Fermat's Last Theorem*, Springer-Verlag, New York. [11]

Ribenboim, P. (1984), Consecutive powers, *Exposit. Math.* **2**, 193–221. [12]

Ribenboim, P. (1986), A note on Catalan's equation, *J. Number Theory*, to appear. [12]
Richter, B. (1982), Ein Sonderfall der diophantischen Gleichung $(x^a-(\pm 1)^a)/(x\mp 1)=z^a$, *Acta Arith.* **40**, 273–88. [12]
Ridout, D. (1957), Rational approximations to algebraic numbers, *Mathematika* **4**, 125–31. [5]
Robba, Ph. (1978), Zéros de suites récurrentes linéaires, Groupe d'Etude d'Analyse Ultramétrique 1977/78, Paris, Exp. 13, 5 pp. [4]
Robbins, N. (1978), On Fibonacci numbers which are powers, *Fibonacci Quart.* **16**, 515–17. [9]
Robbins, N. (1981), Fibonacci and Lucas numbers of the forms w^2-1, $w^3\pm 1$, *Fibonacci Quart.* **19**, 369–73. [9]
Robbins, N. (1983), On Fibonacci numbers which are powers II, *Fibonacci Quart.* **21**, 215–18. [9]
Rosser, J. Barkley and Schoenfeld, L. (1962), Approximate formulas for some functions of prime numbers, *Illinois J. Math.* **6**, 64–94. [Notation, 7]
Roth, K. F. (1955), Rational approximations to algebraic numbers, *Mathematika* **2**, 1–20. Corr. **2**, 168. [5]
Rotkiewicz, A. (1956), Sur l'équation $x^z-y^t=a^t$, où $|x-y|=a$, *Ann. Polon. Math.* **3**, 7–8. [12]
Rotkiewicz, A. (1960), Une remarque sur le dernier théorème de Fermat, *Mathesis* **69**, 135–40. [11]
Rotkiewicz, A. (1961), Sur le problème de Catalan II, *Elem. Math.* **16**, 25–7. [12]
Rumsey, H. and Posner, E. C. (1964), On a class of exponential equations, *Proc. Am. Math. Soc.* **15**, 974–8. [1]
Runge, C. (1887), Ueber ganzzahlige Lösungen von Gleichungen zwischen zwei Veränderlichen, *J. reine angew. Math.* **100**, 425–35. [6]
Sansone, G. (1976), Il sistema diofanteo $N+1=x^2$, $3N+1=y^2$, $8N+1=z^2$, *Ann. Mat. Pura Appl.* (4) **111**, 125–51. [6]
Schäffer, J. J. (1956), The equation $1^p+2^p+3^p+\cdots+n^p=m^q$, *Acta Math.* **95**, 155–89. [10]
Schinzel, A. (1956), Sur l'équation $x^z-y^t=1$, où $|x-y|=1$, *Ann. Polon. Math.* **3**, 5–6. [12]
Schinzel, A. (1962*a*), The intrinsic divisors of Lehmer numbers in the case of negative discriminant, *Ark. Mat.* **4**, 413–16. [3]
Schinzel, A. (1962*b*), On primitive prime factors of a^n-b^n, *Proc. Camb. Phil. Soc.* **58**, 555–62. [3]
Schinzel, A. (1967), On two theorems of Gelfond and some of their applications, *Acta Arith.* **13**, 177–236. Corr. **16** (1969/70), 101. [3, 5, 7, 9]
Schinzel, A. (1968), On primitive prime factors of Lehmer numbers III, *Acta Arith.* **15**, 49–70. Corr. **16** (1969/70), 101. [3]
Schinzel, A. (1969), An improvement of Runge's theorem on diophantine equations, *Comment. Pontific. Acad. Sci.* **2**, No. 20, 9 pp. [6]
Schinzel, A. (1974), Primitive divisors of the expression A^n-B^n in algebraic number fields, *J. reine angew. Math.* **268/269**, 27–33. [3]
Schinzel, A. (1977), Abelian binomials, power residues and exponential congruences, *Acta Arith.* **32**, 245–74, Add. and Corr. **36** (1980), 101–4. [1]
Schinzel, A. and Tijdeman, R. (1976), On the equation $y^m=P(x)$, *Acta Arith.* **31**, 199–204. [10]
Schinzel, A. and Zassenhaus, H. (1965), A refinement of two theorems of Kronecker, *Michigan Math. J.* **12**, 81–5. [A]
Schlickewei, H. P. (1977*a*), On norm form equations, *J. Number Theory* **9**, 370–80. [7]
Schlickewei, H. P. (1977*b*), On linear forms with algebraic coefficients and diophantine equations, *J. Number Theory* **9**, 381–92. [7]
Schlickewei, H. P. (1977*c*), Inequalities for decomposable forms, *Astérisque* 41–2, Soc. Math. France, Paris, pp. 267–71. [7]
Schmidt, W. M. (1970), Simultaneous approximation to algebraic numbers by rationals, *Acta Math.* **125**, 189–201. [5]
Schmidt, W. M. (1971*a*), Linearformen mit algebraischen Koeffizienten II, *Math. Ann.* **191**, 1–20. [5, 7]
Schmidt, W. M. (1971*b*), Approximation to algebraic numbers, *Enseign. Math.* (2) **17**, 187–253. [Notation, 5]

Schmidt, W. M. (1972), Norm form equations, *Ann. Math.* (2) **96**, 526–51. [5, 7]
Schmidt, W. M. (1973), Inequalities for resultants and for decomposable forms, *Diophantine Approximation and its Applications*, Proceedings of the Conference at Washington D.C. 1972, Academic Press, New York, pp. 235–53. [5]
Schmidt, W. M. (1978), Thue's equation over function fields, *J. Austral. Math. Soc.* A **25**, 385–422. [1, 5, 6]
Schmidt, W. M. (1980*a*), Polynomial solutions of $F(x, y) = z^n$, *Proceedings of Queen's Number Theory Conference 1979*, Queen's Papers Pure Appl. Math. No. 54, Kingston, Canada, pp. 33–65. [5]
Schmidt, W. M. (1980*b*), *Diophantine Approximation*, Lecture Notes in Mathematics 785, Springer-Verlag, Berlin. [5]
Schneider, Th. (1934), Transzendenzuntersuchungen periodischer Funktionen I, *J. reine angew. Math.* **172**, 65–9. [B]
Schneider, Th. (1967), Anwendung eines abgeänderten Roth-Ridoutschen Satzes auf diophantische Gleichungen, *Math. Ann.* **169**, 177–82. [1]
Schur, I. (1929), Einige Sätze über Primzahlen mit Anwendung auf Irreduzibilitätsfragen, *Sitzungsber. Preuss. Akad. Wiss.* Phys. Math. Kl. **23**, 1–24. [7]
Scott, S. J. (1960), On the number of zeros of a cubic recurrence, *Am. Math. Monthly* **67**, 169–70. [4]
Senge, H. G. and Straus, E. G. (1973), P.V.-numbers and sets of multiplicity, *Periodica Math. Hung.* **3**, 93–100. [12]
Shafarevich, I. R. (1977), *Basic Algebraic Geometry*, Springer-Verlag, Berlin. [11]
Shanks, D. (1962), *Solved and Unsolved Problems in Number Theory*, Spartan Books, Washington D.C. 2nd edition: Chelsea Publ. Co., New York, 1978. [11]
Shorey, T. N. (1974*a*), Linear forms in the logarithms of algebraic numbers with small coefficients I, *J. Indian Math. Soc.* (N.S.) **38**, 271–84. [B]
Shorey, T. N. (1974*b*), On gaps between numbers with a large prime factor II, *Acta Arith.* **25**, 365–73. [B, 7]
Shorey, T. N. (1976*a*), On linear forms in the logarithms of algebraic numbers, *Acta Arith.* **30**, 27–42. [7]
Shorey, T. N. (1976*b*), Some applications of linear forms in logarithms, Sém. Delange-Pisot-Poitou 1975/76, Paris, Exp. 3, 8 pp. [5]
Shorey, T. N. (1980), On the greatest prime factor of $(ax^m + by^n)$, *Acta Arith.* **36**, 21–5. [10]
Shorey, T. N. (1982), The equation $ax^m + by^m = cx^n + dy^n$, *Acta Arith.* **41**, 255–60. [2, 7]
Shorey, T. N. (1983*a*), Applications of linear forms in logarithms to binary recursive sequences, *Seminar on Number Theory, Paris 1981/82*, Progr. Math. 38, Birkhäuser, Boston, pp. 287–301. [3]
Shorey, T. N. (1983*b*), Divisors of convergents of a continued fraction, *J. Number Theory* **17**, 127–33. [5]
Shorey, T. N. (1983*c*), The greatest square free factor of a binary recursive sequence, *Hardy-Ramanujan J.* **6**, 23–36. [1, 3]
Shorey, T. N. (1984*a*), Linear forms in members of a binary recursive sequence, *Acta Arith.* **43**, 317–31. [2, 3, 7]
Shorey, T. N. (1984*b*), On the equation $a(x^m - 1)/(x - 1) = b(y^n - 1)/(y - 1)$ (II), *Hardy-Ramanujan J.* **7**, 1–10. [12]
Shorey, T. N. (1984*c*), On the ratio of values of a polynomial, *Proc. Indian Acad. Sci.* (Math. Sci.) **93**, 109–16. [6, 12]
Shorey, T. N. (1986*a*), Perfect powers in values of certain polynomials at integer points, *Math. Proc. Camb. Phil. Soc.* **99**, 195–207. [B, 10, 12]
Shorey, T. N. (1986*b*), On the equation $z^q = (x^n - 1)/(x - 1)$, *Indag. Math.*, to appear. [12]
Shorey, T. N. (1986*c*), On the equation $ax^m - by^n = k$, *Indag. Math.*, to appear. [12]
Shorey, T. N., van der Poorten, A. J., Tijdeman, R. and Schinzel, A. (1977), Applications of the Gel'fond–Baker method to diophantine equations, *Transcendence Theory: Advances and Applications*, Academic Press, London, pp. 59–77. [B, 7, 10, 12]
Shorey, T. N. and Stewart, C. L. (1981), On divisors of Fermat, Fibonacci, Lucas and Lehmer numbers II, *J. Lond. Math. Soc.* (2) **23**, 17–23. [3]
Shorey, T. N. and Stewart, C. L. (1983), On the diophantine equation $ax^{2t} + bx^t y + cy^2 = d$ and pure powers in recurrence sequences, *Math. Scand.* **52**, 24–36. [3, 9]

Shorey, T. N. and Stewart, C. L. (1986), Pure powers in recurrence sequences and some related diophantine equations, to appear. [9]

Shorey, T. N. and Tijdeman, R. (1976*a*), New applications of diophantine approximations to diophantine equations, *Math. Scand.* **39**, 5–18. [9, 12]

Shorey, T. N. and Tijdeman, R. (1976*b*), On the greatest prime factors of polynomials at integer points, *Compositio Math.* **33**, 187–95. [7]

Siegel, C. L. (1921), Approximation algebraischer Zahlen, *Math. Z.* **10**, 173–213. [1, 5, 6, 7]

Siegel, C. L. (1926) (under the pseudonym X), The integer solutions of the equation $y^2 = ax^n + bx^{n-1} + \cdots + k$, *J. Lond. Math. Soc.* **1**, 66–8. [6]

Siegel, C. L. (1929), Ueber einige Anwendungen diophantischer Approximationen, *Abh. Preuss. Akad. Wiss.* Phys.-math. Kl., No. 1, 70 pp. [6, 8]

Siegel, C. L. (1969), Abschätzung von Einheiten, *Nachr. Akad. Wiss. Göttingen* Math.-Phys. Kl. **2**, 71–86. [A]

Silverman, J. H. (1982*a*), Integer points and the rank of Thue elliptic curves, *Invent. Math.* **66**, 395–404. [5]

Silverman, J. H. (1982*b*), The Catalan equation over function fields, *Trans. Am. Math. Soc.* **273**, 201–5. [12]

Silverman, J. H. (1983*a*), The Thue equation and height functions, *Diophantine Approximations and Transcendental Numbers*, Proceedings of the Conference at Luminy 1982, Progr. Math. 31, Birkhäuser, Boston, pp. 259–70. [5]

Silverman, J. H. (1983*b*), Representations of integers by binary forms and the rank of the Mordell–Weil group, *Invent. Math.* **74**, 281–92. [1, 5, 7]

Skolem, Th. (1929), Lösung gewisser Gleichungssysteme in ganzen Zahlen oder ganzzahligen Polynomen mit beschränktem gemeinschaftlichen Teiler, Oslo Vid. Akad. Skr. I 1929, No. 12. [6]

Skolem, Th. (1935), Ein Verfahren zur Behandlung gewisser exponentialer Gleichungen und diophantischer Gleichungen, 8. *Skand. Mat. Kongr.* Stockholm 1934, pp. 163–88. [4]

Skolem, Th. (1937), Anwendung exponentieller Kongruenzen zum Beweis der Unlösbarkeit gewisser diophantischer Gleichungen, Avh. Norske Vid. Akad. Oslo 1937, No. 12, 16 pp. [1, 5]

Skolem, Th. (1938), *Diophantische Gleichungen*, Springer-Verlag, Berlin. [1, 5, 6]

Skolem, Th. (1944), Extension of two theorems of C. Størmer (Norwegian), *Norsk Mat. Tidsskr.* **26**, 85–95. [1]

Skolem, Th. (1945*a*), On certain exponential equations, *Norske Vid. Selsk. Forh.* **17**, 126–9. [1]

Skolem, Th. (1945*b*), A method for the solution of the exponential equation $A_1^{x_1} \cdots A_m^{x_m} - B_1^{y_1} \cdots B_n^{y_n} = C$ (Norwegian), *Norsk Mat. Tidsskr.* **27**, 37–51. [1, 7]

Skolem, Th., Chowla, S. D. and Lewis, D. J. (1959), The diophantine equation $2^{n+2} - 7 = x^2$ and related problems, *Proc. Am. Math. Soc.* **10**, 663–9. [7]

Smiley, M. F. (1956), On the zeros of a cubic recurrence, *Am. Math. Monthly* **63**, 171–2. [4]

Sparlinskij, I. E. (1980), Prime divisors of recurrence sequences (Russian), *Izv. Vysš. Učebn. Zaved. Mat.* No. 4, 100–3. [4]

Sprindžuk, V. G. (1968), Effectivization in certain problems of diophantine approximation theory (Russian), *Dokl. Akad. Nauk BSSR* **12**, 293–7. [7]

Sprindžuk, V. G. (1969), Effective estimates in 'ternary' exponential diophantine equations (Russian), *Dokl. Akad. Nauk BSSR* **13**, 777–80. [1, 7]

Sprindžuk, V. G. (1970*a*), A new application of *p*-adic analysis to representation of numbers by binary forms (Russian), *Izv. Akad. Nauk SSSR* Ser. Mat. **34**, 1038–63. English trans.: *Math. USSR Izv.* **4**, 1043–69. [5, 7]

Sprindžuk, V. G. (1970*b*), An effective estimate of rational approximations for algebraic numbers (Russian), *Dokl. Akad. Nauk BSSR* **14**, 681–4. [7]

Sprindžuk, V. G. (1971*a*), Improving the estimate of rational approximations to algebraic numbers (Russian), *Dokl. Akad. Nauk BSSR* **15**, 101–4. [5, 7]

Sprindžuk, V. G. (1971*b*), New applications of analytic and *p*-adic methods in diophantine approximations, *Actes du Congrès International Mathématique, Nice 1970*, Gauthier-Villars, Paris, Vol. 1, pp. 505–9. [7]

Sprindžuk, V. G. (1971*c*), The greatest prime divisor of a binary form (Russian), *Dokl. Akad. Nauk BSSR* **15**, 389–91. [7]

Sprindžuk, V. G. (1971*d*), Rational approximations to algebraic numbers (Russian), *Izv. Akad. Nauk SSSR* Ser. Mat. **35**, 991–1007. English trans.: *Math. USSR. Izv.* **5**, 1003–19. [7]

Sprindžuk, V. G. (1972), On an estimate for solutions of Thue's equation, *Izv. Akad. Nauk SSSR* Ser. Mat. **36**, 712–41. English trans.: *Math. USSR Izv.* **6**, 705–34. [5, 7]

Sprindžuk, V. G. (1973*a*), Squarefree divisors of polynomials and class numbers of ideals of algebraic number fields (Russian), *Acta Arith.* **24**, 143–9. [6]

Sprindžuk, V. G. (1973*b*), The structure of numbers representable by binary forms (Russian), *Dokl. Akad. Nauk BSSR* **17**, 685–8, 775. [7]

Sprindžuk, V. G. (1974*a*), Representation of numbers by the norm forms with two dominating variables, *J. Number Theory* **6**, 481–6. [5]

Sprindžuk, V. G. (1974*b*), An effective analysis of the Thue and Thue–Mahler equations (Russian), *Current Problems of Analytic Number Theory*, Proceedings of the Summer School Minsk 1972, Izdat. Nauka i Tehnika, Minsk, pp. 199–222, 272. [7]

Sprindžuk, V. G. (1976), A hyperelliptic diophantine equation and class numbers (Russian), *Acta Arith.* **30**, 95–108. [6]

Sprindžuk, V. G. (1977), The arithmetic structure of integer polynomials and class numbers (Russian), *Analytic Number Theory, Mathematical Analysis and their Applications*, Trudy Mat. Inst. Steklov, pp. 143, 152–74, 210. [6]

Sprindžuk, V. G. (1980), Achievements and problems in diophantine approximation theory (Russian), *Uspehi Mat. Nauk* **35**, No. 4, 3–68, 248. English trans.: *Russian Math. Surv.* **35**, No. 4, 1–80. [1, 7]

Sprindžuk, V. G. (1982), *Classical Diophantine Equations in Two Unknowns* (Russian), Nauka, Moskva. [1, 5, 6, 7, 10]

Sprindžuk, V. G. and Kotov, S. V. (1973), An effective analysis of the Thue–Mahler equations in relative fields (Russian), *Dokl. Akad. Nauk BSSR* **17**, 393–5, 477. [7]

Sprindžuk, V. G. and Kotov, S. V. (1976), The approximation of algebraic numbers by algebraic numbers from a given field (Russian), *Dokl. Akad. Nauk BSSR* **20**, 581–4, 667. [7]

Stark, H. M. (1973), Effective estimates of solutions of some diophantine equations, *Acta Arith.* **24**, 251–9. [A, 5, 6]

Steiner, R. (1978), On *n*th powers in the Lucas and Fibonacci series, *Fibonacci Quart.* **16**, 451–8. [9]

Steiner, R. (1980), On Fibonacci numbers of the form x^2+1, *A Collection of Manuscripts related to the Fibonacci Sequence*, Fibonacci Association, Santa Clara, California, pp. 208–10. [9]

Stewart, C. L. (1975), The greatest prime factor of a^n-b^n, *Acta Arith.* **26**, 427–33. [3]

Stewart, C. L. (1976), Divisor Properties of Arithmetical Sequences, Ph.D. Thesis, University of Cambridge. [2, 3]

Stewart, C. L. (1977*a*), On divisors of Fermat, Fibonacci, Lucas and Lehmer numbers, *Proc. Lond. Math. Soc.* (3) **35**, 425–47. [3]

Stewart, C. L. (1977*b*), Primitive divisors of Lucas and Lehmer numbers, *Transcendence Theory: Advances and Applications*, Proceedings of the Conference at Cambridge, Academic Press, London, pp. 79–92. [B, 3]

Stewart, C. L. (1977*c*), A note on the Fermat equation, *Mathematika* **24**, 130–2. [11]

Stewart, C. L. (1978), Algebraic integers whose conjugates lie near the unit circle, *Bull. Soc. Math. France* **106**, 169–76. [A]

Stewart, C. L. (1980), On the representation of an integer in two different bases, *J.* reine angew. Math. **319**, 63–72. [12]

Stewart, C. L. (1981), On some diophantine equations and related linear recurrence sequences, *Seminar on Number Theory, Paris 1980/81*, Progr. Math. 22, Birkhäuser, Boston, pp. 317–21. [9]

Stewart, C. L. (1982), On divisors of terms of linear recurrence sequences, *J.* reine angew. Math. **333**, 12–31. [2, 3, 4]

Stewart, C. L. (1983), On divisors of Fermat, Fibonacci, Lucas and Lehmer numbers III, *J. Lond. Math. Soc.* (2) **28**, 211–17. [3]

Stewart, C. L. (1984), A note on the product of consecutive integers, *Topics in Classical Number Theory*, Proceedings of the Conference in Budapest 1981, Colloq. Math. Soc. János Bolyai 34, North-Holland, Amsterdam, pp. 1523–37. [7]

Stewart, C. L. (1986), On the greatest prime factor of terms of a linear recurrence sequence, *Rocky Mountain J. Math.*, to appear. [C]

Stolarsky, K. B. (1974), *Algebraic Numbers and Diophantine Approximation*, Pure Appl. Math. No. 26, Marcel Dekker, New York. [A]

Størmer, C. (1897), Quelques théoremès sur l'équation de Pell $x^2-Dy^2=\pm 1$ et leurs applications, Vid. Skr. I Math. Natur. Kl. (Christiana) 1897 No. 2, 48 pp. [7]

Størmer, C. (1898), Sur une équation indéterminée, *C.r. hebd. Séanc. Acad. Sci., Paris* **127**, 752–4. [1]

Stroeker, R. J. and Tijdeman, R. (1982), Diophantine equations, *Computational Methods in Number Theory*, M.C. Tract 155, Centr. Math. Comp. Sci., Amsterdam, pp. 321–69. [Introduction, 1]

Sylvester, J. J. (1892), On arithmetical series, *Messenger Math.* **21**, 1–19, 87–120. *Collected Mathematics Papers*, Vol. 4, 1912, pp. 687–731. [7]

Szymiczek, K. (1965), On the equations $a^x \pm b^x = c^y$, *Am. J. Math.* **87**, 262–6. [1]

Terjanian, G. (1977), Sur l'équation $x^{2p}+y^{2p}=z^{2p}$, *C.r. hebd. Séanc. Acad. Sci., Paris* **A285**, 973–5. [11]

Thue, A. (1909), Ueber Annäherungswerte algebraischer Zahlen, *J. reine angew. Math.* **135**, 284–305. [Introduction, 1, 5, 7]

Tijdeman, R. (1972), On the maximal distance of numbers with a large prime factor, *J. Lond. Math. Soc.* (2) **5**, 313–20. [7]

Tijdeman, R. (1973), On integers with many small prime factors, *Composition Math.* **26**, 319–30. [1]

Tijdeman, R. (1974), On the maximal distance between integers composed of small primes, *Compositio Math.* **28**, 159–62. [1]

Tijdeman, R. (1975), Some applications of Baker's sharpened bounds to diophantine equations, Sém. Delange-Pisot-Poitou 1974/1975, Paris, Exp. 24, 7 pp. [2, 5]

Tijdeman, R. (1976*a*), Applications of the Gel'fond–Baker method to rational number theory, *Topics in Number Theory*, Proceedings of the Conference at Debrecen 1974, Colloq. Math. Soc. János Bolyai 13, North-Holland, Amsterdam, pp. 399–416. [10]

Tijdeman, R. (1976*b*), On the equation of Catalan, *Acta Arith.* **29**, 197–209. [12]

Tijdeman, R. (1981), Multiplicities of binary recurrences, Sém. Théorie des Nombres Univ. Bordeaux I, Lab. Th. Nombr. C.N.R.S., Exp. 29, 11 pp. [C, 4]

Tijdeman, R. (1985), On the Fermat–Catalan equation, *Jahresber. Deutsche Math. Verein.* **87**, 1–18. [12]

Tijdeman, R. (1986), A note on the Fermat equation, *Colloq. Math.*, to appear. [11]

Townes, S. B. (1962), Notes on the diophantine equation $x^2+7y^2=2^{n+2}$, *Proc. Am. Math. Soc.* **13**, 864–9. [3]

Trelina, L. A. (1977*a*), On algebraic integers with discriminants containing fixed prime divisors (Russian), *Mat. Zametki* **21**, 289–96. English trans.: *Math. Notes* **21**, 161–5. [1]

Trelina, L. A. (1977*b*), The greatest prime divisor of an index form (Russian), *Dokl. Akad. Nauk BSSR* **21**, 975–6, 1051. [7]

Trelina, L. A. (1978), *S*-integral solutions of diophantine equations of hyperelliptic type (Russian), *Dokl. Akad. Nauk BSSR* **22**, 881–4, 955. [8]

Turk, J. (1978), Sets of integers composed of few prime numbers, Sém. Delange-Pisot-Poitou 1977/78, Paris, Exp. 44, 6 pp. [7]

Turk, J. (1979), Multiplicative properties of neighbouring integers, Dissertation, R.U. Leiden. [7]

Turk, J. (1980*a*), Multiplicative properties of integers in short intervals, *Indag. Math.* **42**, 429–36. [7, 12]

Turk, J. (1980*b*), Prime divisors of polynomials at consecutive integers, *J. reine angew. Math.* **319**, 142–52. [7]

Turk, J. (1982), Polynomial values and almost powers, *Michigan Math. J.* **29**, 213–20. [6, 7, 10]

Turk, J. (1983*a*), A note on the Fermat equation, *J. Number Theory* **17**, 76–9. [11]

Turk, J. (1983*b*), The product of two or more neighboring integers is never a power, *Illinois J. Math.* **27**, 392–403. [10]

Turk, J. (1984), Almost powers in short intervals, *Arch. Math.* **43**, 157–66. [6, 12]
Turk, J. (1986), On the difference between perfect powers, *Acta Arith.* **45**, 289–307. [6, 9, 10, 12]
Tzanakis, N. (1983), On the diophantine equation $y^2 - D = 2^k$, *J. Number Theory* **17**, 144–64. [7]
Tzanakis, N. (1984), The complete solution in integers of $X^3 + 2Y^3 = 2^n$, *J. Number Theory* **19**, 203–8. [7]
Tzanakis, N. and Wolfskill, J. (1986), On the diophantine equation $y^2 = 4q^n + 4q + 1$, *J. Number Theory* **23**, 219–37. [7]
Veluppillai, M. (1980), The equations $z^2 - 3y^2 = -2$ and $z^2 - 6x^2 = -5$, *A Collection of Manuscripts Related to the Fibonacci Sequence*, Fibonacci Association, Santa Clara, California, pp. 71–5. [6]
Vinogradov, A. I. and Sprindžuk, V. G. (1968), The representation of numbers by binary forms (Russian), *Mat. Zametki* **3**, 369–76. [7]
Vojta, P. A. (1983), Integral Points on Varieties, Ph.D. Thesis, Harvard University. [1, 5]
Voorhoeve, M., Győry, K. and Tijdeman, R. (1979), On the diophantine equation $1^k + 2^k + \cdots + x^k + R(x) = y^z$, *Acta Math.* **143**, 1–8. [10]
Wagstaff, S. S., Jr (1978), The irregular primes to 125 000, *Math. Comp.* **32**, 583–91. [11]
Wagstaff, S. S., Jr (1979), Solution of Nathanson's exponential congruence, *Math. Comp.* **33**, 1097–100. [1]
Waldschmidt, M. (1980), A lower bound for linear forms in logarithms, *Acta Arith.* **37**, 257–83. [B]
Ward, M. (1954), Prime divisors of second order recurring sequences, *Duke Math. J.* **21**, 607–14. [4]
Ward, M. (1955*a*), The intrinsic divisors of Lehmer numbers, *Ann. Math.* (2) **62**, 230–6. [3]
Ward, M. (1955*b*), The laws of apparition and repetition of primes in a cubic recurrence, *Trans. Am. Math. Soc.* **79**, 72–90. [4]
Warkentin, P. (1984*a*), Die Normformgleichung über dem rationalen Funktionenkörper, *J. Number Theory* **18**, 56–68. [7]
Warkentin, P. (1984*b*), Ueber Normformungleichungen mit beschränkter Lösungsmenge, *J. Number Theory* **18**, 371–90. [7]
Wasén, R. C. (1977), On additive relations between algebraic integers of bounded norm, *Théorie Additive des Nombres*, Université de Bordeaux I, pp. 153–60. [1]
Williams, H. C. (1975), On Fibonacci numbers of the form $k^2 + 1$, *Fibonacci Quart.* **13**, 213–14. [9]
Wyler, O. (1964), Squares in the Fibonacci series, *Am. Math. Monthly* **71**, 220–2. [9]
Zimmer, H. G. (1972), *Computational Problems, Methods, and Results in Algebraic Number Theory*, Lecture Notes in Mathematics 262, Springer-Verlag, Berlin. [A]
Zimmert, R. (1981), Ideale kleiner Norm in Idealklassen und eine Regulatorabschätzung, *Invent. Math.* **62**, 367–80. [A]
Zsigmondy, K. (1892), Zur Theorie der Potenzreste, *Monatsh. Math.* **3**, 265–84. [3]

Index